T0203667

The CRC Press
Laser and Optical Science and Technology Series

Editor-in-Chief: Marvin J. Weber

Handbooks of Laser Science and Technology

Volume I: Lasers and Masers, Edited by Marvin J. Weber

Volume II: Gas Lasers, Edited by Marvin J. Weber

Volume III: Optical Materials, Part 1, Edited by Marvin J. Weber

Volume IV: Optical Materials, Part 2, Edited by Marvin J. Weber

Volume V: Optical Materials, Part 3, Edited by Marvin J. Weber

Supplement I: Lasers, Edited by Marvin J. Weber

Supplement II: Optical Materials, Edited by Marvin J. Weber

Thermodynamic and Kinetic Aspects of the Vitreous State

Sergei V. Nemilov

Optical Constants of Inorganic Glasses

Andrei M. Efimov

Spontaneous Emission and Laser Oscillation in Microcavities

Edited by Hiroyuki Yokoyama and Kikuo Ujihara

Forthcoming Handbook Titles

Handbook of Laser Wavelengths, Edited by Marvin J. Weber

Spontaneous Emission and Laser Oscillation in Microcavities

Edited by
Hiroyuki Yokoyama, Ph.D.
Opto-Electronics Research Laboratories
NEC Corporation
Tsukuba, Japan

Kikuo Ujihara, Ph.D.
Professor
Department of Electronic Engineering
University of Electro-Communications
Tokyo, Japan

CRC Press
Taylor & Francis Group
Boca Raton London New York

CRC Press is an imprint of the
Taylor & Francis Group, an **informa** business

Published in 1995 by
CRC Press
Taylor & Francis Group
6000 Broken Sound Parkway NW, Suite 300
Boca Raton, FL 33487-2742

First issued in paperback 2019

No claim to original U.S. Government works

ISBN 13: 978-0-367-44895-0 (pbk)
ISBN 13: 978-0-8493-3786-4 (hbk)

Library of Congress Card Number 95-4016

Library of Congress Cataloging-in-Publication Data

Spontaneous emission and laser oscillation in microcavities / edited
    by Hiroyuki Yokoyama and Kikuo Ujihara.
        p. cm.—(The CRC Press laser and optical science and technology series)
        Includes bibliographical references and index.
        ISBN 0-8493-3786-0
        1. Quantum electronics. 2. Lasers. 3. Laser beams.
    I. Yokoyama, Hiroyuki. II. Ujihara, Kiluo. III. Series.
    QC688.S66 1995
    621.36'61—dc20   95-4016

PREFACE

Since the advent of the laser, the field of quantum optics and optoelectronics has developed at a great speed and has deeply influenced modern society academically and technologically. The realization of fiber-optical communication with a semiconductor-laser light source is typical of the profound influences. In order to treat a vast amount of information, further innovation in the size and power consumption of laser devices is now being pursued. On the other hand, solid-state electronics with silicon large-scale-integrated circuit (LSI) is approaching its fundamental limit. As the size of the structure decreases, both optical and electronic microstructures are encountering quantum effects, which can be obstacles for conventional devices as well as new mechanisms for novel optical or electronic devices.

We are confronting innovative frontiers, and novel concepts of a new age in quantum-integrated devices are emerging. For an optical device, reducing the size and number of photons leads to inherent quantum fluctuations that must be overcome, for example, by squeezing the photon states. The physics of optical confinement has developed through quantum optical interest in controlling spontaneous emission by an optical microcavity. This stream of research, combined with the stream of miniaturization of the laser device, resulted in the concept of a microcavity laser. The microcavity laser will have a key role in photonics, for example, as an ultralow power-consumption light emitter in an optical interconnect. Research on microcavity lasers represents an interdisciplinary science involving quantum electronics, optics, solid state physics, as well as device physics and device technology. New concepts of electronic devices utilizing quantum confinement and the wave nature of electrons are also being formed. The unification of the optical microcavity with electronic quantum structures will be a challenge for researchers in optical microstructures and electronic microstructures, opening up a new branch of device physics for the future.

A semiconductor microcavity laser involving electronic quantum confinement structures is a forerunner of such devices. Actually, much recent research on a surface emitting semiconductor-quantum-well laser has been carried out in conjunction with controlling spontaneous emission by a microcavity. In the limit of three-dimensional wavelength-size cavity structures, the cavity QED effect will be pronounced and the spontaneous emission probability will be greatly enhanced.

In spite of increasing importance of microcavities, device physics or observable phenomena in optical microcavities, such as enhanced or inhibited spontaneous emission and its relation with the laser oscillation, have not been systematically described. This may be partly because quantum electrodynamics is sometimes necessary for discussing the operating properties of the devices. Another point is that the essential and simple physics of the phenomena are often masked by the quantum electrodynamics view, even though it is well explained classically or semiclassically. The microcavity is obviously not "a small miracle box" that provides a totally new device operation principle based on quantum electrodynamics. However, simultaneously, microcavity devices have many dramatically different features in contrast with the conventional lasers or light-emitting devices. Thus, microcavity effects need further study from both quantum-mechanical and classical approaches to obtain a deeper understanding.

With all of the above background in mind, this book aims to give the basics of optical microcavities rather than describing advanced theoretical topics or the most recent experimental reports or fabrication technologies. Therefore it is usable as a textbook for graduate and postgraduate students, as well as researchers beginning to study the field.

Chapter 1, by E. V. Goldstein and P. Meystre, stresses the fundamentals of controlling spontaneous emission by a microcavity. After a survey of several methods describing free space spontaneous emission, changes in the feature due to the presence of cavity are discussed. As a special topic, spontaneous emission by moving atoms is also included. In Chapter 2, Y. Lee clarifies the effect of optical loss mechanisms in a microcavity that dephases the vacuum field and causes deterioration of the cavity's capability to control spontaneous emission. K. Ujihara, in Chapter 3, gives an analysis of the effect of spectral broadenings of an atom on the spontaneous emission in a microcavity based on a reservoir model and the density matrix method. Chapter 4, by C. Weisbuch, R. Houdré, and R. P. Stanley, describes the splitting in transmission peaks of planar microcavities containing semiconductor quantum wells. Both theoretical and experimental results are given from the viewpoint of transmissive vacuum Rabi splitting for this phenomenon.

Chapters 5 through 10 are devoted to device physics and device applications of optical microcavities. First, in Chapter 5, S. D. Brorson gives a simple but useful way to consider the change in the spontaneous emission rate from the viewpoint of mode density alteration by wavelength sized cavities. Chapter 6, by G. Björk and Y. Yamamoto, presents a concise description of classical calculation methods of lifetime and angular distribution of spontaneous emission in a realistic planar dielectric microcavity with employing reciprocity theorem. In Chapter 7, T. Baba and K. Iga describe classical mode density calculations to obtain the fraction of spontaneous emission coupled into a fundamental laser mode in a pillar-type dielectric microcavity. Effects of electron confinement in semiconductor quantum wells, wires, and boxes are also discussed. Chapter 8, by H. Yokoyama, extends the controlling spontaneous emission phenomenon to laser oscillation. Starting from Fermi's golden rule, the microcavity laser rate equations are derived and the oscillation characteristics are analyzed. In Chapter 9, H. Yokoyama also summarizes the present status of experimental research regarding optical microcavity experiments from the points of view of controlling spontaneous emission and laser oscillation. I. Hayashi gives motivation for microcavity research from the real world application point of view in Chapter 10. The reason why microcavity optical devices are important in ULSI interconnection technologies is described. Furthermore, their applicability in massively optical parallel processing systems, as well as demands for the device performance, are also discussed.

It should be noted that the editors have not included in this volume, topics concerning photonic band-gap structures and squeezed-state photon generation. Although these recent issues are strongly related to the microcavity, they also involve many topics that will be described in different volumes.

The editors would like to acknowledge the close cooperation of the authors. They have mutually reviewed the articles, discussed them, and sometimes made several revisions in the manuscripts. The editors also should like to thank Dr. M. J. Weber, the Laser Science Series Editor, Ms. F. Shapiro, Ms. Carole Sweatman, and Mr. J. Claypool of CRC Press for their help throughout in completing this edition.

H. Yokoyama and K. Ujihara

EDITORS

Hiroyuki Yokoyama, Ph.D., is Research Manager at the Opto-Electronics Research Laboratories for the NEC Corporation in Tsukuba, Japan. He received his Ph.D. degree in 1982 in electronic engineering in the Research Institute of Electrical Communication at Tohoku University, Sendai, Japan and has been with NEC since that time. He also has been a visiting scientist at the Research Laboratory of Electronics, Massachusetts Institute of Technology from 1988 to 1989, and since 1994 has been a visiting professor in the Department of Physics at the University of Tsukuba.

Dr. Yokoyama has been engaged in research of ultrafast opto-electronics and device physics of semiconductor lasers since 1982. He also has done pioneer work in the theoretical and experimental study on microcavity lasers. He is a member of the Optical Society of America, the American Physical Society, and the Japan Society of Applied Physics.

Kikuo Ujihara, Ph.D., is Professor in the Department of Electronic Engineering at the University of Electro-Communications in Chofu, Tokyo, Japan. He received his B.S., M.S. and Ph.D. degrees in electronic engineering from the University of Tokyo, in 1964, 1966, and 1970, respectively. In 1981, he was a visiting scholar at the University of Sussex in Brighton, England; a visiting associate professor at the Institute for Modern Optics at the University of New Mexico in Albuquerque, New Mexico (1981-1982); and a guest researcher at Max-Planck Institut für Quantenoptik in Garching, Germany in 1984-1985.

Dr. Ujihara is a member of the Japan Society of Applied Physics, Physical Society of Japan, Optical Society of Japan, Optical Society of America, and American Association for the Advancement of Science. He has written a textbook on quantum electronics in Japanese and translated a book on quantum optics from English to Japanese. He also has published over 50 technical papers. His research has been on the theory of lasers with output coupling and on optical phase conjugation. His current research interest is on the theory and experiments on microcavity lasers and quantum electrodynamics of open cavities.

CONTRIBUTORS

Toshihiko Baba, Ph.D.
Division of Electrical and
 Computer Engineering
Yokohama National University
Yokohama, Japan

Gunnar Björk, Ph.D.
E. L. Ginzton Laboratory
Stanford University
Stanford, California

Stuart D. Brorson, Ph.D.
Tele Danmark Research
Hørsholm, Denmark

Elena V. Goldstein, Ph.D.
Optical Sciences Center
University of Arizona
Tucson, Arizona

Izuo Hayashi, Ph.D.
Optoelectronic Technology Research
 Laboratory
Tsukuba, Japan

Romuald Houdré, Ph.D.
Institut de Micro- et Optoélectronique
Ecole Polytechnique Fédérale de
 Lausanne
Lausanne, Switzerland

Kenichi Iga, Ph.D.
Precision & Intelligence Laboratory
Tokyo Institute of Technology
Yokohama, Japan

Yong Lee, Ph.D.
Central Research Laboratory
Hitachi, Ltd.
Tokyo, Japan

Pierre Meystre, Ph.D.
Optical Sciences Center
University of Arizona
Tucson, Arizona

Ross P. Stanley, Ph.D.
Institut de Micro- et Optoélectronique
Ecole Polytechnique Fédérale de
 Lausanne
Lausanne, Switzerland

Kikuo Ujihara, Ph.D
Department of Electronic Engineering
The University of Electro-
 Communications
Tokyo, Japan

Claude Weisbuch, Prof.
Laboratoire de Physique de la Matière
 Condensée
Ecole Polytechnique
Palaiseau, France

Hiroyuki Yokoyama, Ph.D.
Opto-Electronics Research Laboratories
NEC Corporation
Tsukuba, Japan

Yoshihisa Yamamoto, Ph.D.
Department of Applied Physics
E. L. Ginzton Laboratory
Stanford University
Stanford, California

CONTENTS

1 Spontaneous Emission in Optical Cavities: A Tutorial Review

Elena V. Goldstein and Pierre Meystre

TABLE OF CONTENTS

I.   INTRODUCTION

In its standard textbook description, spontaneous emission is the irreversible emission of a photon by an atom into the free space modes of the electromagnetic field, accompanied by a transition of the atom from an electronic state of energy E_2 to one of lower energy E_1. The frequency of the emitted light is $(E_2 - E_1)/\hbar$, where $\hbar$ is Planck's constant. The presence of Planck's constant in this frequency clearly indicates that spontaneous emission is an intrinsically quantum mechanical process. Indeed, its proper description requires the quantization of both the atoms and the field. A well-known result of this theory is that the rate of spontaneous emission is proportional to the free space mode density of the electromagnetic field.

In recent years, it has become increasingly clear that this description of spontaneous emission is not general, and that spontaneous emission is not an intrinsic atomic property: rather, it can be modified by tailoring the electromagnetic environment that the atom can radiate into.[1-3b] This was first realized by Purcell,[4] who noted that the spontaneous emission rate can be enhanced for an atom placed inside a cavity with one of its modes resonant with the transition under consideration, and by Kleppner,[5] who discussed the opposite case of inhibited spontaneous emission. It has also been recognized that spontaneous emission need not be an irreversible process. Indeed, the Schrödinger equation always leads to reversible dynamics. It is only when the modes of the electromagnetic field are treated as a Markovian reservoir, that is, a reservoir with infinitely short memory, that irreversibility is achieved. In that case, the upper electronic state of the atom does indeed suffer an irreversible exponential decay. However, qualitatively different types of dynamics can be achieved if the vacuum modes of the electromagnetic field cannot be characterized in such a way. In particular, we shall see that an atom coupled to a single mode of the electromagnetic field undergoes a periodic exchange of excitation between the atom and the field. Remarkably, it has now become possible to observe such reversible behavior, both in the microwave and in the optical regime of cavity quantum electrodynamics (QED).

The purpose of this chapter is to present a tutorial review of spontaneous emission in optical resonators. For simplicity, we restrict our discussion to a single two-level system, which we typically think of as representing a dipole-allowed atomic transition. Hence, we do not address questions such as quantum beats[6] or the superfluorescence from a sample of N atoms,[7] to mention but two examples. Also, our discussion is not specifically directed toward semiconductor systems, whose description forms the bulk of this book. Yet, our results are generic enough that they can be extended to such situations without difficulty.

It is well known that spontaneous emission is normally accompanied by a level shift known as the Lamb shift. In cavities, which are the environment of most interest in this book, the situation is further complicated by the presence of cavity walls, which leads to additional shifts, such as van der Waals shifts, scaling as $1/z^3$ and dominant for small distances z between the radiator and the cavity walls; "retarded Casimir shifts", important for ground state atoms at large z, and scaling as $1/z^4$; and shifts induced by the dipole-wall interaction, relevant for excited atoms at large z, and scaling as $1/z$. A proper description of these shifts, which are of considerable interest in high resolution spectroscopy, requires an accurate description of the atoms under consideration. The interested reader is referred to a recent review by Haroche,[1] and to a treatise by Milonni,[8] who discuss these shifts in great detail.

We first consider the simple problem of a two-level atom dipole-coupled to a single quantized mode of the electromagnetic field. This is the famous Jaynes-Cummings model,[9] which lies at the heart of spontaneous emission, laser and maser theory, and a number of problems in quantum optics. It can be solved exactly within the so-called rotating-wave approximation and yields a simple form of spontaneous emission, which turns out to be reversible: the photon first emitted by the atom can subsequently be reabsorbed, thereby leading to oscillations of the upper level population at the so-called vacuum Rabi frequency.

In order to understand how irreversibility is achieved, it is necessary to consider a multimode vacuum field. Section II.B gives the simplest description of the dynamics of this system in terms of first-order perturbation theory, which leads to the Fermi golden rule.[6] However, this approach, which remarkably predicts the proper spontaneous emission rate, is valid for short times only. Section II.C generalizes this result to long interaction times via the Weisskopf-Wigner theory of spontaneous emission.

An alternative approach to the problem of spontaneous emission is given in Section II.D in terms of reservoir theory, which presents the advantage of yielding a simple physical interpretation of the assumption underlying the derivation of the conventional spontaneous emission decay rate and gives a simple physical feeling for the Born-Markov approximation. We also discuss how the master equation resulting from the reservoir approach may be reinterpreted in terms of an effective, non-Hermitian Hamiltonian which properly describes the evolution of the excited state. In addition to its appeal from an intuitive point of view, this effective Hamiltonian approach opens up the way to the use of Monte Carlo wave function simulations,[10-12] which have recently proven useful to numerically handle large quantum systems whose standard density operator treatment would require excessive amounts of computer memory.

Section III then turns to the description of spontaneous emission in optical resonators. We introduce in Section III.B a generalized master equation description for the combined atom-cavity system and are naturally led to introduce the so-called weak and strong coupling limits.[3a,b] In the weak coupling limit, the atomic evolution is still governed by an exponential decay, but at a rate that can be either higher or lower than in free space. These are the now well-known regimes of enhanced and inhibited spontaneous emission. The strong coupling regime, discussed in Section III.D, is characterized by the fact that the evolution of the upper atomic state population ceases to be exponential.

Instead, a photon initially emitted by the atom may be trapped inside the cavity long enough to have a nonvanishing probability of being reabsorbed. Hence, the dynamics of the system are characterized by a periodic exchange of energy between the atom and the cavity mode, superimposed on an exponential decay. In this case, the spontaneous emission spectrum becomes a doublet that we interpret in terms of dressed states.

In contrast to the free space situation, the coupling of an atom to the vacuum-field modes becomes dependent on its location inside the cavity. This leads to a number of new effects, which become specially apparent in the case of moving atoms. This situation is discussed in Section IV in both the weak and strong coupling regimes. Finally, Section V is a summary and conclusion.

II.   FREE SPACE SPONTANEOUS EMISSION

A.   SINGLE MODE SPONTANEOUS EMISSION

To illustrate the basic physical mechanism underlying spontaneous emission, we first consider the electric dipole interaction between a single two-level system and a single mode of the electromagnetic field. This two-level atom is conveniently described by the Hamiltonian

$$H_a = \hbar\omega_e |e\rangle\langle e| + \hbar\omega_g |g\rangle\langle g|, \tag{1}$$

where $\hbar\omega_e$, $\hbar\omega_g$ are the energies of the excited and the ground level, respectively.

In the absence of interaction, an atom initially in its excited state $|e\rangle$ will remain there for all times. Transitions between the eigenstates of the atom result from the coupling of the atom to some other system. In the case of dipole coupling to a single electromagnetic field mode of frequency ω_c, the total atom-field system is described by the Jaynes-Cummings Hamiltonian[9]

$$H = H_a + H_f + H_{a-f}, \tag{2}$$

where

$$H_f = \hbar\omega_c \, a^\dagger a. \tag{3}$$

and the annihilation and creation operators a and $a^\dagger$ obey the boson commutation relation $[a, a^\dagger] = 1$. The eigenstates of the Hamiltonian (3) are the "number states" $|n\rangle$, where $n = 0, 1, 2,..., \infty$, with $a^\dagger a|n\rangle = n|n\rangle$, $a|n\rangle = \sqrt{n}\,|n - 1\rangle$, $a^\dagger|n\rangle = \sqrt{n+1}\,|n + 1\rangle$; the vacuum state $|0\rangle$ is the eigenstate of the annihilation operator a with zero eigenvalue,

$$a|0\rangle = 0. \tag{4}$$

In the dipole and rotating wave approximations[6] the interaction Hamiltonian H_{a-f} between this field mode and the two-level atom is

$$H_{a-f} = -\hat{\mathbf{d}} \cdot \hat{\mathbf{E}} \tag{5}$$

where $\hat{\mathbf{d}} = \wp(|g\rangle\langle e| + h.c.)$ is the dipole moment operator of the transition, and the electric field operator is

$$\hat{\mathbf{E}}(\mathbf{R}) = \mathscr{E}[\boldsymbol{\varepsilon} f(\mathbf{R})a + h.c.]. \tag{6}$$

Here, $\boldsymbol{\varepsilon}$ is the polarization of the field mode. For standing-wave quantization, the spatial variation of the field mode is $f(\mathbf{R}) = \sin(\mathbf{kR})$ and the "electric field per photon" $\mathscr{E} = \sqrt{\hbar\omega_c/\varepsilon_0 V}$. For running-wave quantization, in contrast, $f(\mathbf{R}) = \exp(i\mathbf{k} \cdot \mathbf{R})$ and $\mathscr{E} = \sqrt{\hbar\omega_c/2\varepsilon_0 V}$. While both quantization schemes are equivalent in principle, the choice of one vs. the other is normally dictated by the geometry of the problem at hand.

The atom-field dipole coupling constant is

$$\hbar g(\mathbf{R}) = \mathscr{E}f(\mathbf{R})(\wp \cdot \boldsymbol{\varepsilon}) \tag{7}$$

where g is half the so-called vacuum Rabi frequency. Invoking the rotating-wave approximation, the electric dipole interaction Hamiltonian (5) then takes the form

$$H_{a-f} = \hbar(g(\mathbf{R})a^\dagger|g\rangle\langle e| + h.c.). \tag{8}$$

The vacuum Rabi frequency $2g(\mathbf{R})$ normally depends on the location $\mathbf{R}$ of the atom. However, this spatial dependence can be ignored if considering only a single atom at rest. Introducing the Pauli pseudospin operators $\sigma_z = (|e\rangle\langle e| - |g\rangle\langle g|)/2, \sigma_+ = |e\rangle\langle g|, \sigma_- = (\sigma_+)^\dagger$, one can re-express the Hamiltonian (2) as

$$H = E_0 + \hbar\omega_0\sigma_z + \hbar\omega_c a^\dagger a + \hbar(g(\mathbf{R})a^\dagger\sigma_- + h.c.), \tag{9}$$

where $\omega_0 \equiv \omega_e - \omega_g$ is the atomic transition frequency.

The interaction Hamiltonian (8) is physically quite transparent: the atom can either absorb a photon and undergo a transition from the ground state to the excited state, or emit a photon while undergoing a transition from the excited to the ground state.

Within the framework of the Jaynes-Cummings model, a state of the combined atomic system, with the atom in its excited state $|e\rangle$ and the field in the number state $|n\rangle$, $|\psi\rangle = |e, n\rangle$, is coupled only to the state $|\psi'\rangle = |g, n + 1\rangle$, which corresponds to the atom in the ground state and the field in the number state $|n + 1\rangle$. Hence, the atom-field interaction can be considered for each manifold of levels $\mathscr{E}_{n+1} = \{|e, n\rangle, |g, n + 1\rangle\}$ independently. Physically, this property indicates that the total number of excitations in the atom-field system is conserved, the excitations being merely exchanged between the atom and the cavity mode. Technically, this also means that for each manifold $\mathscr{E}_n$, the Jaynes-Cummings problem reduces to a simple two-level problem that can be solved exactly. The full dynamics are solved by an appropriate sum over the dynamics of the system within each manifold. For our present purpose, which is merely to elucidate the basic mechanism at the origin of spontaneous emission, it is sufficient to assume that the atom is initially in its excited state $|e\rangle$, while the field mode is in the vacuum state $|0\rangle$,

$$|\psi(0)\rangle = |e, 0\rangle. \tag{10}$$

The atom-field system therefore remains in the one-quantum manifold $\mathscr{E}_1$ for all time, and its general state is given by a superposition of the states $|e,0\rangle$ and $|g,1\rangle$. On resonance $\omega_c = \omega_0$, the time-dependent state of the atom-field system is

$$|\psi(t)\rangle = \cos(gt)|e, 0\rangle - i\sin(gt)|g, 1\rangle, \tag{11}$$

so that the probability $P_g(t)$ for the atom to be in its ground electronic state $|g\rangle$ is

$$P_g(t) = |\langle g,1 |\psi(t)\rangle|^2 = \sin^2(gt). \tag{12}$$

This result is the simplest form of "spontaneous emission": in contrast to a classical field of zero amplitude, a quantized field in the vacuum induces transitions from the upper to the lower electronic state. This difference between the quantum and classical descriptions arises because even though the expectation value of the electric field vanishes in the vacuum state, its fluctuations, proportional to $\langle(a + a^\dagger)^2\rangle$, do not. It is these "vacuum fluctuations" that induce the atom to undergo spontaneous emission. Note also that if the atom is initially in its ground state and the field is in a vacuum, $|\psi(0)\rangle = |g, 0\rangle$, the atom will remain in its ground state for all times.[6] This fundamental asymmetry is a direct consequence of the fact that the operators a and $a^\dagger$ do not commute.

This simple example of spontaneous emission may appear puzzling, as $P_g(t)$ exhibits an oscillatory behavior at the "vacuum Rabi frequency" $2g$, in contrast to the irreversible exponential decay that we are accustomed to. This is because in our simple model, once a photon is emitted, it can subsequently be reabsorbed by a standard stimulated emission process: the dynamics of the atom-field system are characterized by a periodic exchange of a quantum of excitation between the atom and the cavity mode. This is in contrast to the conventional situation, where once an excitation is transferred from the atom to the multimode vacuum field, it is *de facto* irretrievably lost. We shall see in Section III how recent experimental progress has made it possible to reach the oscillatory regime of spontaneous emission. However, we first discuss how spontaneous emission effectively becomes an irreversible process in conventional situations.

B. FERMI GOLDEN RULE

The major "flaw" of the model considered so far is that it considers a single mode of the electromagnetic field only. Such a simple description is not adequate in general: rather, atoms in free space interact with a continuum of field modes, so that the atom-field interaction Hamiltonian takes the form

$$H = H_a + \sum_k \hbar \omega_k a_k^\dagger a_k + \sum_k \hbar\left(g_k a_k^\dagger \sigma_- + h.c.\right) \tag{13}$$

where $[a_k, a_{k'}^\dagger] = \delta_{k,k'}$, and g_k is the dipole matrix element between the atom and the k-th mode of the field. In the Hamiltonian (13), we use a compact notation where the mode-labeling index $k = \{\mathbf{k}, \lambda\}$ stands for both the wavevector $\mathbf{k}$ and the polarization $\lambda = 1,2$ of the mode.

As before, we assume that the field is initially in the vacuum state and the atom is in its excited state, so that $|\psi(0)\rangle = |e, \{0\}\rangle$, where $|\{0\}\rangle$ labels the multimode vacuum, with $a_k|\{0\}\rangle = 0$ for all k. In analogy to the single mode situation, the state of the atom-field system at time t is therefore of the form

$$|\psi(t)\rangle = a(t)e^{-i\omega_0 t}|e,\ 0\rangle + \sum_k b_k(t)e^{-i\omega_k t}|g,\ 1_k\rangle, \tag{14}$$

with $a(0) = 1, b_k(0) = 0$. Here we have extracted the rapid time dependence of the various probability amplitudes for convenience, and we have reset the origin of atomic energies so that $\omega_g = 0$ and $\omega_e = \omega_0$. The state $|1_k\rangle$ is given by

$$|1_k\rangle = a_k^\dagger |0\rangle. \tag{15}$$

The equations of motion for the probability amplitudes $a(t)$ and $b_k(t)$ are readily found as

$$\frac{da(t)}{dt} = -i\sum_k g_k e^{-i(\omega_k - \omega_0)t} b_k(t), \tag{16}$$

$$\frac{db_k(t)}{dt} = -ig_k^* e^{i(\omega_k - \omega_0)t} a(t). \tag{17}$$

To first order in perturbation theory, which is valid for times short enough that the various probability amplitudes do not change appreciably, we can approximate $a(t)$ in the equation for $b_k(t)$ by its initial value $a(0) = 1$. This yields readily

$$|b_k(t)|^2 = |g_k|^2 \frac{\sin^2\left[(\omega_k - \omega_0)t/2\right]}{(\omega_k - \omega_0)^2/4}. \tag{18}$$

The probability for the atom to be in its upper state at time t is

$$P_e(t) = 1 - \sum_k |b_k(t)|^2 .$$

(19)

Going to a continuous limit and replacing the sum over k by an integral, we have

$$P_e(t) = 1 - \int dk |g_k|^2 \frac{\sin^2[(\omega_k - \omega_0)t/2]}{[(\omega_k - \omega_0)/2]^2},$$

(20)

Note that the integral over k actually involves a three-dimensional integral over all wavevectors $\mathbf{k}$ of the continuum of field modes, as well as a sum over two orthogonal polarizations. This integral is carried out in a number of texts for the case of a free space spontaneous emission, (see, e.g., Reference 6). One finds

$$P_e(t) = 1 - \frac{1}{6\varepsilon_0 \pi^2 \hbar c^3} \int d\omega_k \omega_k^3 |d|^2 \frac{\sin^2[(\omega_k - \omega_0)t/2]}{[(\omega_k - \omega_0)/2]^2},$$

(21)

where d is the magnitude of the atom-field dipole moment. In general, of course, this integral will lead to different results for different mode densities. Hence, $P_e(t)$ depends explicitly on the specific geometry of the problem. This is one of the key observations leading to cavity QED.

The last step in the evaluation of $P_e(t)$ is achieved by noting that as time increases, the 4 $\sin^2[(\omega_k - \omega_0)t/2]/(\omega_k - \omega_0)^2$ term is significantly different from zero only for $\omega_k \simeq \omega_0$. In other words, it acquires the character of a Dirac delta function, which selects only the energy-conserving terms in the atom-field interaction. For such times, and with

$$\lim_{t \to \infty} \frac{\sin^2((\omega_k - \omega_0)t/2)}{(\omega_k - \omega_0)^2/4} = 2\pi t \delta(\omega_k - \omega_0),$$

(22)

we find that in the free space the excited atomic state population decays at the rate

$$\gamma_f = \frac{dP_e}{dt} = \frac{\omega_0^3 |d|^2}{3\pi \varepsilon_0 \hbar c^3},$$

(23)

where $k_0 = \omega_0/c$.

This result, which is an example of Fermi's golden rule, predicts that *for times large enough that energy conservation is established, yet short enough for first-order perturbation theory to remain valid*, the excited atomic state decays at the (time-independent) rate γ_f of Equation 23. More precisely, the prediction of this calculation is that for this range of times, the evolution of the upper state population is given by

$$\frac{dP_e(t)}{dt} = -\gamma_f \simeq -\gamma_f P_e(t),$$

(24)

the second approximate equality relying on the fact that $P_e(t) \simeq 1$ in the regime of validity of first-order perturbation theory. Remarkably, it turns out that this result can be readily extended to the form

$$\frac{dP_e(t)}{dt} = -\gamma_f P_e(t),$$

(25)

that is, the spontaneous emission rate γ_f is valid for all times and any degree of depletion of the upper state population — except of course for very short times, where a quadratic dependence of $P_e(t)$ on time is to be expected! (There are also very long time departures from the exponential decay law, but we won't discuss them in this review.) Clearly, first-order perturbation theory is not in a position to predict the result (25). The derivation of this equation is the subject that we now turn to.

C. WEISSKOPF-WIGNER THEORY

A better way to approximately solve Equations 16 and 17 is given by the Weisskopf-Wigner theory of spontaneous emission.[13] Here, one proceeds by formally integrating Equation 17 and inserting the result in Equation 16 to get

$$\frac{da(t)}{dt} = -\sum_k |g_k|^2 \int_0^t dt' e^{-i(\omega_k - \omega_0)(t-t')} a(t').$$

(26)

As was the case in the Fermi golden rule discussion, we replace the sum over k by an integral. Following steps similar to those of the preceding discussion, we find readily that

$$\frac{da(t)}{dt} = -\frac{1}{6\varepsilon_0 \hbar c^3} \int d\omega_k \omega_k^3 |d|^2 \int_0^t dt' e^{-i(\omega_k - \omega_0)(t-t')} a(t').$$

(27)

If we assume that $a(t')$ varies sufficiently slowly compared to the exponential factor in the time integral that it can be evaluated at time t and removed from that integral — an assumption which will have to be checked for self-consistency against the final result — and consider as in the Fermi golden rule approach times long enough that this exponential factor actually selects energy-conserving terms via

$$\lim_{t \to \infty} \int_0^t dt' e^{-i(\omega_k - \omega_0)(t - t')} = \pi \delta(\omega_k - \omega_0) - \mathcal{P} \left[\frac{i}{\omega_k - \omega_0} \right] \tag{28}$$

we find

$$\frac{da(t)}{dt} = -\frac{\gamma_f}{2} a(t), \tag{29}$$

or

$$\frac{dP_e(t)}{dt} = -\gamma_f P_e(t), \tag{30}$$

which is precisely the advertised Equation 25. Note that in arriving at this expression, we have neglected a frequency shift due to the principal part in Equation 28. This shift is a caricature of the Lamb shift, for two-level atoms and in the dipole approximation. An evaluation of the spontaneous decay rate γ_f for a given transition will readily show that it is indeed slow compared to an optical frequency, typical values being of the order of 10^9 sec^{-1} for γ_f and 10^{14} sec^{-1} for ω_0. This justifies *a posteriori* taking $a(t')$ outside the time integral in Equation 27.

The basic lesson of the Weisskopf-Wigner theory of spontaneous emission is that the Fermi golden rule result is fine, provided that perturbation theory is carried out for the instantaneous value of the excited state population $P_e(t)$. The validity of this observation actually transcends the problem at hand, and can be applied whenever a small system is coupled to a large system that remains in thermal equilibrium and has a broadband response. Because the coupling of small systems to such "reservoirs" plays an important role in cavity QED, we outline their general theory in the next subsection. This will permit us to recover the spontaneous decay rate γ_f by yet another method, and, in a second example, to determine the dissipation of the field inside an optical resonator.

D. RESERVOIR THEORY APPROACH

Consider a small system, described by the Hamiltonian H_s and coupled to a large system described by the Hamiltonian H_r via an interaction Hamiltonian V, so that the Hamiltonian of the whole system is $H = H_s + H_r + V$. For instance, in the case of spontaneous emission the small system could be a two-level atom and the large system the continuum of modes of the electromagnetic field. We assume that the large system always remains in thermal equilibrium at some temperature T, independent of what happens to the small system. In other words, it acts as a thermal reservoir, described at all times by the thermal equilibrium density operator

$$\rho_r = \frac{1}{Z} e^{-H_r/k_B T},$$
(31)

where k_B is Boltzmann's constant, and $Z = Tr_r \exp(-\beta H_r)$ is the partition function of the reservoir. The symbol Tr_r stands for "trace over the reservoir."

We are interested in the evolution of the small system only, i.e., in the evolution of its *reduced density operator* ρ_s

$$\rho_s = Tr_r \rho_{sr},$$
(32)

where ρ_{sr} is the density operator describing the state of the full system. It is convenient to treat this problem in the interaction picture, where the interaction Hamiltonian V becomes

$$V_I(t - t_0) = e^{iH_0(t-t_0)/\hbar} V e^{-iH_0(t-t_0)/\hbar},$$
(33)

and $H_0 = H_s + H_r$. Correspondingly, the interaction picture density matrix of the system $P_{sr}(t)$ is given by

$$\rho_{sr}(t) = e^{-iH_0(t-t_0)/\hbar} P_{sr}(t) e^{iH_0(t-t_0)/\hbar},$$
(34)

which gives, by differentiation,

$$\frac{\partial \rho_{sr}}{\partial t} = -\frac{i}{\hbar} e^{-iH_0(t-t_0)} \left([H_0, P_{sr}(t)] + \frac{\partial P_{sr}}{\partial t} \right) e^{iH_0(t-t_0)}.$$
(35)

This equation relates the motion of the density operator in the Schrödinger and interaction pictures and yields the interaction picture density operator equation of motion

$$\frac{\partial P_{sr}}{\partial t} = -\frac{i}{\hbar} [V_I(t - t_0), P_{sr}(t)].$$
(36)

Assuming that the small system and the reservoir are decorrelated at time $t = t_0$, the interaction picture density matrix P_{sr} of the complete system is given to second order in perturbation theory by

$$P_{sr}(t) = P_{sr}(t_0) - \frac{i}{\hbar} \int_{t_0}^{t} dt' [V_I(t'-t_0), P_{sr}(t_0)]$$

$$- \frac{1}{\hbar^2} \int_{t_0}^{t} dt' \int_{t_0}^{t'} dt'' [V_I(t'-t_0), [V_I(t''-t_0), P_{sr}(t_0)]] + \cdots$$
(37)

Tracing this expression over the reservoir yields the equation of motion for the interaction picture reduced density operator

$$\rho(t) = e^{iH_s(t-t_0)/\hbar} \rho_s e^{-iH_s(t-t_0)/\hbar}. \tag{38}$$

The next step consists in introducing the so-called coarse-grained derivative

$$\dot{\rho}(t) \approx \frac{\rho(t) - \rho(t-\tau)}{\tau}, \tag{39}$$

where τ is a time interval long compared to the correlation time of the reservoir — we shall discuss this in more detail shortly — but short compared to the characteristic time over which the small system evolves, for instance γ_f in the case of spontaneous emission. In this coarse-grained sense, Equation 37 yields readily

$$\dot{\rho}(t) = -\frac{i}{\hbar\tau} \int_0^\tau d\tau' Tr_r\big(V_I(\tau')P_{sr}(t)\big)$$

$$-\frac{1}{\hbar^2\tau} \int_0^\tau \int_0^{\tau'} d\tau' d\tau'' Tr_r\big(V_I(\tau')V_I(\tau'')P_{sr}(t) - V_I(\tau')P_{sr}(t)V_I(\tau'')\big) + \text{adj.} \tag{40}$$

In many problems, and in particular in all situations of interest in this chapter, the first term on the right-hand side of this equation is equal to zero. The second term consists of a sum of two-time correlation functions of reservoir operators. For instance, in the case of the dipole-coupling Hamiltonian (13) we have, with Equation 33,

$$V_I(t) = \hbar\sigma_+ \hat{F}(t) + \text{adj.}, \tag{41}$$

where

$$\hat{F}(t) = \sum_k g_k a_k e^{i(\omega_0 - \omega_k)t}, \tag{42}$$

and the trace over the reservoir involves two-time correlation functions of the forms $\langle \hat{F}(t')\hat{F}(t'')\rangle$, $\langle \hat{F}^\dagger(t')\hat{F}^\dagger(t'')\rangle$, $\langle \hat{F}(t')\hat{F}^\dagger(t'')\rangle$, and $\langle \hat{F}^\dagger(t')\hat{F}(t'')\rangle$. For example, we have

$$\left\langle \hat{F}(t')\hat{F}^\dagger(t'')\right\rangle_r = \sum_{k,k'} g_k g_{k'}^* \left\langle a_k a_{k'}^\dagger\right\rangle_r e^{i\omega_0(t'-t'')} e^{-i(\omega_k t' - \omega_{k'} t'')}, \tag{43}$$

where $\langle\ \rangle_r$ means "average over the reservoir". For the reservoir thermal equilibrium density operator (31), we find readily

$$\left\langle a_k a_{k'}^\dagger \right\rangle_r = \left(\bar{n}_k + 1\right)\delta_{kk'}, \tag{44}$$

where $\bar{n}_k$ is the number of the thermal photons in the mode k of the field, and $\bar{n}_k = 0$ at zero temperature. Hence, Equation 43 reduces to

$$\left\langle \hat{F}(t')\hat{F}^\dagger(t'') \right\rangle_r = \sum_k \left|g_k\right|^2 \left(\bar{n}_k + 1\right)e^{i(\omega_0 - \omega_k)(t'-t'')}. \tag{45}$$

Similar calculations can be carried out for the other two-time correlation functions appearing in Equation 40. Since the various correlations depend only on the time difference $\Delta t = t' - t''$ — a consequence of the fact that the bath is assumed to be in thermal equilibrium and hence its statistical properties are stationary — a typical term in Equation 40 may be re-expressed in the form

$$\int_0^\tau \int_0^{\tau'} d\tau' d\tau'' \left\langle \hat{F}(\tau')\hat{F}^\dagger(\tau'') \right\rangle_r = \int_0^\tau d\tau' \int_0^{\tau'} d\Delta t \sum_k \left|g_k\right|^2 \left(\bar{n}_k + 1\right)e^{i(\omega_0 - \omega_k)\Delta t}. \tag{46}$$

The time dependence of this integral is due to the two-time correlation functions of the bath. For a broadband reservoir, they decay in a time that is short compared to the time characterizing the evolution of the small system, and hence the upper limit of integration τ' can be extended to infinity. Then

$$\int_0^\tau \int_0^{\tau'} d\tau' d\tau'' \left\langle \hat{F}(\tau')\hat{F}^\dagger(\tau'') \right\rangle_r = \int_0^\tau d\tau' \int_0^\infty d\Delta t \sum_k \left|g_k\right|^2 \left(\bar{n}_k + 1\right)e^{i(\omega_0 - \omega_k)\Delta t}$$

$$= \frac{\tau}{6\varepsilon_0 \pi^2 \hbar c^3} \int d\omega_k \omega_k^3 |d|^2 \left(\bar{n}_k + 1\right)\int_0^\infty dt\, e^{i(\omega_0 - \omega_k)t}, \tag{47}$$

where the sum over modes is replaced as in Fermi's gold rule or the Weisskopf-Wigner approach by an integral. The latter expression has the same structure as Equation 27. Using the representation (28) of the delta function, ignoring the associated frequency shifts, re-combining the different contributions from Equation 40, and considering explicitly the case of a two-level atom described by the Hamiltonian (1) finally yields the interaction picture master equation

$$\dot{\rho}_A(t) = -\frac{\gamma_f}{2}(\bar{n}+1)\left[\sigma_+\sigma_-\rho_A(t) - \sigma_-\rho_A(t)\sigma_+\right]$$

$$-\frac{\gamma_f}{2}\bar{n}\left[\rho_A(t)\sigma_-\sigma_+ - \sigma_+\rho_A(t)\sigma_-\right] + \text{adj.} \tag{48}$$

This master equation describes the decay of a two-level atom coupled to a reservoir of modes of the electromagnetic field (or more generally of harmonic oscillators) at temperature $T, \bar{n}$ being the corresponding number of reservoir thermal quanta at the atomic transition frequency ω_0. The reader can easily convince himself that for a reservoir at zero temperature, the upper state population $P_e(t) = Tr[|e\rangle\langle e|\rho_A(t)] = Tr[(\sigma_z + 1/2)\rho_A(t)]$ is governed by the equation of motion

$$\frac{dP_e(t)}{dt} = Tr_e\big[|e\rangle\langle e|\dot{\rho}(t)\big] = -\gamma_f P_e(t),$$

(49)

which is once more the Weisskopf-Wigner result.

In this same zero temperature limit, and returning to the Schrödinger picture, the two-level system master equation becomes

$$\dot{\rho}_A(t) = -\frac{i}{\hbar}\big[H_a, \rho_A\big] - \frac{\gamma_f}{2}\big[\sigma_+\sigma_-\rho_A(t) + \rho_A(t)\sigma_+\sigma_- - 2\sigma_-\rho_A(t)\sigma_+\big].$$

(50)

It is instructive to note that this equation may be recast in the form

$$\dot{\rho}_A(t) = -\frac{i}{\hbar}\big[H_{eff}\rho_A(t) - \rho_A(t)H_{eff}^\dagger\big] + \gamma_f\sigma_-\rho_A(t)\sigma_+,$$

(51)

where the non-Hermitian effective Hamiltonian H_{eff} is given by

$$H_{eff} = H_a - i\hbar\frac{\gamma_f}{2}\sigma_+\sigma_-.$$

(52)

This indicates that as far as the decay of the *unnormalized excited atomic state*

$$\big|\phi_e(t)\rangle = a(t)|e\rangle$$

(53)

is concerned, the master equation (50) may be replaced by the effective Schrödinger equation

$$i\hbar\frac{d|\phi_e(t)\rangle}{dt} = H_{eff}|\phi_e(t)\rangle.$$

(54)

Such an equation is usually easier to handle than the corresponding master equation, since it permits us to deal with state vectors rather than density operators. We shall make extensive use of this formal simplification in the remainder of this paper. We also note that such an effective Schrödinger equation forms the basis of the Monte Carlo wave function simulation techniques which have been enjoying considerable success recently.[10-12]

There are several advantages to the reservoir theory approach, as compared to the standard Weisskopf-Wigner theory. First, the approximation of the time integral in Equation 47 has a simple physical interpretation, since the integrand is just one of the two-time reservoir correlation functions. Approximating such correlation functions as delta functions means that the reservoir is assumed to lose its memory in a short time, compared to the time scale over which the dynamics of the small system occurs, the Markov approximation. The meaning of the coarse-grained derivative (39) then also becomes clear: it describes the evolution of the small system over large time scales, compared to the correlation time of the reservoir. The combination of this Markov approximation and the second-order perturbation theory used in the derivation of the master equation is called the Born-Markov approximation.

A further advantage of reservoir theory is that it does not rely on the explicit forms of the small system and of the reservoir. Indeed, the detailed structure of the reservoir turns out not to be important, provided that it remains in thermal equilibrium at all times and that its memory is so short that it decorrelates itself from the small system faster than any other time scale in the problem. In particular, this approach permits us to readily obtain the master equation of a simple harmonic oscillator with Hamiltonian $H_f = \hbar \Omega a^\dagger a$ coupled to a thermal reservoir of harmonic oscillators. Following the steps outlined in this section one finds

$$\dot{\rho}_f(t) = -\frac{i}{\hbar}\left[H_f, \, \rho_f\right] - \frac{\kappa}{2}(\bar{n}+1)\left[a^\dagger a \rho_f(t) - a \rho_f(t) a^\dagger\right]$$

$$-\frac{\kappa}{2}\bar{n}\left[\rho_f(t) a a^\dagger - a^\dagger \rho_f(t) a\right] + \text{adj}, \tag{55}$$

where $\bar{n}$ is now the number of thermal excitations at the frequency Ω of the harmonic oscillator. At zero temperature, this equation predicts that the mean number of quanta $\langle a^\dagger a \rangle$ in the damped oscillator decays at the rate κ, while the expectation value of the annihilation operator a — which, we recall, is proportional to the positive frequency part of the electric field in the case of a single mode electromagnetic field — decays at the rate $\kappa/2$. We shall use this master equation later on to describe the damping of a single mode field in a high-Q cavity.

III. SPONTANEOUS EMISSION IN CAVITIES

A. PHENOMENOLOGY

The essential message of our discussion so far is that spontaneous emission is *not* an intrinsic atomic property, but rather results from the coupling of the atom to the vacuum modes of the electromagnetic field. As such, it can be modified by tailoring the electromagnetic environment that the atom can radiate into.[1-3] This was already realized by Purcell,[4] who noted that the spontaneous emission rate can be enhanced if an atom at rest inside a cavity is resonant with one of the cavity modes, and by Kleppner,[5] who discussed the opposite case of inhibited spontaneous emission. It has also been recognized that spontaneous emission need not be an irreversible process. Indeed, the Schrödinger equation always leads to reversible dynamics. Spontaneous decay only appears to be irreversible when the electromagnetic field modes are treated as a Markovian reservoir. If the vacuum modes cannot be approximated in this way, qualitatively different types of dynamics can be achieved. In particular, we have seen that an atom coupled to a single mode of the electromagnetic field undergoes a periodic exchange of excitation between the atom and the field. Remarkably, it has now become possible to observe such reversible behavior, both in the microwave and in the optical regime of cavity QED.[14,15]

In a generic cavity QED experiment, an atom is coupled to one, or possibly a few, cavity mode(s). These modes are in turn coupled to the external world via mirror losses and diffraction. In addition, the atom might also be coupled to a continuum of vacuum modes, in the case where the cavity is open-sided. Consequently, such an experiment is typically characterized by three coupling constants: the first one is the dipole coupling constant g between the atom and the cavity mode, the second one is the decay rate κ of this mode, and the third is the rate γ of spontaneous emission into the vacuum modes of the electromagnetic field, which differs from the free space spontaneous emission rate γ_f due to the reduction of the solid angle in which the atoms see the vacuum.[1-3] (More coupling constants would be needed in situations involving, say, several cavity modes or multilevel atoms, but such complications will not be addressed in this review.) To gain a better feeling for the respective roles of these coupling constants, consider a linear cavity which consists of a pair of opposing mirrors separated by a distance L (see Figure 1). In the absence of mirror and diffraction losses, such a cavity supports a discrete set of longitudinal standing-wave modes of frequencies $\omega_{c,n} = ck_{c,n} = c\pi n/L$, n integer, along the cavity axis, as well as transverse modes that we ignore in this discussion. We ignore as well the spatial variation of the longitudinal modes transverse to the cavity axis (the x-axis).

Assume now that these modes are well separated in frequency, so that only one of them, of frequency ω_c, is nearly resonant with the atomic transition under consideration. In the presence of mirror and diffraction losses, this mode acquires a width proportional to κ. If this width is small compared to the longitudinal mode separation, the effective density of modes $\mathcal{D}_e(\omega)$ that the atom effectively interacts with can be approximated by the Lorentzian

$$\mathcal{D}_e(\omega) = \left(\frac{\kappa}{2\pi V}\right) \frac{1}{\left(\frac{1}{2}\kappa\right)^2 + \left(\omega_c - \omega\right)^2},$$

(56)

where $\omega = ck$, and the resonator damping rate κ is related to the quality factor Q of the cavity by

$$Q = \omega_c/\kappa.$$

(57)

One can estimate the spontaneous emission rate in such a geometry by follow-ing the arguments of the Weisskopf-Wigner theory, except that the free space density of modes is now replaced by $\mathcal{D}_c(\omega)$. For a cavity tuned near the atomic resonance frequency ω_0, one finds readily

$$\gamma_c \simeq \gamma_f \left(\frac{\lambda_0^3}{V} \right) \varrho,$$

$$(58)$$

while for a cavity detuned from the atomic transition

$$\gamma_c \simeq \gamma_f \left(\frac{\lambda_0^3}{V} \right) \varrho^{-1}.$$

$$(59)$$

These approximate results clearly show the appearance of enhanced, respec-tively inhibited spontaneous emission, since λ_0^3 is less than or of the order of V. However, they do not treat properly the geometrical factors in the scalar product between the atomic dipole moment and the field polarization, which will be discussed in the next section. In addition, they assume that the coupling between the atom and the cavity modes can be treated to second order in perturbation theory, consistently with the Born-Markov approximation. Fi-nally, they do not take into account the fact that in case the cavity is not closed, the atom also interacts with a continuum of vacuum modes, leading to an additional decay rate γ'. We now proceed to put these results on a firmer basis by introducing a master equation that describes both cavity damping and the coupling of the atom to a reduced set of continuum modes.

B. GENERALIZED MASTER EQUATION FOR THE ATOM-CAVITY SYSTEM

We assume as before that one single atom is at rest inside a cavity of dimensions sufficiently small that the axial mode frequency separation $c/2L$ is large compared to the atomic transition frequency. Under these conditions, and neglecting the effects of axial modes, the atom effectively interacts with a single cavity mode of frequency $\omega_c \approx \omega_0$. In the spirit of Section II.D, we treat the subsystem consisting of the atom coupling to this single mode field as a small system, described by the Hamiltonian $H_s = H_a + H_f + H_{a-f}$ of Section II.A. We consider a standing-wave quantization scheme, so that $g(x) = g \cos(qx)$, where $q = \omega_c/c$ and

$$g = \sqrt{\hbar \omega_c / \varepsilon_0 V} \left(\wp \cdot \varepsilon \right)$$

$$(60)$$

and using a system of reference which rotates at the cavity mode frequency, this Hamiltonian may be re-expressed as

$$H_s = -\hbar \delta \sigma_z + H_{a-f},$$

(61)

where

$$H_{a-f} = \hbar g \, \cos(qx)\left(a^\dagger \sigma_- + \sigma_+ a\right),$$

(62)

and $\delta \equiv \omega_c - \omega_0$ is the atom-cavity detuning. Recall that in general, the atom-field coupling is location dependent, as already indicated in Section II.A. We assume for now an atom at rest at position $x = 0$, but shall return to this important point in Section IV.

In addition to its Hamiltonian evolution, the small atom-cavity mode system is influenced by two dissipative processes. The first one is associated with the coupling of the atom to the free space electromagnetic field background, and the second one to the coupling of the cavity mode to the outside world via mirror losses and diffraction — see Figure 1. The first process is particularly important in open cavities, and results in an incoherent decay of the excited electronic state of the atom á la Weisskopf-Wigner, while mirror losses and diffraction lead to the irreversible escape of cavity photons. Due to the additive nature of these dissipation mechanisms, the reservoir theory summarized in Section II.D allows us to treat them straightforwardly: the small atom-cavity mode system is coupled to two thermal reservoirs that model the electromagnetic background and the mirror losses as continua of harmonic oscillators. In the Born-Markov approximation, the coupling of the small system to these two reservoirs is described by a master equation whose non-Hermitian part is given by the sum of two terms given by Equations 48 and 55, respectively. Since the atom interacts only with a subset of the free space modes, determined by the solid angle over which the atom "sees" that background, the free space spontaneous emission rate γ_f is now replaced by the smaller γ'. In analogy with Equation 51, the atom-cavity mode master equation is then given at zero temperature by

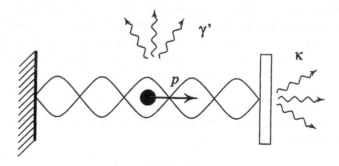

FIGURE 1 The Fabry-Pérot cavity, a standing-wave mode of frequency ω_c, and the atom with momentum p along the cavity axis. The rates γ' and κ account for atomic losses into the free space background and cavity losses due to imperfect mirrors, respectively.

$$\dot{\rho}_s = -\frac{i}{\hbar}\left[H_{eff}\rho - \rho H_{eff}^\dagger\right] + \kappa a \rho_s a^\dagger + \gamma' \sigma_- \rho_s \sigma_+, \tag{63}$$

where

$$H_{eff} = H_s + H_{loss} \tag{64}$$

and

$$H_{loss} = -i\hbar\frac{\gamma'}{2}\sigma_+\sigma_- - i\hbar\frac{\kappa}{2}a^\dagger a. \tag{65}$$

Spontaneous emission involves the decay of an initially excited atom, with no photon in the cavity mode, so that the relevant initial condition is as before

$$|\psi(0)\rangle = |e,\, 0\rangle. \tag{66}$$

The Hamiltonian part of the small system evolution involves the exchange of excitation between the atom and the cavity mode, but as we have seen, the total number of excitations in the system, one in the present case, remains constant. In contrast, the coupling to the reservoirs involves an irreversible loss of excitation from the small system. Consequently, there are only three relevant states involved in its dynamics, the "one-quanta" states $|e, 0\rangle$ and $|g, 1\rangle$, and the "zero-quanta" state $|g, 0\rangle$: the total evolution of the atom-cavity mode system is restricted to the one- and zero-quanta subspaces of the associated Hilbert space. Since the coupling between these two subspaces irreversibly populates the zero-quanta subspace, we can follow an argument similar to that of Section II.D, Equation 54, and introduce a general *unnormalized* one-quantum state

$$|\psi(t)\rangle = c_e(t)e^{\frac{1}{2}\delta t}|e,\, 0\rangle + c_g(t)e^{\frac{1}{2}\delta t}|g,\, 1\rangle. \tag{67}$$

The evolution of this state is readily seen from the master Equation 63 to be governed by the effective Schrödinger equation

$$i\hbar\frac{d|\psi(t)\rangle}{dt} = H_{eff}|\psi(t)\rangle, \tag{68}$$

H_{eff} being given by Equation 64. The corresponding equations of motion for the probability amplitudes $c_e(t)$ and $c_g(t)$ are

$$\frac{dc_e(t)}{dt} = -(\gamma'/2)c_e(t) - igc_g(t), \tag{69}$$

$$\frac{dc_g(t)}{dt} = -(i\delta + \kappa/2)c_g(t) - igc_e(t).$$

(70)

For the closed cavities typically used in microwave experiments, $\gamma' \approx 0$.

At this point, it is useful to distinguish between two qualitatively different regimes. The first one is characterized by the fact that the irreversible decay rates κ and γ' dominate over the Hamiltonian dipole interaction between the atom and the cavity mode, whose strength is given by g. This is traditionally called the *weak coupling regime,* or bad cavity regime. In contrast, the *strong coupling regime,* or good cavity regime, is characterized by the fact that the coherent interaction between the atom and the cavity mode dominates over the irreversible decay mechanisms. In the closed cavities often used in microwave experiments, $\gamma' \approx 0$, so that the strong coupling regime corresponds to $g \gg \kappa$ and the weak coupling regime to $g \ll \kappa$. In contrast, most optical cavities encompass only a small fraction of the free-space solid angle 4π, so that $\gamma' \approx \gamma_f$, the free space spontaneous emission rate. In this case, the strong coupling regime is defined by $g \gg \gamma',\kappa$, and the weak coupling regime by $\gamma',\kappa \gg g$.

C.   WEAK COUPLING REGIME

Consider first the weak coupling regime, characterized by $g \ll \gamma',\kappa$ for open cavities and $g \ll \kappa$ and $\gamma' \approx 0$ for closed microwave cavities. Formally integrating Equation 70 yields

$$c_g(t) = -ig \int_0^t dt' c_e(t') e^{-(i\delta + \kappa/2)(t-t')}.$$

(71)

Equations 69 and 70 show that $c_e(t)$ will be a slow variable provided that g and γ' are small compared to $|\delta| + |\kappa|/2$. Under these conditions, we can remove $c_e(t')$ from the integral and evaluate it at time t. The remaining integral gives, for $t \gg \kappa^{-1}$,

$$c_g(t) = \frac{-ig}{i\delta + \kappa/2} c_e(t),$$

(72)

and after substitution into Equation 69,

$$\frac{dc_e(t)}{dt} = -\left[(\gamma'/2) + \frac{g^2(\kappa/2 - i\delta)}{\delta^2 + \kappa^2/4}\right] c_e(t).$$

(73)

Hence, the upper electronic state population $P_e(t) = |c_e(t)|^2$ undergoes a simple exponential decay at the rate

$$\gamma = \gamma' + \gamma_0, \tag{74}$$

where

$$\gamma_0 = \left(\frac{2g^2}{\kappa}\right)\frac{1}{1 + 2(2\delta/\kappa)^2}. \tag{75}$$

The decay rate of the upper atomic level is therefore exponential, and is given by the sum of a "vacuum field" component and a "cavity" component. In the free space limit, $\kappa \to \infty$ and $\gamma' \to \gamma$, so that γ reduces as it should to the Weisskopf-Wigner result γ_f.

1.   Enhanced Spontaneous Emission

Consider in particular the case of a closed microwave cavity, with $\gamma' = 0$ and atoms at resonance with the field mode, $\delta = 0$. Γ_0 then reaches its maximum value $\Gamma_{max} = 2g^2/\kappa$. Expressing the dipole matrix element g in terms of the free space spontaneous emission rate γ_f of Equation 23 with the help of Equations 60 and 57, this rate may be re-expressed as[46]

$$\Gamma_{max} = \frac{3Q}{4\pi^2}\left(\frac{\lambda_0^3}{V}\right)\gamma_f, \tag{76}$$

where $\lambda_0 = 2\pi c/\omega_0$. This result has the same functional dependence as the approximate expression (58), but with the difference that by re-expressing g in terms of γ_f, we automatically take into account the geometrical factors resulting from the scalar product between the atomic dipole moment and the field polarization, a step which was ignored in the approximate expression (58).

For sufficiently high quality factors and transition wavelengths comparable to the linear dimensions of the cavity, this expression predicts a considerable *enhancement of the spontaneous emission rate* as compared to its free space value.

2.   Inhibited Spontaneous Emission

Equation 75 also predicts an *inhibition of spontaneous emission*[5] for atoms far detuned from the mode frequency ω_c. For instance, for $|\delta| = \omega_0$, we have for $Q \gg 1$

$$\Gamma_0 \simeq \Gamma_{max}\left(\frac{1}{4Q^2}\right) = \frac{3}{16\pi^2 Q}\left(\frac{\lambda_0^3}{V}\right)\gamma_f. \tag{77}$$

For large quality factors, it is therefore possible to switch off spontaneous emission almost completely.

It may be useful at this point to make a general comment about the dependence of the spontaneous emission rate on the cavity density of modes. The mode structure depends on the boundary conditions imposed by the cavity, and one might wonder how the atom can initially "know" that it is inside a cavity, rather than in free space. Is there some action at a distance involved, and if not, what is the mechanism through which the atom learns of its environment? The single mode theory presented so far does not permit to answer this question, as it does not permit to describe the propagation of wave packets along the cavity axis. Using a proper multimode theory, Parker and Stroud[16] and Cook and Milonni[17] showed that there is a simple answer to this question. In a real cavity, the excited atom starts to spontaneously decay while radiating a multimode field that propagates away from it. Eventually, this field encounters the cavity walls, which reflect it. The reflected field acts back on the atom, carrying information about the cavity walls as well as about the state of the atom itself at earlier times. Depending on the phase of this field relative to the phase of the atomic polarization, it will then either accelerate or prevent further atomic decay. An alternative way to think of this problem, first discussed in Reference 18, is the image method, which replaces the mirror cavity by a string of virtual images. This method, which is valid for cavities of dimensions $L \ll c/\gamma_p$, leads to the same results as the mode expansion results. Clearly, such argumentation indicates that the initial stages of the atomic decay are not exponential, in contrast to the result (75). However, recall that our approximations are not valid for very short times: a more careful treatment is necessary to describe the early emission stages where the atom "learns" about its environment.

Equations 76 and 77 might indicate that a transition wavelength comparable to the cavity size is necessary to obtain significant enhancement or inhibition of spontaneous emission. This turns out to be incorrect, however, and results from an oversimplified description of the cavity modes neglecting transverse effects. In particular, in the case of a confocal resonator of length L, and for Gaussian modes of waist $\omega_0 = \sqrt{L\lambda/\pi}$, the possible wavelengths are given by $L = (q + 1/4)\lambda/2$, where q is an integer, and the mode volume is $V = \pi\omega_0^2 L/4 = (q + 1/4)^2\lambda^3/16$.[19] In this case, the wavelength dependence in Equations 76 and 77 largely disappears, demonstrating that wavelength size cavities are not required in general to observe enhanced and inhibited spontaneous emission.

D. STRONG COUPLING REGIME

When the coupling between the atom and the cavity mode is so strong that a photon emitted into the cavity is likely to be reabsorbed before it escapes, a perturbative analysis of the coupling between the atom and the cavity mode ceases to be justified. The general solution of Equations 69 and 70 for arbitrary g, γ', and κ is of the form

$$c_e(t) = c_{e1}e^{\alpha_1 t} + c_{e2}e^{\alpha_2 t} \tag{78}$$

where

$$\alpha_{1,2} = -\frac{1}{2}\left(\frac{\gamma'}{2}+\frac{\kappa}{2}+i\delta\right)\pm\frac{1}{2}\left[\left(\frac{\gamma'}{2}+\frac{\kappa}{2}+i\delta\right)^2-4g^2\right]^{1/2}$$

(79)

and the constants c_{e1}, c_{e2} are determined from the initial conditions $c_e(0) = 1$, $c_g(0) = 0$. In the strong coupling regime $g \gg \gamma'$, κ, and δ, these exponents are

$$\alpha_{1,2} = -\frac{1}{2}\left(\frac{\gamma'}{2}+\frac{\kappa}{2}+i\delta\right)\pm ig.$$

(80)

The amplitude of the imaginary part of the exponents (80) is much larger than that of the real part, so that the evolution of the upper state population will consist of oscillations at the vacuum Rabi frequency which slowly decay in time. This dependence is illustrated in Figure 2. In this regime, the spectrum of spontaneous emission ceases to be a simple Lorentzian, as is the case in the weak coupling regime or in free space. Rather, a simple Fourier transform shows that this spectrum now consists of a doublet of Lorentzian lines of equal widths $(\gamma' + \kappa)/4$ and split by the vacuum Rabi frequency $2g$.[20] Under such conditions one should observe reversible Rabi oscillations slowly decaying in time. Figure 2 shows the time dependence of the probability of atom to be in excited electronic state for both weak coupling (a) and strong coupling (b) regimes.

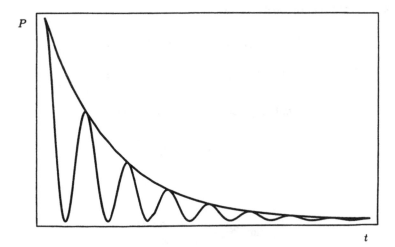

FIGURE 2 Atomic excited state probability. The exponentially decaying curve is for the weak coupling regime, and the damped oscillations correspond to the strong coupling regime.

1.   Dressed States Interpretation

A simple physical interpretation of the spontaneous emission spectrum can be obtained if one adopts the point of view that the atom-cavity mode small system represents a single quantum system, and by analyzing its eigenenergy spectrum. This small system is described by the Jaynes-Cummings Hamiltonian (2), and is sometimes referred to as a "Jaynes-Cummings atom", or a "Jaynes-Cummings molecule". We have previously mentioned that the Jaynes-Cummings Hamiltonian conserves the total number of excitations in the system, that is, it couples only the states $|e, n\rangle$ and $|g, n + 1\rangle$. Hence, we can consider the atom-field interaction for each manifold $\mathscr{E}_{n+1} = \{|e, n\rangle, |g, n + 1\rangle\}$ independently and decompose H into the sum

$$H = \sum_n H_n,$$

(81)

where H_n acts only in the $(n + 1)$-quanta space. In the $(|e, n\rangle, |g, n + 1\rangle)$ basis, H_n takes the form of a 2×2 matrix which can readily be diagonalized to yield the eigenenergies

$$E_{2n} = \hbar\left(n + \frac{1}{2}\right)\omega_c - \hbar R_n$$

$$E_{1n} = \hbar\left(n + \frac{1}{2}\right)\omega_c + \hbar R_n,$$

(82)

and corresponding eigenvectors

$$|2n\rangle = -\sin\theta_n|e, n\rangle + \cos\theta_n|g, n+1\rangle$$

$$|1n\rangle = \cos\theta_n|e, n\rangle + \sin\theta_n|g, n+1\rangle.$$

(83)

Here R_n is the n-photon generalized Rabi frequency

$$R_n = \frac{1}{2}\sqrt{\delta^2 + 4g^2(n+1)},$$

(84)

and

$$\tan 2\theta_n = -\frac{2g\sqrt{(n+1)}}{\delta}.$$

(85)

The eigenstates (83) are the so-called dressed states of the atom, that is, the eigenstates of the atom dressed by the cavity mode. In contrast, the states $|e, n\rangle$,

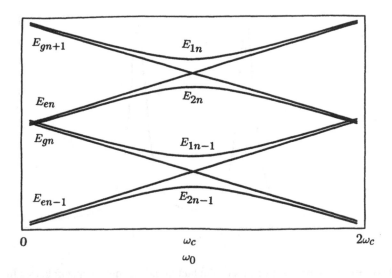

FIGURE 3 Two manifolds of eigenstates of the Jaynes-Cummings model, as a function of the atomic transition frequency ω_0. The anticrossings correspond to the resonance condition $\omega_0 = \omega_c$, and the straight lines correspond to the bare states of the system.

$|g, n + 1\rangle$ are called bare states. The zero-quanta manifold $\mathscr{E}_0$ has a single eigenstate $|g, 0\rangle$, with eigenenergy $E_0 = 0$.

The dressed state eigenvalues, or, in other words, the energy spectrum of the Jaynes-Cummings molecule, are illustrated as solid lines in Figure 3 as a function of the atom-field detuning. The dashed lines are the corresponding bare state energies. The energies of the bare levels $|e, n\rangle$ and $|g, n + 1\rangle$ cross at $\delta = 0$, but the atom-field interaction removes this degeneracy, causing the dressed states $|1, n\rangle$ and $|2, n\rangle$ to repel each other, or anticross. The energy separation between dressed states belonging to the same manifold is $E_{1n} - E_{2n} = R_n$, which has its minimum value $\left|2g\sqrt{(n+1)}\right|$ at resonance $\delta = 0$.

Since the dressed states are eigenstates of the Jaynes-Cummings Hamiltonian, it follows from the master Equation 63 that in the dressed states basis, the only transitions induced by dissipation occur between manifolds with n excitations and manifolds with $n - 1$ excitations; there are no transitions inside a given manifold, nor between manifolds differing by more than one quantum. In addition, at zero temperature there are no upward transitions from an n-quantum to an $(n + 1)$-quantum manifold. For $n > 1$, four transitions are possible between the n-quantum and $(n - 1)$-quantum manifold, corresponding to $|1, n\rangle \rightarrow |1, n-1\rangle$, $|2, n\rangle \rightarrow |2, n-1\rangle$, $|1, n\rangle \rightarrow |2, n-1\rangle$, and $|2, n\rangle \rightarrow |1, n-1\rangle$. The corresponding energies are $\hbar\omega_c$, $\hbar\omega_c$, $\hbar(\omega_c + R_n)$ and $\hbar(\omega_c - R_n)$. As is well known, these four transitions correspond to the triplet structure observed in the resonance fluorescence of two-level atoms resonantly driven by a classical field for which $n \simeq n-1$, with a central peak at ω_c and two sidebands at $\omega_c \pm R_n$.

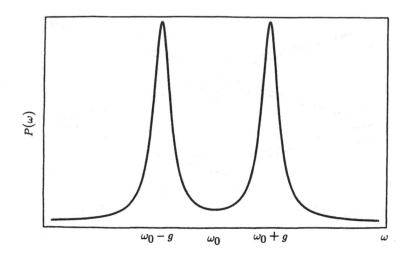

FIGURE 4   Spontaneous emission doublet in the one-quanta manifold as a function of frequency.

The situation is of course different in the case of spontaneous emission, since the Jaynes-Cummings molecule is now initially in the one-quantum manifold $\mathscr{E}_1$, see Figure 4. There are now only two allowed transitions, $|1, 0\rangle \rightarrow |g, 0\rangle$ and $|2, 0\rangle \rightarrow |g, 0\rangle$, with frequencies $-\delta/2 + R_0$ and $-\delta/2 - R_0$, consistently with Equation 80. On resonance $\delta = 0$, the frequency separation between these two transitions is precisely the vacuum Rabi frequency $2g$.

Atomic and field relaxation processes can be expressed in the dressed states basis in a particularly simple way, provided that the strong coupling condition is satisfied. It is then possible to decouple the evolution of the dressed states population from that of the coherences between states of a given manifold, the so-called secular approximation.[21] The relaxation of the dressed state populations then occurs with decay rates depending on γ', κ as well as on the angle θ_0, and may be expressed in the simple form

$$\Gamma_{1,0} = \gamma' \cos^2\theta_0 + \kappa \sin^2\theta_0$$

$$\Gamma_{2,0} = \gamma' \sin^2\theta_0 + \kappa \cos^2\theta_0. \tag{86}$$

Hence, the Jaynes-Cummings molecule relaxes either because the atom decays via emission of a photon into the continuum of vacuum modes at rate γ' (γ'-contribution), or via the escape of a cavity photon (κ-contribution). The precise relationship between these contributions depends upon the mixing angle θ_0, i.e., upon the atom-cavity detuning. At resonance ($\theta_0 = \theta_n = \pi/4$), the decay rate of the dressed states is just the average of the cavity and atomic decay rates.

2.   Vacuum Doublet

It is interesting to note that one can perform the spectroscopy of the Jaynes-Cummings molecule in precisely the same way that one would proceed for a conventional atom or molecule. A standard technique is the use of probe spectroscopy, which is achieved by coupling the molecule to a weak, tunable monochromatic field. We assume that the probe can be treated as a classical field $E_p \exp(-i\omega t)$, so that the total system is described with a time-dependent Hamiltonian

$$H(t) = H_a + H_f + H_{a-f} + V(t), \tag{87}$$

where

$$V(t) = -\hbar g_p \left(\sigma_+ e^{-i\omega t} + h.c. \right) \tag{88}$$

is a time-dependent perturbation describing the coupling of the atom to the probe field, and $g_p = d_{eg} E_p / \hbar$ is the Rabi frequency associated to the probe. The matrix elements of $V(t)$ between the zero- and one-quantum manifolds of the Jaynes-Cummings molecule are

$$\langle g, \, 0 | V(t) | 1, \, 0 \rangle = -\hbar g_p e^{i\omega t} \cos \theta_0,$$

$$\langle g, \, 0 | V(t) | 2, \, 0 \rangle = -\hbar g_p e^{i\omega t} \sin \theta_0. \tag{89}$$

Assuming that the atom-cavity system is now prepared in its ground state $|g, 0\rangle$ — a situation corresponding to absorption spectroscopy, rather than the spontaneous emission problem discussed so far — one can readily evaluate the power $P(\omega)$ absorbed by the atom-cavity system. Obviously, $P(\omega)$ exhibits two resonance lines at the frequencies corresponding to the energy differences between $|g, 0\rangle$ and the two levels of the first excited manifold, their intensities being proportional to $\cos^2\theta_0$ and $\sin^2\theta_0$, respectively, and their widths being equal to the relaxation rates $\Gamma_{1,0}$ and $\Gamma_{2,0}$. On resonance, this probe absorption doublet is symmetrical, with line splitting given as before by the vacuum Rabi frequency $2g$.

While on resonance, the atomic and field characters of the Jaynes-Cummings molecule carry equal weights in their contributions to the fluorescence (or absorption) doublet, the situation is quite different far off resonance, ($|\delta/g| \gg 1$).[1] Consider first the case $-\delta/g \gg 1$, i.e., $\tan\theta_0 \simeq \theta_0 = -g/\delta$, or $\cos\theta_0 \simeq 1$ and $\sin\theta_0 \simeq -g/\delta$. In this case the dressed states of the system almost coincide with its bare states. From (83) one readily finds

$$|1, 0\rangle \simeq |e, 0\rangle - \frac{g}{\delta}|g, 1\rangle$$

$$|2, 0\rangle \simeq \frac{g}{\delta}|e, 0\rangle + |g, 1\rangle \tag{90}$$

with corresponding eigenenergies

$$E_{10} = \hbar\omega_0 - \frac{\hbar g^2}{\delta}$$

$$E_{20} = \hbar\omega_c + \frac{\hbar g^2}{\delta} \tag{91}$$

and damping rates

$$\Gamma_{10} \simeq \gamma', \ \Gamma_{20} \simeq \kappa. \tag{92}$$

From these expressions, it follows that the state $|1, 0\rangle$ is predominantly atomic in nature, with an eigenenergy close to that of the bare excited atom, and accordingly suffers a decay rate given almost exactly by the spontaneous atomic decay rate γ'. In contrast, the eigenstate $|2, 0\rangle$ is predominantly cavity-like, with an eigenenergy close to $\hbar\omega_c$ and a damping rate close to the cavity damping rate κ. In that limit, the absorption spectrum of the Jaynes-Cummings molecule consists of a strong "atomic" line centered at a frequency close to ω_0, and a weak "field" line centered close to ω_c. The strong line is essentially the atomic absorption line, slightly shifted by the atomic coupling to the cavity mode by an amount $\Delta = -g^2/\delta$. Clearly, the situation is just the reverse for $\delta/g \gg 1$.

This discussion is summarized in Figure 5, which shows the dependence of the probe absorption spectrum on the detuning δ. As $-\delta > 0$ decreases, the atomic absorption and cavity absorption lines first move toward each other, the splitting reaching the minimum value at exact resonance. The lines then move apart as δ becomes positive.

IV. VELOCITY-DEPENDENT SPONTANEOUS EMISSION

Spontaneous emission by moving atoms in cavities presents a number of new features, as compared to the situation of atoms at rest.[22-29] Intuitively, one can understand why this is so by recalling that as we have seen in Section II, the dipole coupling constant $g = g \cos(qx)$ between the atom and the cavity mode is spatially dependent. This has a number of consequences on the radiative properties of moving atoms. The most immediate of these consequences results from the simple fact that one can switch from the weak to the strong coupling regime by changing the location of the atom inside the cavity. Hence, an atom moving along the axis of the resonator will alternatively experience both these situations. In addition to this modulation of the dipole coupling constant, the Doppler shift can play an important role: for instance, an atom whose transition frequency is below the cavity cutoff, and would be decoupled from the cavity field when at rest, may be brought into resonance by a suitable choice of velocity, thereby greatly enhancing its coupling to the cavity field. Before proceeding to a discussion of these effects, though, we return for a moment to the case of spontaneous emission in free space to determine whether motional effects are possible there. In addition, this discussion will allow us to set the stage for the case of moving atoms in cavities.

A. FREE SPACE SPONTANEOUS EMISSION
AND ATOMIC MOTION

We have seen in Section II that an excited two-level atom traveling in free space through the electromagnetic vacuum spontaneously undergoes a transition to its ground state, thereby emitting a photon of energy $\hbar\omega_0$ and of wave number $k = |\mathbf{k}| = \omega_0/c$, where ω_0 is the atomic transition frequency. There is, however, another aspect to this problem that we have neglected so far: As a photon is emitted by the atom, momentum conservation requires that the corresponding photon momentum $\hbar\mathbf{k}$ be compensated by a recoil of the atomic center-of-mass motion. In order to properly account for energy *and* momentum conservation, the master Equation (50) describing spontaneous emission in free space must therefore be modified to take into account the center-of-mass motion of the atom. The goal of this section is to derive this master equation, using general conservation and invariance arguments.

If atomic recoil is neglected, the master Equation (50) implies that the effects of spontaneous emission assume the simple form

$$\dot{\rho}_{ee} = -\gamma_f \rho_{ee}, \tag{93}$$

$$\dot{\rho}_{eg} = -\frac{\gamma_f}{2} \rho_{eg}, \tag{94}$$

$$\dot{\rho}_{ge} = -\frac{\gamma_f}{2} \rho_{ge}, \tag{95}$$

$$\dot{\rho}_{gg} = \gamma_f \rho_{ee}. \tag{96}$$

where ρ_{ee} and ρ_{gg} are the populations of the excited and ground state, respectively, and ρ_{eg}, ρ_{ge} are the coherences between electronic levels. When atomic recoil is included, however, the quantities ρ_{ab} become *operator-valued* instead of complex numbers, due to the inclusion of the atomic center-of-mass degrees

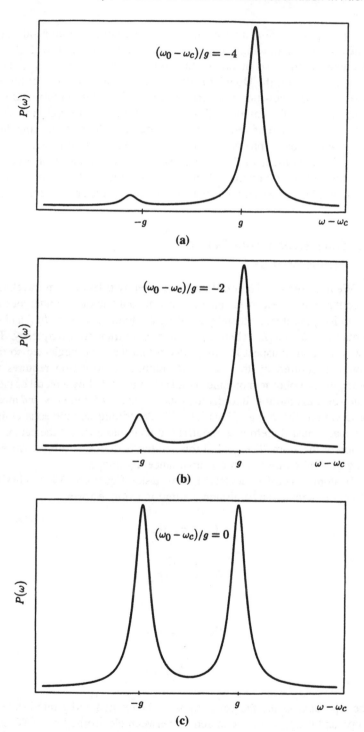

FIGURE 5   Dependence of the probe absorption spectrum on the atom-field detuning $\delta = \omega_0 - \omega_c$. The five curves are labeled by the detuning in units of g.

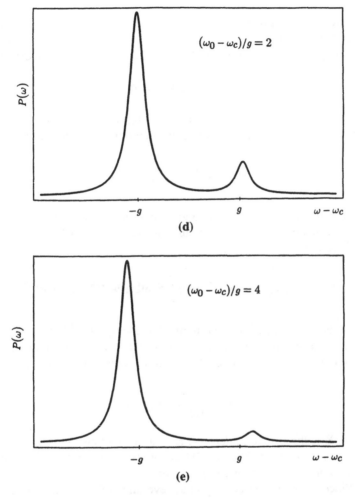

FIGURE 5 (continued)

of freedom. Nonetheless, the depletion of the excited state population ρ_{ee} and the decay of the electronic coherences ρ_{eg} and ρ_{ge} are still described by Equations 93 to 95. This is because the decay of the excited state neither depends on the motional state of the atom (Galilean invariance), nor can it change the momentum of the excited state (momentum conservation). The only modification appears in the equation for the electronic ground state population, Equation 96, as atomic recoil leads to transitions between electronic states of different center-of-mass momenta.

Indeed, an excited atom with momentum $\mathbf{P}$ decays into its ground state with shifted momentum $\mathbf{P} - \hbar\mathbf{k}$. Such a momentum change is most conveniently described by the momentum shift operator $e^{-i\mathbf{k}\hat{\mathbf{R}}}$, which acts on the center-of-mass state vector as $e^{-i\mathbf{k}\hat{\mathbf{R}}}|\mathbf{P}\rangle = |\mathbf{P} - \hbar\mathbf{k}\rangle$, where $|\mathbf{P}\rangle$ is an eigenstate of the momentum operator. Writing $\mathbf{k} = k\mathbf{n}$, the contribution of spontaneous emission in the **n**-direction to the increase in population of the electronic ground state is then given by $d\gamma_n e^{-ik\mathbf{n}\hat{\mathbf{R}}} \rho_{ee} e^{ik\mathbf{n}\hat{\mathbf{R}}}$, where $d\gamma_n$ is the differential rate of spontaneous emission in the **n** direction and may be expressed as

$$d\gamma_n = \gamma_f \Phi(\mathbf{n}) d^2\mathbf{n}, \tag{97}$$

Here, we have introduced the probability $\Phi(\mathbf{n})d^2\mathbf{n}$ of emission into the infinitesimal solid angle $d^2\mathbf{n}$ in direction **n**. Integrating over all directions yields

$$\dot{\rho}_{gg} = \int d\gamma_n e^{-ik\mathbf{n}\hat{\mathbf{R}}} \rho_{ee} e^{ik\mathbf{n}\hat{\mathbf{R}}}. \tag{98}$$

Inserting Equations 93 to 95 and 98 into Equation 50, we finally obtain the master equation describing spontaneous emission by a freely traveling two-level atom as[30-32]

$$\dot{\rho}_A = -\frac{i}{\hbar}\left[H_{eff}\rho_A - \rho_A H_{eff}^\dagger\right] + \gamma \int d^2\mathbf{n}\Phi(\mathbf{n})e^{-ik\mathbf{n}\hat{\mathbf{R}}}\sigma_-\rho_f\sigma_+ e^{ik\mathbf{n}\hat{\mathbf{R}}}, \tag{99}$$

where the non-Hermitian "Hamiltonian" $H_{eff} = H_a - i\hbar\frac{\gamma_f}{2}\sigma_+\sigma_-$ now accounts for the center-of-mass atomic motion as well as for the irreversible decay of the excited electronic state and of the electronic coherences, that is,

$$H_a = \frac{\hat{\mathbf{P}}^2}{2M} + \hbar\omega_0\sigma_+\sigma_- \tag{100}$$

where $\hat{\mathbf{P}}$ is the atomic center-of-mass momentum and M is the atomic mass. The integral term in Equation 99 accounts for the irreversible increase in population of the electronic ground state.

Since we are interested only in the evolution of the excited electronic state, we can as before introduce an effective Schrödinger equation for the unnormalized excited state ket-vector in the center-of-mass Hilbert space of the atom,

$$i\hbar\left|\dot{\phi}_e\right\rangle = \left[\hat{T} + \hbar\left(\omega_0 - i\frac{\gamma_f}{2}\right)\right]\left|\phi_e\right\rangle, \tag{101}$$

where $\hat{T} = \hat{\mathbf{P}}^2/2M$. The formal solution of Equation 101,

$$\left|\phi_e(t)\right\rangle = \exp\left\{-\frac{i}{\hbar}\left(\hat{T} + \hbar\omega_0\right)t - \frac{\gamma_f}{2}t\right\}\left|\phi(0)\right\rangle \tag{102}$$

readily yields the decay of the upper electronic state population as $P(t) \equiv \langle\phi_e(t)|\phi_e(t)\rangle = e^{-\gamma_f t}\langle\phi_e(0)|\phi_e(0)\rangle$. As expected, this decay is purely exponential and independent of the center-of-mass state of the atom.

B. ANGULAR PHOTON DISTRIBUTION

In the preceding discussion we applied general symmetry and invariance arguments to obtain the master equation describing spontaneous emission by moving atoms in free space. Alternatively, one might attempt to derive such a master equation by treating the center-of-mass position **R** as a true dynamical variable. The dipole interaction then plays the role of a potential energy, giving rise to the dipole force in the atomic center-of-mass equation of motion $M\ddot{\mathbf{R}} = \nabla_{\mathbf{R}}[\mathbf{d} \cdot \mathbf{E}(\mathbf{R},t)]$. This approach has its flaws, however. In a recent paper[33] Wilkens showed that the naive dipole interaction leads to a nonphysical atomic velocity dependence in the angular distribution of photons spontaneously emitted by a moving atom. Somewhat disturbingly, the nonphysical terms of first order in the atomic velocity, i.e., of the same order as the linear Doppler shift of frequencies. The usage of the dipole interaction for moving atoms was also criticized by Baxter et al.,[34] who showed that it violates energy-mentum conservation, as well as the gauge invariance of radiation induced mechanical forces. However, as these authors demonstrated, these deficiencies are cured by including the effects of the so-called Röntgen interaction $H_{int}^{r\ddot{o}} = -\mathbf{d} \cdot \left(\dot{\mathbf{R}} \times \mathbf{B} \right)$ of the moving atom with the magnetic component of the radiation field. This interaction results from the fact that the electromagnetic field experienced by the moving atom differs from the fields as measured in the laboratory frame, both fields being related by a suitable Lorentz transformation.

Taking into account atomic motion, the interaction of a two-level atom and the electromagnetic field is therefore properly described by the Hamiltonian

$$H = H_a + H_f + H_{a-f},\tag{103}$$

where H_a is given by Equation 100, the free field Hamiltonian

$$H_f = \sum_{k,\lambda} \hbar\omega_k a_{k\lambda}^\dagger a_{k\lambda},\tag{104}$$

and the atom-field interaction is

$$H_{a-f} = -\hat{\mathbf{d}} \cdot \hat{\mathbf{E}}_\perp\left(\hat{\mathbf{R}}\right)$$

$$-\frac{1}{2M}\left\{\hat{\mathbf{P}} \cdot \left[\hat{\mathbf{B}}\left(\hat{\mathbf{R}}\right) \times \hat{\mathbf{d}}\right] + \left[\hat{\mathbf{B}}\left(\hat{\mathbf{R}}\right) \times \hat{\mathbf{d}}\right] \cdot \hat{\mathbf{P}}\right\},\tag{105}$$

where $\hat{\mathbf{E}}_\perp$ is the transverse electric field operator. The interaction involves, beside the familiar dipole contribution, a Röntgen term which results from the convective atomic motion. (Note that in Equation 104 we now include explicitly both the sum over wavevectors and the sum over polarizations.)

We consider for concreteness a $\Delta m = 0$ transition in free space. In this case it is convenient to expand the electromagnetic field in terms of linearly polarized plane waves with periodic boundary conditions, so that

$$\hat{\mathbf{E}}_\perp(\hat{\mathbf{R}}) = \sum_{k\lambda} \mathscr{E}_k \mathbf{e}_{k\lambda} \left[a_{k\lambda} e^{ik\hat{\mathbf{R}}} + h.c. \right]$$

(106)

and

$$\hat{\mathbf{B}}(\hat{\mathbf{R}}) = \frac{1}{c} \sum_{k\lambda} \mathscr{E}_k (\mathbf{n}_k \times \mathbf{e}_{k\lambda}) \left[a_{k\lambda} e^{ik\hat{\mathbf{R}}} + h.c. \right],$$

(107)

where $\mathbf{n}_k = \mathbf{k}/|\mathbf{k}|$ is a unit vector in the direction of propagation of the k-th mode, and $\mathbf{e}_{k1}$, $\mathbf{e}_{k2}$ with $(\mathbf{e}_{k1}\mathbf{e}_{k2}) = 0$ and $\mathbf{e}_{k1} \times \mathbf{e}_{k2} = \mathbf{n}_k$ are the orthogonal unit vectors associated with the two polarizations of that mode.

Using the vector-relation $(\mathbf{n}_k \times \mathbf{e}_{k\lambda}) \times \wp = (\wp \cdot \mathbf{n}_k)\mathbf{e}_{k\lambda} - (\wp \cdot \mathbf{e}_{k\lambda})\mathbf{n}_k$, and making use of the operator identity $\hat{\mathbf{P}}e^{\pm i k\hat{\mathbf{R}}} = e^{\pm i k\hat{\mathbf{R}}}[\hat{\mathbf{P}} \pm \hbar\mathbf{k}]$, the atom-field interaction Hamiltonian (105) assumes the form in the rotating-wave approximation,

$$H_{a-f} = \hbar \sum_{k\lambda} \left(g_{k\lambda}(\hat{\boldsymbol{\beta}}) e^{ik\hat{\mathbf{R}}} \sigma_+ a_{k\lambda} + h.c. \right),$$

(108)

where $\hat{\boldsymbol{\beta}} = \hat{\mathbf{P}}/(Mc)$, and $\hbar g_{k\lambda}(\hat{\boldsymbol{\beta}})$ is the operator-valued coupling

$$\hbar g_{k\lambda}(\hat{\boldsymbol{\beta}}) = \mathscr{E}_k \wp \left\{ (\mathbf{e}_\wp \cdot \mathbf{e}_{k\lambda}) \left[1 - (\mathbf{n}_k \cdot \hat{\boldsymbol{\beta}}) + \hbar\omega_k/(2Mc^2) \right] \right.$$

$$\left. - (\mathbf{e}_\wp \cdot \mathbf{n}_k)(\mathbf{e}_{k\lambda} \cdot \hat{\boldsymbol{\beta}}) \right\},$$

(109)

and $\mathbf{e}_\wp \equiv \wp/\wp$ is the unit vector in direction of the atomic dipole transition moment. We note that the $\hat{\boldsymbol{\beta}}$-dependence of the coupling (109), which is in contrast to the standard model of nonrelativistic QED, and the presence of the recoil term $\omega_k/(2Mc^2)$, are entirely due to the Röntgen interaction in Equation 105. In addition, it is worth noting that this interaction also couples the moving atom to those modes whose polarizations are perpendicular to the atomic dipole transition moment, as indicated by the last term in Equation 109.

To derive the angular distribution of spontaneously emitted photons, we consider an initially excited atom which travels with momentum $\mathbf{P}$ through the electromagnetic vacuum. Due to the atom-field interaction, the atom will eventually end up in its ground state, traveling with some momentum $\mathbf{P} - \hbar\mathbf{k}$, where $\mathbf{k}$ is the wave vector of the emitted photon. The differential count rate of photons spontaneously emitted into an infinitesimal solid angle $d^2\mathbf{n}$, centered around the direction of observation $\mathbf{n}$, can be obtained via the Fermi golden rule as

$$dy_\beta(\mathbf{n}) = 2\pi\hbar \sum_{k\lambda} |g_{k\lambda}(\boldsymbol{\beta})|^2 \delta\left\{\hbar\omega_k + \frac{(\mathbf{P} - \hbar\mathbf{k})^2}{2M} - \left[\hbar\omega_0 + \frac{\mathbf{P}^2}{2M}\right]\right\} \delta(\mathbf{n} - \mathbf{n}_k) d^2\mathbf{n}$$

(110)

where the argument of the first delta distribution accounts for energy conservation. Expanding this argument, we observe that the only contributions to the count rate Equation 110 result from frequencies which obey

$$\omega_k = \omega_0 + \omega_k \mathbf{n}_k \boldsymbol{\beta} - \frac{\hbar\omega_k^2}{2Mc^2},$$

(111)

instead of $\omega_k = \omega_0$ as was the case when atomic motion and recoil were neglected. The second term on the right-hand side stems from the nonrelativistic Doppler shift, while the third term accounts for the nonrelativistic recoil shift.

For simplicity, we restrict ourselves to the case of infinitely heavy atoms, $M \to \infty$, which allows us to drop the recoil part in Equation 111. In this limit, the coupling constant assumes the form

$$g_{k\lambda}(\boldsymbol{\beta}) = \wp\mathscr{E}_k\left\{\left(\mathbf{e}_\wp \cdot \mathbf{e}_\lambda\right)\left[1 - (\mathbf{n} \cdot \boldsymbol{\beta})\right] + \left(\mathbf{e}_\wp \cdot \mathbf{n}\right)\left(\mathbf{e}_\lambda \cdot \boldsymbol{\beta}\right)\right\}, \quad \lambda = 1, 2.$$

(112)

As in Section II, we evaluate the sum (110) in the continuum limit. Due to the presence of the directional delta function, the solid angle integration is trivial. Taking into account the frequency dependence of the coupling, the remaining frequency integral reads $I = \int \omega_k^3 \delta[\omega_k - \omega_0 - \omega_k \mathbf{n} \cdot \boldsymbol{\beta}] d\omega_k$ with the result $I = \omega_0^3[1 - \mathbf{n} \cdot \boldsymbol{\beta}]^{-4}$. The differential count rate Equation 110 becomes then

$$dy_\beta(\mathbf{n}) = \gamma_f \Phi_\beta(\mathbf{n}) d^2\mathbf{n},$$

(113)

where γ_f is the free space spontaneous decay rate of Section II, and

$$\Phi_\beta(\mathbf{n}) = \frac{3}{8\pi} \frac{1}{[1 - \mathbf{n} \cdot \boldsymbol{\beta}]^4} \times \sum_\lambda \left\{\left(\mathbf{e}_\wp \cdot \mathbf{e}_\lambda\right)\left[1 - (\mathbf{n} \cdot \boldsymbol{\beta})\right] + \left(\mathbf{e}_\wp \cdot \mathbf{n}\right)\left(\mathbf{e}_\lambda \cdot \boldsymbol{\beta}\right)\right\}^2$$

(114)

is the angular distribution density (dipole radiation pattern). Its β-dependence accounts for the aberration effects of light emitted by a moving source.

Due to the nonrelativistic nature of the underlying Hamiltonian model, expression 114 is only valid to first order in $\beta = v/c$. Expanding the radiation pattern to lowest order in β, we observe the summand in the polarization sum is proportional to $(\mathbf{e}_\wp \cdot \mathbf{e}_\lambda)$. We exploit the freedom left in the transversality relation $\mathbf{e}_1 \times \mathbf{e}_2 = \mathbf{n}$ to choose $\mathbf{e}_2$ orthogonal to the dipole moment. The polarization sum in Equation 114 then becomes

$$\sum_\lambda \{\cdots\} = \left(\mathbf{e}_\wp \cdot \mathbf{e}_1\right)^2 + 2\left(\mathbf{e}_\wp \cdot \mathbf{e}_1\right)\left(\mathbf{e}_\wp \cdot \mathbf{n}\right)\left(\mathbf{e}_1\boldsymbol{\beta}\right) - 2\left(\mathbf{e}_\wp \cdot \mathbf{e}_1\right)^2(\mathbf{n} \cdot \boldsymbol{\beta}).$$

(115)

The last two terms on the right-hand side of this expression are due to the Röntgen interaction. With $(e_\wp \cdot e_1)^2 = 1 - (e_\wp \cdot n)^2$, $(e_\wp \cdot e_1)(e_1 \cdot \beta) = (e_\wp \cdot \beta) - (e_\wp \cdot n)(n \cdot \beta)$, and combining the last two terms of Equation 115 as

$$2\left(e_\wp \cdot e_1\right)\left(e_\wp \cdot n\right)\left(e_1 \cdot \beta\right) - 2\left(e_\wp \cdot e_1\right)^2 (n \cdot \beta)$$

$$= 2\left(n \cdot e_\wp\right)\left(e_\wp \cdot \beta\right) - 2(n \cdot \beta) = 2\left(\beta \times e_\wp\right) \cdot \left(e_\wp \times n\right). \tag{116}$$

Inserting Equation 115 with Equation 116 into Equation 114, and expanding the remaining prefactor, we finally obtain

$$d\gamma_\beta(n) = \gamma_f \Phi_\beta(n) d^2 n, \tag{117}$$

where

$$\Phi_\beta(n) = \frac{3}{8\pi}\left[1 - \left(e_\wp \cdot n\right)^2\right]\left[1 + 4(n \cdot \beta)\right] + \frac{6}{8\pi}\left(\beta \times e_\wp\right) \cdot \left(e_\wp \times n\right) \tag{118}$$

up to corrections of order $(v/c)^2$.

The angular distribution density of Equation 118 consists of two parts. The first part stems from the $d \cdot E$ interaction, while the second part results from the Röntgen interaction. The velocity dependence of the first part is entirely due to the Doppler shift of frequencies, Equation 111, which modifies the mode density in the frequency integral. In contrast, the velocity-dependent contribution of the Röntgen interaction is related to the modified field strengths as experienced by a moving atom. The Röntgen contribution vanishes if either the atomic dipole transition moment and the atomic velocity are parallel or the atom is observed in direction of the dipole transition. In all other cases, the Röntgen contribution is of the same magnitude as the modification of the ordinary $d \cdot E$ contribution due to the Doppler effect. Wilkens has shown that the inclusion of the Röntgen interaction leads to the Lorentz invariance of the angular distribution density of spontaneously emitted photons to first order in v/c.

C. SPONTANEOUS EMISSION BY A MOVING ATOM INSIDE A CAVITY

We now return to the problem of spontaneous emission in cavities, and consider the situation of atoms moving at some velocity v along the cavity axis. Equation 109 shows that the Röntgen part of the atom-field interaction leads to the dynamical coupling $g_{k\lambda}(\beta)$. For thermal atoms, the ensuing dynamical effect is quite small, leading primarily to a velocity dependent change in the Rabi frequency $\Delta g/g = v/c \propto 10^{-5}$. In this case, the secular Doppler shift in the denominator of the optical response function certainly dominates, and the Röntgen interaction can safely be ignored. With this approximation, the Hamiltonian of the moving atom has the form

$$H = \frac{\hat{p}_r^2}{2M} - \hbar\delta\sigma_z + H_{a-f} \tag{119}$$

with $H_{a-f} = \hbar g \cos(q\,\hat{x})(a^\dagger\sigma_+ + h.c.)$, and we consider the atomic motion along the cavity axis only. The coupling of the atom to the vacuum radiation modes, and the inclusion of the mirror and diffraction losses, lead then to the master equation for the atom-cavity mode system Equation 63

$$\dot{\rho} = -\frac{i}{\hbar}\left[H_{eff}\rho - \rho H_{eff}^\dagger\right] + \kappa a\rho a^\dagger + \gamma'\int_{-1}^{1}d\eta w(\eta)e^{-ik\hat{x}\eta}\sigma_-\rho\sigma_+e^{ik\hat{x}\eta}, \tag{120}$$

where

$$H_{eff} = H + H_{loss} \tag{121}$$

and H_{loss} is given by Equation 65. Here, we have used Equation 55 with $\bar{n} = 0$ to describe cavity damping, and spontaneous emission, including the recoil effects, is described by Equation 99. The weight function in the fill-up integral in that part of the Liouvillian describing spontaneous emission into the vacuum modes results from the reduction of Equation 99 from three to one dimension,

$$w(\eta) = \int d^2\mathbf{n}\Phi'(\mathbf{n})\delta(\eta - n_x), \tag{122}$$

where n_x is the component of the unit vector $\mathbf{n}$ along the x-direction and the prime accounts for the modification of the dipole radiation pattern due to the presence of the cavity boundaries.

We assume that the atom enters the cavity in its excited electronic state, while the cavity field is in its vacuum state. As is the case for atoms at rest, the evolution of the partial state vector that describes the electronic state of the atom and the degree of excitation of the cavity mode is restricted to the one- and zero-quanta subspaces of the associated Hilbert space. Since the coupling between these two subspaces irreversibly populates the zero-quanta subspace, we can follow an argument similar to that of Section III to describe the dynamics of the one-quantum subspace by the effective Schrödinger equation

$$i\hbar\frac{d}{dt}|\Psi\rangle = H_{eff}|\Psi\rangle, \tag{123}$$

where the one-quantum state vector is

$$|\Psi(t)\rangle = |\phi_e(t)\rangle \otimes |e,\,0\rangle + |\phi_g(t)\rangle \otimes |g,\,1\rangle. \tag{124}$$

Here $|\phi_e(t)\rangle$ and $|\phi_g(t)\rangle$ specify the center-of-mass state of the atom in the excited and ground state, respectively.

Being primarily interested in the impact of the atomic velocity on the decay of the excited electronic state, we transform the Schrödinger equation into a frame moving with the initial average velocity of the atom, $v = \langle \phi_e(0)| \hat{p}/M|\phi_e(0)\rangle$. In terms of the transformed kets

$$\left|\phi'_{e,g}\right\rangle = G_v\left|\phi_{e,g}\right\rangle, \tag{125}$$

where $G_v \equiv \exp\left\{\dfrac{i}{\hbar}\mathbf{v}\cdot\left(\hat{\mathbf{p}}t - \hat{\mathbf{x}}M\right)\right\}$ is a Galilean boost, we obtain, dropping the primes for notational clarity,[32]

$$i\hbar\left|\dot{\phi}_e\right\rangle = \left[\hat{T} - \hbar\left(\delta + i\frac{\gamma'}{2}\right)\right]\left|\phi_e\right\rangle - i\hbar g\ \cos\!\left(q\hat{x} + qvt\right)\left|\phi_g\right\rangle, \tag{126}$$

$$i\hbar\left|\dot{\phi}_g\right\rangle = \left[\hat{T} - i\hbar\frac{\kappa}{2}\right]\left|\phi_g\right\rangle + i\hbar g\ \cos\!\left(q\hat{x} + qvt\right)\left|\phi_e\right\rangle, \tag{127}$$

where $qv = \omega_c v/c$ is the Doppler shift of the cavity frequency as seen from the moving frame. Our choice of initial conditions implies $|\phi_g(0)\rangle = 0$ and $P_e(0) \equiv \langle\phi_e(0)|\phi_e(0)\rangle = 1$.

D. WEAK COUPLING REGIME

To derive an equation for the excited electronic state amplitude in the weak coupling limit, it is convenient to proceed by temporarily moving to an interaction picture defined by the unitary transformation

$$\left|\tilde{\phi}_{e,g}\right\rangle \equiv \exp\left\{\frac{i}{\hbar}\hat{T}t - i\delta t\right\}\left|\phi_{e,g}(t)\right\rangle, \tag{128}$$

in terms of which the Schrödinger equations 126 and 127 become

$$i\hbar\left|\dot{\tilde{\phi}}_e\right\rangle = -i\hbar\frac{\gamma'}{2}\left|\tilde{\phi}_e\right\rangle - i\hbar g\ \cos\!\left(q\hat{x} + \hat{\Omega}t\right)\left|\tilde{\phi}_g\right\rangle, \tag{129}$$

$$i\hbar\left|\dot{\tilde{\phi}}_g\right\rangle = \hbar\left(\delta - i\frac{\kappa}{2}\right)\left|\tilde{\phi}_g\right\rangle + i\hbar g\ \cos\!\left(q\hat{x} + \hat{\Omega}t\right)\left|\tilde{\phi}_e\right\rangle, \tag{130}$$

where

$$\hat{\Omega} = \frac{\omega_c}{c}\left(v + \frac{\hat{p}_x}{M}\right) \tag{131}$$

is the detuning operator in the moving frame. We formally integrate Equation 130 to obtain

$$\left|\tilde{\phi}_g(t)\right\rangle = g\int_0^t dt' \; \exp\left\{-i\left(\delta - i\frac{\kappa}{2}\right)(t - t')\right\}\cos\left(q\hat{x} + \hat{\Omega}t'\right)\left|\tilde{\phi}_e(t')\right\rangle. \tag{132}$$

Equation 129 shows that $\left|\tilde{\phi}_e\right\rangle$ is a slow variable provided that g and γ' are small compared with $|\delta| + |\kappa|/2$. Under these conditions, we can perform the Markov approximation in Equation 132, replacing $\left|\tilde{\phi}_e(t')\right\rangle$ by $\left|\tilde{\phi}_e(t)\right\rangle$. However, we leave the kernel cosine function intact to account for the dynamic Doppler correction to the detuning. Using the Baker-Campbell-Haussdorff formula to disentangle the operator-valued argument of the cosine, $q\hat{x} + \hat{\Omega}t'$, the integration over t' is easily carried out. The result, valid for $\kappa \gg g, \gamma'$, is

$$\left|\tilde{\phi}_g(t)\right\rangle \approx -i\frac{g}{2}\left(e^{i\left(q\hat{x} + \hat{\Omega}t\right)}\frac{1}{\delta' + \hat{\Omega} - i\kappa/2} + e^{-i\left(q\hat{x} + \hat{\Omega}t\right)}\frac{1}{\delta' - \hat{\Omega} - i\kappa/2}\right)\left|\tilde{\phi}_e(t)\right\rangle, \tag{133}$$

where

$$\delta' \equiv \omega_c - \omega_0 + \omega_r \tag{134}$$

is an effective detuning taking into account the recoil frequency $\omega_r \equiv \hbar q^2/2M$. Upon inserting Equation 133 into Equation 129 and using Equation 128 to transform back to the Schrödinger picture, we obtain[25]

$$i\hbar\left|\dot{\phi}_e(t)\right\rangle = \left[\mathcal{T} + \mathcal{V} - \hbar\left(\delta + i\frac{\gamma'}{2}\right)\right]\left|\phi_e(t)\right\rangle, \tag{135}$$

where

$$\mathcal{T} = \frac{\hat{p}_x^2}{2M} - \hbar\frac{|g|^2}{4}\left(\frac{1}{\delta' + \hat{\Omega} - i\kappa/2} + \frac{1}{\delta' - \hat{\Omega} - i\kappa/2}\right) \tag{136}$$

is a (non-Hermitian) generalized kinetic energy operator, and

$$\mathcal{V} = \hbar\frac{|g|^2}{4}\left(e^{2iq(\hat{x}+vt)}\frac{1}{\delta' + \hat{\Omega} - i\kappa/2} + e^{-2iq(\hat{x}+vt)}\frac{1}{\delta' - \hat{\Omega} - i\kappa/2}\right) \tag{137}$$

is an "optical" potential that depends on the atomic momentum through the detuning operator $\hat{\Omega}$.

The effective Schrödinger Equation 135 describes the dynamics of the upper state probability amplitude in the bad cavity limit $\kappa \gg g, \gamma'$. The separation of the effective Hamiltonian in Equation 135 into a kinetic and a potential part reveals the different roles that they play for the dynamics of the spontaneous emission. In particular, dissipation is given by their imaginary part. The imaginary part of $\mathcal{T}$ is spatially uniform but yields different loss rates for different velocities as a result of the linear Doppler effect. In contrast, the imaginary part of the optical potential $\mathcal{V}$ is spatially modulated and gives different loss rates at different positions within the cavity. In addition, the optical potential leads to the scattering of atoms into different velocity states, which in turn have different Doppler induced decay rates.

Under the assumption that the momentum spread remains small in the atomic rest frame, we may perform the Raman-Nath approximation, which consists in setting $\hat{p}_x = 0$ in the Schrödinger equation 135 for $|\phi_e\rangle$. With $\hat{x}_t \equiv \hat{x} + vt$, this gives

$$i\hbar|\dot{\phi}_e\rangle = -\hbar\left\{\delta + i\frac{\gamma'}{2} + \frac{|g|^2}{4}\left(\frac{1+e^{2iq\hat{x}_t}}{\delta' + qv - i\kappa/2} + \frac{1+e^{-2iq\hat{x}_t}}{\delta' - qv - i\kappa/2}\right)\right\}|\phi_e\rangle. \tag{138}$$

Being diagonal in the position representation, this equation is easily integrated and yields the overall upper state probability

$$P(t) = e^{-(\gamma' + \gamma_v)t}\int_{-\infty}^{\infty} dx\, e^{-\eta(x,t)}|\phi(x,\,0)|^2, \tag{139}$$

where

$$\gamma_v = \frac{2|g|^2}{\kappa}\left\{\frac{1}{1+4(\delta'+qv)^2/\kappa^2} + \frac{1}{1+4(\delta'-qv)^2/\kappa^2}\right\} \tag{140}$$

and

$$\eta(x,\,t) = \frac{\gamma_v \sin(qvt)}{qv}$$

$$\left\{\cos(2qx+qvt) + \frac{2qv}{\kappa}\frac{1-4(\delta'^2 - q^2v^2)/\kappa^2}{1+4(\delta'^2 + q^2v^2)/\kappa^2}\sin(2qx+qvt)\right\} \tag{141}$$

accounts for the local variation of the spontaneous decay due to the spatial modulation of the resonant mode function.

If the atom is initially well localized around some position x_0, the excited state probability becomes $P(t) = \exp\{-[(\gamma' + \gamma_v(t - \eta(x_0 t))]\}$. It displays an oscillatory behavior superimposed to an exponential decay. If the atom enters the cavity at a node of the field, it will not decay until its velocity brings it into a region of nonzero field strength. If, on the other hand, it enters the cavity at an antinode of the cavity mode, for example at $x_0 = 0$, its decay is described by

$\gamma' + \gamma_v$. For an atom at rest and resonant with the field mode in a closed cavity, $\gamma' = 0$, we readily recover Equation 76

$$\gamma_{v=\delta'=0} = \frac{3Q}{4\pi^2} \frac{\lambda^3}{V} \gamma_f.$$
(142)

For arbitrary velocities, the decay rate γ_v is a superposition of two Lorentzians centered around $v = \pm\delta'/q$. These two Lorentzians account for the interaction of the atom with two symmetrically Doppler-shifted counterpropagating cavity modes. The FWHM of these Lorentzians is given by $v_{cav} = c/Q$. If the atom at rest is in resonance with the cavity field, $\delta' = 0$, an increase in atomic velocity leads to a noticeable decrease in the spontaneous decay rate for velocities which are close to or exceed the characteristic cavity velocity v_{cav}. For $v = v_{cav}$ the decay due to atom-cavity coupling is half as fast as given by Expression 142. Conversely, if the cavity is strongly detuned from the atoms at rest, $\delta' > qv_{cav}$, the Doppler shift of moving atoms can tune them into resonance with one of the two counterpropagating waves, thereby increasing their spontaneous decay rate.

In general, the effect of the spatial mode variation on the decay of the atom results is an oscillation with period $\lambda/2v$. This period has to be larger than the lifetime $(\gamma' + \gamma_v)^{-1}$ in order to lead to observable effects.

E.   STRONG COUPLING REGIME

To simplify the analysis of the strong coupling regime,[24,28] we adopt again the Raman-Nath approximation, i.e., we set $\hat{p}_x \to 0$ in the Hamiltonian underlying Equations 126 and 127, and we replace the position operator by a c-number x_0, which gives the position of the atom as it enters the cavity. Under these circumstances the atom effectively moves at constant velocity along the axis of the cavity, to which it is coupled by the time-dependent coupling constant

$$g(t) = g \, \cos(qx_0 + \Omega t),$$
(143)

where $\Omega = qv$ is the linear Doppler shift induced by the atomic motion.

Having effectively eliminated the center-of-mass degrees of freedom of the atom, the evolution of the atom-cavity system in the one-quanta subspace is described by a closed set of equations for the expectation values of linear and quadratic operator products. From Equations 126 and 127 one obtains

$$\langle \dot{\sigma}_+ \rangle = -(\gamma'/2 + i\delta)\langle \sigma_+ \rangle - g(t)\langle a^\dagger \rangle,$$
(144)

$$\langle \dot{a}^\dagger \rangle = -\kappa/2\langle a^\dagger \rangle + g(t)\langle \sigma_+ \rangle,$$
(145)

and

$$\langle \dot{\sigma}_+ \sigma_- \rangle = -\gamma'\langle \sigma_+ \sigma_- \rangle - g(t)\langle a^\dagger \sigma_- + \sigma_+ a \rangle,$$
(146)

$$\langle a^\dagger a \rangle = -\kappa \langle a^\dagger a \rangle + g(t)\langle a^\dagger \sigma_- + \sigma_+ a \rangle, \tag{147}$$

$$\langle a^\dagger \sigma_- \rangle = -(\gamma'/2 + \kappa/2 - i\,\delta)\langle a^\dagger \sigma_- \rangle + g(t)\langle \sigma_+ \sigma_- - a^\dagger a \rangle. \tag{148}$$

Here, $\langle \sigma_+ \rangle$ and $\langle a^\dagger \rangle$ measure the negative frequency part of the atomic polarization and cavity field, while $\langle \sigma_+ \sigma_- \rangle$ and $\langle a^\dagger a \rangle$ measure the excited state population and cavity photon number, respectively.

The assumption that the atom enters the cavity in its excited state with the cavity field in its vacuum state implies the initial conditions $\langle \sigma_+(0) \rangle = \langle a^\dagger(0) \rangle = 0$, $\langle a^\dagger a \rangle(0) = \langle a^\dagger \sigma_- \rangle(0) = 0$, and $\langle \sigma_+ \sigma_- \rangle(0) = 1$. These initial conditions yield in turn the trivial solutions $\langle \sigma_+(t) \rangle = \langle a^\dagger(t) \rangle = 0$, and we are only left with the task of having to solve Equations 146 to 148 for the quadratic operator products.

In general, these equations are not amenable to an analytical solution. At resonance $\delta = 0$, and for $\gamma' = \kappa$, however, we readily find

$$P(t) \equiv \langle \sigma_+ \sigma_- \rangle(t) = e^{-\gamma' t} \cos^2(A(t)) \tag{149}$$

where $A(t) = \int_0^t g(t')dt'$ is the "vacuum area" associated with the atom-cavity coupling,

$$A(t) = \frac{g}{\Omega}\left(\sin(qx_0 + \Omega t) - \sin(qx_0)\right). \tag{150}$$

the "pulse area" of conventional coherent atom-field interactions measures the effective strength of the atom-field coupling, the vacuum area is a measure of the effective coherent coupling between the atom and the vacuum cavity mode as the atom flies along the resonator.

Expressing the cosine in Equation 149 in terms of Bessel functions as

$$\cos(A(t)) = \cos(r\,\sin\,qx_0)J_0(r)$$

$$+\, 2\,\cos(r\,\sin\,qx_0)\sum_{n=1}^{\infty} J_{2n}(r)\,\cos\left[2n(qx_0 + \Omega t)\right]$$

$$+\, 2\,\sin(r\,\sin\,qx_0)\sum_{n=0}^{\infty} J_{2n+1}(r)\,\sin\left[(2n+1)(qx_0 + \Omega t)\right], \tag{151}$$

where $r = g/\Omega$ shows that the upper electronic state population develops an infinite number of sidebands separated by the Doppler shift Ω. These sidebands are most visible for slow atoms such that the dynamic detuning Ω is smaller than or comparable to the vacuum Rabi frequency g. For fast atoms such that $v/c \gg g/\omega_c$, the contributions of the Bessel functions of order larger than zero become increasingly negligible: the atom is effectively decoupled from the cavity field, and would remain in its upper state for all times, were it not for

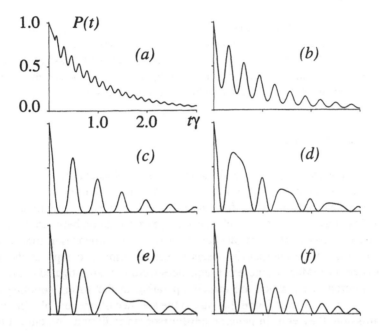

FIGURE 6 Probability $P(t)$ of finding an atom in the excited state and no photon in the Fabry-Pérot cavity in the strong coupling regime $g/\gamma' = g/\kappa = 10$ and at resonance $\delta' = 0$. The velocities $v = \Omega c/\omega_c$ are $\Omega/g = 2$ (a), $\Omega/g = 1$ (b), $\Omega/g = 2/\pi$ (c), $\Omega/g = 1/\pi$ (d), $\Omega/\gamma' = 1$ (e), and $\Omega/\gamma' = 0.1$ (f).

the decay due to spontaneous emission into the free space background. The evolution of the upper state probability $P(t)$ is displayed in Figure 6a–f for $g/\gamma' = g/\kappa = 10$ and for various values of g/Ω.

Another limit that can be handled analytically is that of large and approximately equal detuning and Doppler shift, $\Omega \simeq \Delta \gg g$ with $\kappa = \gamma'$. In this regime, Equations 146 to 148 reduce approximately to

$$\langle \sigma_+ \sigma_- \rangle = -\gamma' \langle \sigma_+ \sigma_- \rangle - (g/2)e^{-iqx_0 - i\Omega t}\langle a^\dagger \sigma_- \rangle - (g/2)e^{iqx_0 + i\Omega t}\langle \sigma_+ a \rangle, \quad (152)$$

$$\langle a^\dagger a \rangle = -\kappa \langle a^\dagger a \rangle + (g/2)e^{-iqx_0 - i\Omega t}\langle a^\dagger \sigma_- \rangle + (g/2)e^{iqx_0 + i\Omega t}\langle \sigma_+ a \rangle, \quad (153)$$

$$\langle a^\dagger \sigma_- \rangle = -(\gamma' - i\delta)\langle a^\dagger \sigma_- \rangle + ge^{iqx_0 + i\Omega t}\langle \sigma_+ \sigma_- - a^\dagger a \rangle, \quad (154)$$

up to corrections of order g/Ω and g/δ. This set of equations is easily brought into a form reminiscent of a two-level atom interacting in the rotating-wave approximation with a monochromatic field, with a Doppler-shifted detuning $\Delta = \delta - \Omega$. The upper state population is readily found to be given by

$$\langle \sigma_+ \sigma_- \rangle(t) = e^{-\gamma t}\frac{1}{2}\left[1 + s^2 + c^2 \cos(\Re t)\right] \quad (155)$$

where $\mathcal{R} = [g^2 + \Delta^2]^{1/2}$ is an effective Rabi frequency, $s^2 = \Delta^2/\mathcal{R}^2$, and $c^2 = g^2/\mathcal{R}^2$. Hence, the upper state population undergoes damped Rabi-like oscillations. The major difference between this situation and the usual case of atoms at rest is that the Doppler shift can now compensate exactly for the atom-field detuning, so that the effective detuning becomes $\Delta = 0$.

V.   CONCLUSION

In this chapter, we have discussed the basic physical processes underlying spontaneous emission in optical cavities. We found that it can be enhanced or reduced as compared to its free space value, or can even cease to be exponential, taking instead the form of an exchange of excitation between the atom and the cavity mode. These results have important consequences in applications, in particular in the microcavity lasers whose description comprises the bulk of this volume. Many of these consequences were first investigated in the context of micromasers, which are masers operating in the strong coupling regime $g \gg \kappa, \gamma' = 0$. It was found that due to the absence of irreversible spontaneous emission, they emit in general sub-Poissonian radiation, in contrast to conventional single mode lasers and masers operating well above threshold. Indeed, the study of micromasers showed that it is the noise associated with pump and relaxation mechanisms that leads to their Poissonian photon statistics.

While the role of pump noise in conventional lasers was also recognized quite early, the effects of spontaneous emission were not explicitly discussed in much detail, as it was not clear in the early days how to build cavities which would permit tailoring the vacuum field in the optical regime. The advent of MBE has changed the situation dramatically, and it is now possible to engineer microcavity semiconductor lasers where spontaneous emission is significantly different from its free space value. In particular, it is now possible to reduce γ' significantly, to the point where the fraction of spontaneous emission into the cavity mode, the so-called β-factor, is of the order of 10^{-1} or higher. Microcavity lasers operating in this regime are characterized by low threshold currents — these lasers are sometimes referred to, somewhat inaccurately, as "thresholdless" — and may produce a sub-Poissonian, or intensity squeezed output. However, their linewidth, which is dominated by spontaneous emission and nonequilibrium carrier fluctuations, does not display the usual Schalow-Townes inverse power dependence in the case of large β,[35] and is typically substantial. Although, in contrast to micromasers, these microcavity lasers still operate in the weak coupling regime, there is every reason to believe that improvements in cavity design, and the possible use of "photonic bandgap structures", will change the situation in the near future.

ACKNOWLEDGMENTS

This work was supported by the U.S. Office of Naval Research contract N00014–91-J205, by the National Science Foundation Grants PHY-8902548 and PHY-9213762, and by the Joint Services Optics Program.

REFERENCES

1. S. Haroche, in *Fundamental Systems in Quantum Optics*. Dalibard, J., Raimond, J.M., and Zinn-Justin, J., editors, Elsevier Science, North Holland, Amsterdam, 1992.
2. D. Meschede, *Phys. Rep.*, **211**, 201 (1992).
3a. P. Meystre, in *Progress in Optics*, Vol. XXX, E. Wolf, editor, p. 261, 1992.
3b. P. Meystre, *Opt. Commun.*, **90**, 41 (1992).
4. E. M. Purcell, *Phys. Rev.* **69**, 681 (1946).
5. D. Kleppner, *Phys. Rev. Lett.* **47**, 233 (1981).
6. P. Meystre and M. Sargent III. *Elements of Quantum Optics*. Springer-Verlag, Berlin, Heidelberg, 1991. Second Edition.
7. S. Haroche and J. M. Raimond. *Advances in Atomic and Molecular Physics*, Vol. 20. Academic Press, New York, 1985. pp. 350–411.
8. P. W. Milonni, *The Quantum Vacuum: An Introduction to Quantum Electrodynamics*. Academic Press, San Diego, 1994.
9. E. T. Jaynes and F. W. Cummings, *Proc. IEEE* **51**, 89 (1963).
10. K. Moelmer, Y. Castin, and J. Dalibard, *J. Opt. Soc. Am. B*, **10**, 524 (1993).
11. H. J. Carmichael, *An Open Systems Approach to Quantum Optics*. Lectures presented at l'Université Libre de Bruxelles, Bruxelles, Belgium, 1991.
12. N. Gisin and I. Percival, *Phys. Rev. A*, **46**, 4382 (1992).
13. V. Weiskopf and E. P. Wigner, *Z. Phys.* **63**, 54 (1930).
14. M. G. Raizen, R. J. Thompson, R. J. Brecha, H. J. Kimble, and H. J. Carmichael, *Phys. Rev. Lett.* **63**, 240 (1989).
15. F. Bernardot, P. Nussenzweig, M. Brune, J. M. Raimond, and S. Haroche, *Europhys. Lett.* **17**, 33 (1992).
16. J. Parker and C. R. Stroud, *Phys. Rev. A* **35**, 4226 (1987).
17. R. G. Cook and P. M. Milonni, *Phys. Rev. A* **35**, 5071 (1987).
18. P. W. Milonni and P. L. Knight, *Opt. Commun.* **9**, 119 (1973).
19. D. J. Heinzen, J. J. Childs, J. E. Thomas, and M. S. Feld, *Phys. Rev. Lett.* **58**, 1320 (1987).
20. H. J. Carmichael, R. J., Brecha, M. G. Raizen, H. J. Kimble, and P. R. Rice, *Phys. Rev. A* **40**, 5516 (1989).
21. C. Cohen-Tannoudji, J. Dupont-Roc, and G. Grynberg, *Atom-Photon Interactions: Basic Processes and Applications*. Wiley-Interscience, New York, 1992.
22. R. R. Schlicher, *Opt. Commun.* **70**, 97 (1989).
23. W. Ren and H. J. Carmichael, *OSA Annual Meeting Technical Digest*, 80 (1990).

24. W. Ren, J. D. Cresser, and J. H. Carmichael, *Phys. Rev. A* **46**, 7162 (1992).
25. Z. Bialynicka-Birula, P. Meystre, E. Schumacher, and M. Wilkens, *Opt. Commun.* **85**, 315 (1991).
26. T. W. Mossberg, M. Lewenstein, and D. J. Gauthier, *Phys. Rev. Lett.* **67**, 1723 (1991).
27. M. Wilkens and P. Meystre, *Opt. Commun.* **94**, 66 (1992).
28. P. Meystre, *Opt. Commun.* **90**, 41 (1992).
29. G. S. Agarwal and Y. Zhu, *Phys. Rev. A* **46**, 479 (1992).
30. A. P. Kazantsev, G. J. Surdovich, and V. P. Yakovlev, *Mechanical Action of Light on Atoms.* World Scientific, Singapore, 1990.
31. J. Dalibard and C. Cohen-Tannoudji, *J. Phys. B,* **18**, 1661 (1985).
32. P. Meystre and M. Wilkens. Spontaneous Emission by Moving Atoms, in *Cavity QED: Advances in Atomic, Molecular and Optical Physics,* P. R. Berman, editor, Academic Press, New York, Suppl. 2, p. 301, 1994.
33. M. Wilkens, *Phys. Rev. A* **47**, 671 (1993).
34. C. Baxter, M. Babiker, and R. Loudon, *Phys. Rev. A* **47**, 1278 (1993).
35. R. E. Slusher, M. Mohideen, F. Jahnke, and S. W. Koch, *Opt. Phot. News* **4**, 27 (1993).

2 A Simple Theory on the Effect of Dephasing of Vacuum Fields on Spontaneous Emission in a Microcavity

Yong Lee

TABLE OF CONTENTS

I. INTRODUCTION

In recent years, one of the most interesting subjects in quantum optics and cavity quantum electrodynamics (cavity QED) is alteration of the spontaneous emission (SE) using a "microcavity" with dimensions comparable to the transition wavelength. One might be able to explain this phenomenon in the following way. Imagine an excited atom sitting inside a microcavity. If the atom emits a photon which constructively interferes with itself due to multiple reflections inside the cavity, then the SE is enhanced. On the other hand, if the emitted photon destructively interferes with itself, then the SE is inhibited. However, it is an open question how the atom can predict what will happen to itself in the future.[1] This question can be solved by adopting the quantum mechanical viewpoint for the origin of the SE, which is that the SE occurs as a result of an interaction between an excited atom and vacuum-field fluctuations (VFFs) surrounding the atom, in other words, the SE can be viewed as "stimulated emission", stimulated by "zero-photon states", i.e., VFFs.[2] In a microcavity, the VFFs are squeezed in both spatial and frequency domains by "self-interference". Consequently, the strength of the interaction between the atom and the VFFs is modified, resulting in the enhancement or inhibition of the SE.

Because of the word "fluctuations" in VFFs, one might view the vacuum field (VF) as a wave whose amplitude and phase randomly change in time (Figure 1(a)). If this is true, how can such an incoherent wave be altered by self-interference caused by multiple reflections inside a microcavity? In the quantum theory of light, the two quadrature components of an optical field amplitude, which represent the real (a_1) and imaginary (a_2) parts of the field phasor, are noncommuting operators that satisfy the Heisenberg uncertainty principle: $\langle(\Delta a_1)^2\rangle \langle(\Delta a_2)^2\rangle \geq 1/16$, like the position and momentum of a harmonic oscillator.[3] Coherent light has quadrature components with equal uncertainties and a minimum uncertainty product, i.e., $\langle(\Delta a_1)^2\rangle \langle(\Delta a_2)^2\rangle = 1/16$ (Figure 1(b)). The minimum-uncertainty product for the quadrature components can also be converted into that for the photon number, n (or amplitude), and the phase, ϕ, of light, which is $\langle(\Delta n)^2\rangle \langle(\Delta\phi)^2\rangle = 1/4$. Since a vacuum field can in general be regarded as a coherent state with no real photon field, the VF also obeys the same minimum uncertainty for n and ϕ as a coherent state does. Thus, the "fluctuations" of the VFFs means such inherent quantum mechanical uncertainty for the amplitude and phase of the VF, and each VF, as illustrated by a sinusoidal wave in Figure 1(c), is a wave with a definite amplitude and phase. Therefore, it is possible for the VF to be altered by self-interference.

The most sophisticated model for the spontaneous emission was developed by Dailbard et al. based on reservoir fluctuations and self-reaction.[4] The basic idea of this model is that a system (small system) is fluctuated and polarized by the reservoir (large system) surrounding it (reservoir fluctuations), and simultaneously the reservoir is also fluctuated and polarized by the system as back-action (Figure 2) (self-reaction). Applying this idea to the case of the SE processes, VFFs (large system) first interact with an atom (small system) to induce an atomic dipole moment (effects of reservoir fluctuations), and then the created atomic fields fluctuate and polarize the VFs, which in turn react back on the dipole moment (effects of self-reaction). With the help of such a model for the SE, the Einstein A coefficient, i.e, the SE rate, was successfully derived from the first principle. Also, the model successfully explained why an excited atom can emit a spontaneous photon but an unexcited atom cannot.

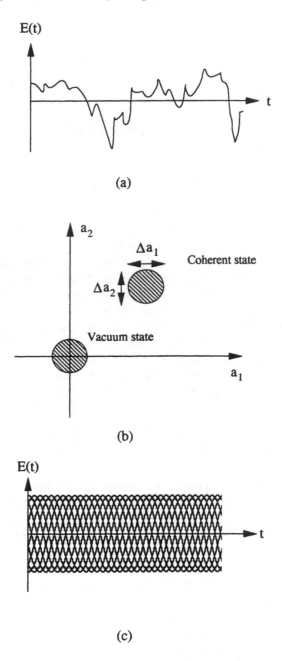

FIGURE 1 (a) A fluctuating electric field: E(t). (b) Coherent state and vacuum state in (a_1, a_2)-plane. (c) Time-varying electric field of vacuum field fluctuations: E(t). Uncertainty of the amplitudes is not depicted to avoid making the figure complicated.

A number of experimental demonstrations on controlled SE in solid-state materials have recently been carried out,[2,5-12] and the realization of the controlled SE, especially in a semiconductor system,[5-11] is expected to shed light on a

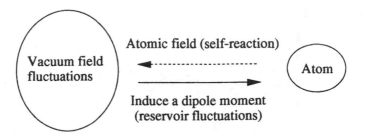

FIGURE 2   Model for spontaneous emission adapted from Dailbard et al.[4]

great improvement on conventional semiconductor optical devices, such as a
low threshold (even thresholdless) lasing operation, an extremely wideband
modulation and the suppression of shot noise, etc.[13,14] In a solid-state system,
unlike a vacuum system, a number of electrons are embedded in a thermal
phonon reservoir. In such a system, the induced dipole moments are randomly
destroyed by phonon collisions, resulting in dipole-moment fluctuations. As
mentioned in the previous paragraph, not only do vacuum fields affect electrons, electrons affect the VFs. Therefore, the vacuum fields would be dephased
by suffering dipole-moment fluctuations as the back-action from the phonon
reservoir and self-interference of the VFs inside a solid-state microcavity
would be weakened. As a result, the alteration of the SE in the cavity would
be affected. In the cavity-QED theory, such an effect of dephasing (or phase-
breaking) of the VFs has not been taken into account so far. Thus, the objective
of this chapter is to present two simple theoretical models which take into
account the effect of dephasing of the VFs on the SE in a microcavity. This
would be necessary to evaluate the limit of the capability of microcavity-based
solid-state optical devices.

The usual recipe to quantize the electromagnetic field in a dielectric is a
canonical quantization scheme with use of its effective linear or nonlinear
susceptibilities.[15] However, it has been known that such a quantization scheme
leads to inconsistency in the canonical equal-time commutation relation between the field operator and its conjugate, that is, a decay of the commutator
if the medium is dispersive.[16-18] This is because only the electromagnetic
energy is taken into account in this scheme. Indeed, the dispersion results from
a continuous exchange of energy between the field and medium.[19] Therefore,
it is clear that the total Hamiltonian for the coupled system has to be employed
in the quantization of light fields.[16-18]

The dephasing of the quantized fields, i.e., the vacuum fields, which is our
main concern, results from the energy exchange between fields and medium
polarizations interacting with phonon reservoirs, that is, from the dispersion
(see Figure 3). Therefore, in our problem, we have to deal with the whole
system (system + reservoir) together in order to keep the quantum mechanical
consistency in the field quantization. The most popular and relatively simpler
way to do it is a reservoir approach independently developed by Lax[20]

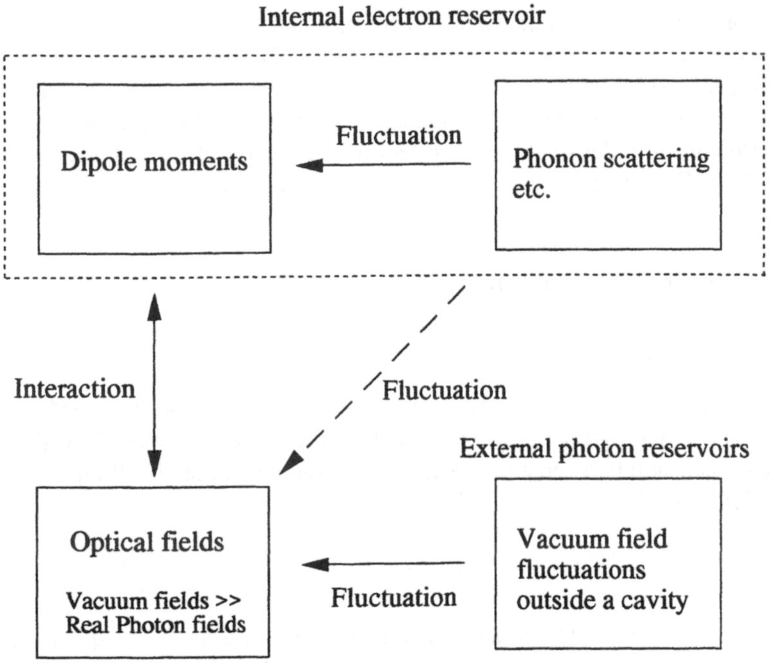

Internal electron reservoir

FIGURE 3   Schematic representation showing mechanism for dephasing of vacuum fields.

and Haken[21] based on the quantum mechanical Langevin equation. In this approach, fluctuation forces as a back-action from a reservoir are incorporated to hold quantum mechanical consistency and their magnitudes are related via the quantum mechanical fluctuation-dissipation theorem to a damping constant.

We present two simple theoretical models based on the reservoir approach[20,21] to take into account the effect of dephasing of VFs on the SE in a microcavity. In the two models, called model I and model II, the energy exchange between fields and medium polarizations interacting with phonon reservoirs is explicitly taken into account so that the quantum mechanical consistency in the field quantization holds. In addition, the energy conservation of fields (vacuum fields) is satisfied. Although an application of model I is restricted to the case where the interaction between fields and the reservoirs is spatially localized, model I has the advantage of being able to obtain analytical solutions. Model II is the model extended from model I so as to be able to deal with continuously distributed interactions between fields and the reservoirs. However, computational simulations are in general necessary in model II.

This chapter is organized as follows. Model I is presented in Section II. In Section II.A, model I is described. In Section II.B, we provide a microscopic picture for model I in order to know under what conditions model I is valid.

Some numerical results obtained by model I are presented in Section II.C. Model II is presented in Section III. Model II is described in Section III.A. Section III.B shows the validity of model II by calculating the SE rate in a homogeneous dielectric. Then, model II is applied to a more practical microcavity structure in order to study the effect of dephasing of VFs on the SE (Section III.C). Finally, a summary is given in the last section.

II. THEORETICAL MODEL I [22]

A. DESCRIPTION OF MODEL I

Let us consider a light wave incident towards an absorbing medium. In the classical description of this problem, the amplitudes of the fields, which are the complex numbers, entering (b_{in}) and leaving (b_{out}) the absorbing medium are related by

$$b_{out} = \sqrt{A} \cdot b_{in} \qquad (1)$$

where A is the field extinction parameter ($0 \leq A \leq 1$). In the quantum theory of radiation, field amplitudes of light are replaced by the creation (b^+) and annihilation (b) operators. These operators obey the boson commutation relations $[b, b^+] = 1$ and $[b, b] = [b^+, b^+] = 0$. As is easily seen, in the quantum mechanical treatment of the same problem by using Equation 1, the quantum mechanical consistency, i.e., the conservation of the commutation relation is broken for the operator b_{out} because $[b_{out}, b_{out}^+] = A \neq 1$ (where the commutator for b_{in}, $[b_{in}, b_{in}^+] = 1$, was assumed to hold). This results from the fact that the back-action on the field, b_{out}, from a reservoir is neglected as mentioned previously. Taking the back-action into account leads to the following relation between b_{in} and b_{out} in the Heisenberg picture:

$$b_{out} = \sqrt{A} \cdot b_{in} + \sqrt{1-A} \cdot f \qquad (2)$$

where f represents the back-action from the reservoir. We focus on the reservoir composed of an electron system surrounded by thermally fluctuating phonons and therefore f is the field operator for the dipole-moment fluctuations, as is shown later in Section II.B (see Figure 3). With the use of $[b_{in}, b_{in}^+] = [f, f^+] = 1$ and $[b_{in}, f^+] = 0$, simple algebra shows that the proper commutation relation for b_{out}, $[b_{out}, b_{out}^+] = 1$, is recovered. In quantum mechanics, $1 - A$ in Equation 2 is the average absorption rate of a photon, and the fluctuation of the absorption rate around its average value, originating from the dipole moment fluctuations, is expressed by the second term on the right-hand side (RHS) of Equation 2. This is easily shown by calculating the expectation value of the photon number in the output light: b_{out},

$$\langle 0, n | b_{out}^+ \, b_{out} | n, 0 \rangle = A \langle n | b_{in}^+ \, b_{in} | n \rangle + (1 - A) \langle 0 | f^+ f | 0 \rangle = An \qquad (3)$$

where $|n,0\rangle$ $(= |n\rangle |0\rangle)$ is the coupled state of the n-photon state, $|n\rangle$, and the ground state of dipole moments, and $\langle n | b_{in}^+ b_{in} | n \rangle = n$ (n is the mean photon number) and $\langle 0 | f^+ f | 0 \rangle = 0$ were used.

Suppose, for simplicity, a one-dimensional planar microcavity with a very thin electron-phonon system inside it, which we henceforth refer to as an internal "electron-reservoir (ER)" as depicted in Figure 4(a). The ER serves not

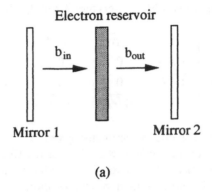

(a)

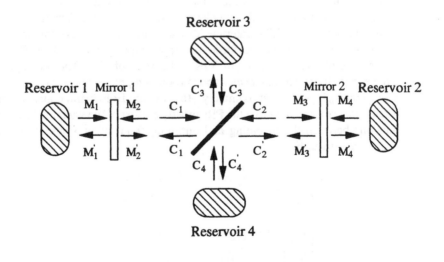

(b)

FIGURE 4 (a) One-dimensional planar microcavity with an internal thin electron reservoir. (b) Model of (a).

only as a source of dephasing of VFs but also as that of the SE. By restricting ourselves to the case of low excitation where very few real photons are present, we can assume that the vacuum field intensity dominates over the real photon field intensity: $|E|^2 \cong |E_{vac}|^2$. Next, we proceed in calculating the spatial field distributions of the VFs inside the microcavity by taking into account their self-interferences due to the cavity walls as well as the destruction of the self-interferences due to the interaction with the localized ER and external "photon reservoirs (PRs)". The system of interest is modeled as shown in Figure 4(b). A coupling of the VFs and the ER is introduced through a virtual beam-splitter (BS) connecting to two ERs, reservoirs 3 and 4, which is described by

$$\begin{bmatrix} C_1' \\ C_2' \\ C_3' \\ C_4' \end{bmatrix} = \begin{bmatrix} 0 & \sqrt{A} & \sqrt{1-A} & 0 \\ \sqrt{A} & 0 & 0 & \sqrt{1-A} \\ \sqrt{1-A} & 0 & 0 & -\sqrt{A} \\ 0 & \sqrt{1-A} & -\sqrt{A} & 0 \end{bmatrix} \begin{bmatrix} C_1 \\ C_2 \\ C_3 \\ C_4 \end{bmatrix} \qquad (4)$$

where A is the absorption parameter given by $A = \exp(-\alpha \cdot W)$ (α is the absorption coefficient and W is the width of the ER). It is readily confirmed that the commutation relations for the outgoing field operators from the BS, C_i' s (i = 1 to 4), preserve their proper relations: $[C_i', C_i'^+] = 1$, with use of $[C_i, C_j^+] = \delta_{i,j}$ ($\delta_{i,j}$ is the Kronecker delta: $\delta_{i,j} = 1$ for i = j, $\delta_{i,j} = 0$ for i≠j). Here, we assume an incoherent directional coupling of the back-action from the ERs with an incident field; that is, the fluctuating fields are assumed to propagate in the same direction as the incident field does after suffering its phase randomization in the ERs. Internal VFs also lose their phase memories by coupling with the external PRs through output coupling mirrors. The PRs serve as a source of both dephasing of VFs and supply of VFs. The mean photon number n with optical frequencies and at room temperature is nearly zero (n(T) = $[\exp(\hbar\omega/k_BT) - 1]^{-1}$ approximately ~0; k_B is the Boltzmann constant and T is temperature), and therefore the PRs mostly consist of the vacuum fields. The relations between the incoming and outgoing fields at two output coupling mirrors are given by

$$\begin{bmatrix} M_1' \\ M_2' \end{bmatrix} = \begin{bmatrix} j\sqrt{R_1}e^{-j\phi_1} & \sqrt{1-R_1} \\ \sqrt{1-R_1} & j\sqrt{R_1}e^{j\phi_1} \end{bmatrix} \begin{bmatrix} M_1 \\ M_2 \end{bmatrix}$$

$$\begin{bmatrix} M_3' \\ M_4' \end{bmatrix} = \begin{bmatrix} j\sqrt{R_2}e^{j\phi_2} & \sqrt{1-R_2} \\ \sqrt{1-R_2} & j\sqrt{R_2}e^{-j\phi_2} \end{bmatrix} \begin{bmatrix} M_3 \\ M_4 \end{bmatrix} \qquad (5)$$

where R_1 and R_2 are the mirror reflectivities for the left-hand side (LHS) and RHS mirrors, respectively, and ϕ_1 and ϕ_2 are the phase shifts due to the mirror reflections. The commutation relations for M_i''s (i = 1,2) are easily shown to be proper by $[M_i, M_j^+] = \delta_{i,j}$ ((i,j) = (1,2), (3,4)). Björk et al. have explicitly taken into account the effect of dephasing of internal VFs due to a coupling with the external PRs.[2] In their model, when an internal field escapes from a cavity, a field, which has no mutual phase relation to the internal field, is injected into the cavity from outside so that no net power flows. This "no-net-power-flow" condition comes from the fact that the vacuum fields carry no net power[2] (or the energy of the vacuum fields is conserved and is not attenuated[23]).

Let us calculate the spatial distributions of VFs inside the microcavity with the help of Equations 4 and 5. Assuming an infinite cross-sectional area of the system, the problem is reduced to a one-dimensional problem. Suppose plane waves incident from each reservoir described by

$$\hat{E}_i^{(+)}(z,t) = j V_i b_{\bar{k}} e^{-j\left(\omega t - \bar{k}\cdot\bar{z}\right)} \qquad \text{for } i = 1,3$$

$$\hat{E}_i^{(+)}(z,t) = j V_i b_{\bar{k}} e^{-j\left(\omega t + \bar{k}\cdot\bar{z}\right)} \qquad \text{for } i = 2,4 \tag{6}$$

where $b_{\bar{k}}$ are the boson field operator and satisfy $\left[b_{\bar{k}}, b_{\bar{k}}^{\pm}\right] = \delta_{\bar{k}}$, $k_{\bar{k}}$, ω is the angular frequency, $\bar{k}$ is the wavevector, and V_i is the magnitude of the electric field incident from the reservoir i. Here, we assume, for simplicity, that field polarizations are conserved in interacting with the reservoirs, so, we suppress them for a while. The plane wave, for example, from reservoir 1 is modified by its self-interference as follows,

$$\hat{E}_{M,1}^{(+)}(z,t) = j V_1 \left[c_{\bar{k}}^M e^{-j\left(\omega t - \bar{k}\cdot\bar{z}\right)} + d_{\bar{k}}^M e^{-j\left(\omega t + \bar{k}\cdot\bar{z}\right)} \right]$$

$$= j V_1 \left[F_1 b_{\bar{k}} e^{-j\left(\omega t - \bar{k}\cdot\bar{z}\right)} + F_2 b_{\bar{k}} e^{-j\left(\omega t + \bar{k}\cdot\bar{z}\right)} \right]$$

$$= j V_1 \chi_1(z) b_{\bar{k}} e^{-j\omega t} \tag{7}$$

where

$$F_1 = \frac{\sqrt{T_1}}{1 - \sqrt{R_1 R_2}\, e^{j(\phi_1 + \phi_2)} e^{j 2 k_z L}}$$

$$F_2 = \frac{\sqrt{T_1 R_1}\, e^{j 2 k_z L}}{1 - \sqrt{R_1 R_2}\, e^{j(\phi_1 + \phi_2)} e^{j 2 k_z L}}, \tag{8}$$

$\chi_1(z)\left(= F_1 e^{j\bar{k}\cdot\bar{z}} + F_2 e^{-j\bar{k}\cdot\bar{z}}\right)$ represents the spatial modification of $b_{\bar{k}}$, k_z is the z-component (z // the cavity normal) of the wave number, L is the cavity-length, and $T_{1(2)} = 1 - R_{1(2)}$. Ueda and Imoto[24] have recently pointed out that field operators in free space, that is, those for plane-wave modes, are modified inside a microcavity by self-interference, resulting in anomalous commutation relation and the alteration of the SE is a direct manifestation of the anomaly. In our case, such anomaly appears in the commutators for $c_{\bar{k}}^{M}$ and $d_{\bar{k}}^{M}$ in Equation 7, which are

$$\left[c_{\bar{k}}^{M},\ c_{\bar{k}}^{M+}\right]=\left|F_{1}\right|^{2}\left[b_{\bar{k}},\ b_{\bar{k}}^{+}\right]=\left|F_{1}\right|^{2}$$

$$\left[d_{\bar{k}}^{M},\ d_{\bar{k}}^{M+}\right]=\left|F_{2}\right|^{2}\left[b_{\bar{k}},\ b_{\bar{k}}^{+}\right]=\left|F_{2}\right|^{2}.$$

$$(9)$$

It is clear that the commutation relations in Equation 9 depend on the property of a microcavity. Although the commutation relations for $c_{\bar{k}}^{M}$ and $d_{\bar{k}}^{M}$ are not conserved, the commutation relation for the plane-wave mode, $b_{\bar{k}}$ has to be conserved. Note that the conservation of the commutation relation, which is the basis of our theoretical model, is the conservation of the commutation relation for plane-wave modes, which is normal. Assuming that fields incident from different reservoirs are completely uncorrelated to each other in phase, that is, the fields coupled to the reservoirs are completely phase-randomized, the operator for the square of the electric fields is obtained by

$$\hat{E}_{M}^{2}(z,\ t)=\sum_{i=1}^{4}\hat{E}_{M,i}^{2}(z,\ t)$$

$$(10)$$

where $\hat{E}_{M,i}(z,\ t)=\hat{E}_{M,i}^{(+)}(z,\ t)+\hat{E}_{M,i}^{(-)}(z,\ t)$ $(\hat{E}_{M,i}^{(-)}(z,\ t)=\hat{E}_{M,i}^{(+)^{*}}(z,\ t))$, where $*$ denotes the Hermitian adjoint). By taking the expectation value of $\hat{E}_{M}^{2}$ for a vacuum state $\left|0_{\bar{k}}>\right.$, we have

$$E_{M}^{2}(z)=\left\langle 0\left|\hat{E}_{M}^{2}(z,\ t)\right|0\right\rangle=\sum_{i=1}^{4}\left\langle 0\left|\hat{E}_{M,i}^{2}(z,\ t)\right|0\right\rangle$$

$$=\sum_{i=1}^{4}\left|V_{i}\right|^{2}\left|\chi_{i}(z)\right|^{2}\left(2\left\langle 0\left|b_{\bar{k}}^{+}b_{\bar{k}}\right|0\right\rangle+1\right)$$

$$=\sum_{i=1}^{4}\left|V_{i}\right|^{2}\left|\chi_{i}(z)\right|^{2}.$$

$$(11)$$

The magnitudes of the V_{i}'s can be determined in the following manner based on the fact that VFs carry no net power.[2,23] The net electromagnetic power flow from the reservoir i, which propagates in the direction normal to the mirror surfaces, is given, in analogy to the quantum mechanical electron transport equation,[25] by

$$\bar{P}_{i}=\sum_{j\neq i}\left\{T_{j,i}\bar{I}_{i}-T_{i,j}\bar{I}_{j}\right\}$$

$$(12)$$

where $\bar{I}_{i}$ is the power flow incident on the cavity from the reservoir i, expressed by $\bar{I}_{i}=\varepsilon_{i}\left|V_{i}\right|^{2}v_{i}\cos\theta_{i}$ (ε_{i} is the dielectric constant, v_{i} is the speed of light, and

θ_i is the incident angle relative to the cavity normal) and $T_{i,j}$ is the probability that a field propagates coherently from the reservoir j to i. It is readily shown by time-reversal symmetry: $T_{i,j} = T_{j,i}$ that $\vec{P}_i = 0$ at all the reservoirs if all $\bar{I}_i$'s are equal. Therefore, by using $|V_1|^2$, the other $V_i|2$ (i ≠1) are obtained by

$$|V_i|^2 = \frac{n_1 \cos\theta_1}{n_i \cos\theta_i} |V_1|^2 \ (i \neq 1)$$

(13)

where n_i is the refractive index in the reservoir i and V_1 is the electric field amplitude in free space given by $V_1 = (\hbar\omega/2\varepsilon_1 V)^{1/2}$ (V is a quantization volume for the reservoir 1). Imposing the boundary conditions, Equations 4 and 5, and using Equations 11 and 13, we obtain the following expression for the internal spatial distribution of VFs. For brevity, we do not present this calculation in detail but give the final result:

$$E_M^2(z) = |V_1|^2 [T_1\{1 + A^2R_2 + 2A\sqrt{R_2}\cos\phi_1^L\}$$

$$+ \frac{n_1 \cos\theta_1}{n_2 \cos\theta_2} AT_2\{1 + R_1 + 2\sqrt{R_1}\cos\phi_2^L\}$$

$$+ \frac{n_1 \cos\theta_1}{n_3 \cos\theta_3} (1 - A)\{1 + R_1 + 2\sqrt{R_1}\cos\phi_3^L\}$$

$$+ \frac{n_1 \cos\theta_1}{n_4 \cos\theta_4} A(1 - A)R_2\{1 + R_1 + 2\sqrt{R_1}\cos\phi_4^L\}$$

for $0 \leq z \leq \ell$ (14a)

where $\phi_1^L = 2k_z(L - z) + \phi_2$, $\phi_2^L = 2k_zz + \phi_1$, $\phi_3^L = 2k_zz + \phi_1$, $\phi_4^L = 2k_zz + \phi_1$, $|Z|^2 = 1 - 2\sqrt{R_1}\sqrt{R_2} A \cos\zeta + A^2R_1R_2$, $\zeta = 2k_zL + \phi_1 + \phi_2$, ℓ is the location of the ER, and ϕ_1 and ϕ_2 are the phase shifts by mirror reflections. For $\ell < z \leq$ L,

$$E_M^2(z) = |V_1|^2 [AT_1\{1 + R_2 + 2\sqrt{R_2}\cos\phi_1^R\}$$

$$+ \frac{n_1 \cos\theta_1}{n_2 \cos\theta_2} T_2\{1 + A^2R_1 + 2A\sqrt{R_1}\cos\phi_2^R\}$$

$$+ \frac{n_1 \cos\theta_1}{n_3 \cos\theta_3} A(1 - A)R_1\{1 + R_2 + \sqrt{R_2}\cos\phi_3^R\}$$

$$+ \frac{n_1 \cos\theta_1}{n_4 \cos\theta_4} (1 - A)\{1 + R_2 + 2\sqrt{R_2}\cos\phi_4^R\}]/|Z|^2$$

for $\ell < z \leq$ L   (14b)

where $\phi_1^R = 2k_z(L-z) + \phi_2$, $\phi_2^R = 2k_zz + \phi_1$, $\phi_3^R = 2k_z(L-z) + \phi_2$, and $\phi_4^R = 2k_z$ $(L-z) + \phi_2$. The electric fields calculated here are macroscopic transverse fields and local field correction[26] is not included.

The above-described model is analogous to the well-known theoretical model for quantum electron transport developed by Büttiker to take into account the effect of inelastic scattering events on coherent electron transport.[25]

B.   VALIDITY OF MODEL I

In this section, we provide a microscopic picture for model I described in the previous section in order to know on what condition this model is valid.

Let us consider a situation depicted in Figure 5. For simplicity, an internal electron reservoir (ER) is sitting at the center of the cavity with two identical mirrors. Starting from Equation 2 and following Yamamoto and Imoto,[27] we have the following set of equations for one-way time-evolution of the field operator, b,

$$b(t+\tau_0) = e^{jv\tau_0}\sqrt{A}\,b(t) + \sqrt{1-A}\,f_A(t+\tau_0) \tag{15a}$$

$$b\left(t+\frac{\tau+\tau_0}{2}\right) = e^{jv\left(\frac{\tau-\tau_0}{2}\right)}b(t+\tau_0) \tag{15b}$$

$$b\left(t+\frac{\tau+\tau_0}{2}\right) = \sqrt{R}\cdot b\left(t+\frac{\tau+\tau_0}{2}\right) + \sqrt{1-R}\,f_R\left(t+\frac{\tau+\tau_0}{2}\right) \tag{15c}$$

$$b(t+\tau) = e^{jv\left(\frac{\tau-\tau_0}{2}\right)}\cdot b\left(t+\frac{\tau+\tau_0}{2}\right) \tag{15d}$$

where R is the mirror reflectivity, f_A and f_R are the operators for the fluctuating forces from the internal ER and the external photon reservoirs (PRs), respectively, τ_0 is the time taken for a field to traverse the ER, and τ is the time taken

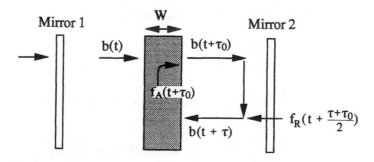

FIGURE 5   Time-dependent field operator in a one-dimensional planar microcavity with an internal thin electron reservoir.

for the field to reenter the ER after entering the ER. It has been assumed in Equation 15c that fluctuating forces from the external photon reservoir are instantaneously added. A symmetric structure of the system allows us to use the one-way time-evolution of field b, in order to obtain a motion equation of the field, which is in general derived from a round-trip time-evolution of the field. So, combining Equations 15a to 15d immediately leads us to

$$b(t+\tau) - b(t) = \left(e^{j v \tau}\sqrt{RA} - 1\right) \cdot b(t)$$

$$+ e^{j v \tau}\sqrt{R}\sqrt{1-A} \cdot f_A(t)$$

$$+ e^{j v \left(\frac{\tau - \tau_0}{2}\right)}\sqrt{1-R} \cdot f_R\left(t + \frac{\tau}{2}\right) \tag{16}$$

and after some calculations, we obtain the quantum mechanical Langevin equation, given by

$$\frac{d}{dt}b(t) = -\frac{v}{2Q}b(t) - \frac{\alpha c_0}{2n}b(t) + \sqrt{\frac{\alpha c_0}{n}}F_A(t) + \sqrt{\frac{v}{Q}}F_R(t) \tag{17}$$

where in deriving Equation 17 from Equation 16, we have used $A \equiv \exp(-\alpha c_0 \tau_0/n)$ (α is the absorption coefficient, c_0 is the speed of light in vacuum, and n is the refractive index) and $R \equiv \exp(-v/Q)$ (v is the angular frequency of the field and Q is the quality factor of the cavity), and have introduced $F_i(t) \equiv f_i(t)/\sqrt{\tau}$ (i = A,R) which are the average random field fluxes over a time interval of τ. If the fluctuations are Markovian, the correlation functions for the $F_i(t)$'s at different times take the form of $\langle F_i(t) F_i^\dagger(t')\rangle = \delta(t-t')$. The δ-function expresses the fact that the reservoirs ER and PR have a very short memory.

Next, we derive the Langevin equation given by Equation 17 in a more rigorous way.[28] We begin by Heisenberg's equations of motion for two decoupled systems: one is a photon system in a cavity and another is a two-level atomic system surrounded by a phonon heat bath:

$$\frac{d}{dt}b_p(t) = -\left[\frac{v}{2Q} + j(\Omega - v)\right]b_p(t) + F(t) \tag{18}$$

$$\frac{d}{dt}\Sigma_k(t) = [-\gamma + j(\omega - v)]\Sigma_k(t) + F_\Sigma(t) \tag{19}$$

where Σ_k is the atomic dipole moment operator ($\Sigma_k = a_{ex,k}^+ a_{g,k}$: a^+ and a are the creation and annihilation operators for electrons, ex and g indicate the excited and ground states, respectively, and k is the wave number), b_p is the photon field operator (p is the mode index), Ω is the resonant angular frequency, v is the angular frequency of the photon field, ω is the optical transition frequency, and γ is the relaxation rate of the dipole moments. F(t) and $F_\Sigma(t)$ represent fluctuating forces from external photon reservoirs and from an internal phonon heat bath, respectively. Assuming that the above fluctuation operators are described as the Markovian noise source, the corresponding correlation functions at different times have the following properties:

$$\langle F(t)F^+(t')\rangle = \frac{v}{Q}\delta(t-t') \tag{20a}$$

$$\langle F_\Sigma(t)F_\Sigma^+(t')\rangle = \frac{2\gamma}{N}\delta(t-t') \tag{20b}$$

where N is the number of atoms which interact with the field b_p.

We now introduce the interaction Hamiltonian to couple the above two systems, which is given by

$$H_I = \hbar\sum_{k',k''}\left(a_{ex,k'}^+ a_{g,k''} b_p \cdot g_{ex,g} + a_{g,k'}^+ a_{ex,k''} b_p^+ \cdot g_{g,ex}\right) \tag{21}$$

where $g_{ex,g}$ and $g_{g,ex}$ are the coupling coefficients. This interaction leads Equations 18 and 19 into the following coupled equations:

$$\frac{d}{dt}b(t) = -\left[\frac{v}{2Q} + j(\Omega - v)\right]b(t) + jg_cN\Sigma(t) + F(t) \tag{22}$$

$$\frac{d}{dt}\Sigma(t) = \left[-\gamma + j(\omega - v)\right]\Sigma(t) + jg_c\left(\sigma_g - \sigma_{ex}\right)b(t) + F_\Sigma(t) \tag{23}$$

where σ_{ex} and σ_g are the number operators for excited and unexcited atoms ($\sigma_{ex\,(g)} = a_{ex\,(g)}^+ a_{ex\,(g)}$), respectively, and for simplicity, we put $g_{ex,g} = g_{g,ex} = g_c$ and suppress the mode indices. By assuming that the dipole moment relaxes before field b escapes from the cavity, i.e., $\gamma \gg v/Q$ ($=1/\tau_{ph}$; τ_{ph} is the photon lifetime), and by putting $\frac{d}{dt}\Sigma = 0$ we have

$$\frac{d}{dt}b(t) = -\frac{v}{2Q}b(t) + \frac{g_c^2N}{\gamma}\left(\sigma_{ex} - \sigma_g\right)b(t) + j\frac{g_cN}{\gamma}F_\Sigma(t) + F(t) \tag{24}$$

where $v = \Omega = \omega$ was assumed. The typical value of γ is the order of 10^{13} sec[-1] in a solid-state system at room temperature[29] so that Equation 24 is approximately valid for a microcavity with $\tau_{ph} \geq 10^{-12}$ sec (or R ≥ 0.99).

By a comparison between Equations 17 and 24, it has been found that the fluctuating force f_A originates from dipole moment fluctuations and model I is valid only for the case where $\gamma \gg v/Q$.

C. NUMERICAL EXAMPLES

We present some numerical results obtained by model I in order to show the effect of dephasing of VFs on the spontaneous emission (SE) characteristics in a microcavity.

Suppose an enhancement-type half-wavelength microcavity containing a thin electron reservoir (ER) serves both as a source of dephasing of VFs and as that of the SE. The mirror reflectivity, R, is set to be 0.999. By assuming that all dipole moments are along the ER plane (this assumption has been known to be reasonable in a quantum-well (QW) layer[30]) and with the help of Equations 14a and 14b, we can obtain the SE rates as a function of emission angle θ relative to the cavity normal by[2]

$$dI^{s,p}(r,\phi,\theta) \cong \frac{1}{r^2} E_M^2(\ell;\omega) \begin{cases} \times \sin^2\phi & \text{for S - wave} \\ \times \cos^2\phi \cos^2\theta & \text{for P - wave} \end{cases} \qquad (25)$$

where (r,ϕ,θ) is the polar coordinate. Here, we restrict ourselves to a single frequency optical transition by assuming that the optical line broadening is negligibly small. This assumption will also be used in model II. Calculated radiation patterns for three different values of the dephasing parameter, κ (=1 – A), which express the strength of dephasing of the VFs due to the ER, are shown in Figure 6. As κ is increased, especially when $\kappa \gg 1 - R$, the peak is significantly reduced and broadens. This is because the cavity effect weakens due to the dephasing. Figure 7 shows κ-dependence of the SE lifetime, τ_{sp} obtained by integrating Equation 25 over all solid angles.[2] The

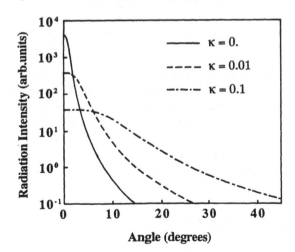

FIGURE 6 Spontaneous radiation patterns for three different dephasing parameters: $\kappa = 0$, 0.01, and 0.1. The mirror reflectivity R is 0.999.

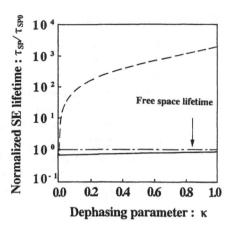

FIGURE 7 Spontaneous emission lifetime as a function of κ. The $\kappa = 0$ and $\kappa = 1$ correspond to no dephasing and the complete dephasing at the electron reservoir, respectively. The solid and dashed lines were obtained based on model I and Equation 1, respectively.

SE lifetime is normalized by that in free space, τ_{sp0}. As κ is increased, the SE lifetime tends to slightly increase. On the other hand, the result shown by a dashed line, which is obtained based on Equation 1 where the back-action from the ER is ignored, exceeds τ_{sp0} by far.

In spite of $\kappa = 1$, which means a complete dephasing of the VFs at the ER, the SE lifetime is not equal to its free space value, τ_{sp0}. This is because the ER is localized and self-interference of the VFs is slightly present between the ER and the mirrors. In order to completely destroy the self-interference of the VFs in the cavity, we take the spatial average of E_M^2 and obtain

$$\overline{E_M^2} = |V_1|^2 \big[T_1 \big(1 + A^2 R_2 \big) + A T_2 \big(1 + R_1 \big)$$

$$+ (1 - A)(1 + R_1) + A(1 - A)R_2 \big(1 + R_1 \big) \big] \big/ |Z|^2. \tag{26}$$

Substituting $A = 0$ ($\kappa = 1$) leads to $\overline{E_M^2} = 2|V_1|^2$, which is the intensity of the VFs in free space, resulting in a free space lifetime. Therefore, model I provides the reasonable result that the spontaneous emission rate in a microcavity with a complete destruction of self-interference of VFs equals that in free space.

Although we have here presented only the results on how the dephasing of VFs affects an enhancement of the SE in a microcavity, the dephasing would also affect an inhibition of the SE in a qualitatively similar way.

III. THEORETICAL MODEL II

A. Description of Model II

We extend model I so as to be able to deal with the case where electron reservoirs (ERs), which cause the destruction of self-interference of vacuum fields (VFs), are continuously distributed. We will henceforth refer to the extended model as model II.

By using the same recipe as was used in extending the multiprobe Landauer-Büttiker formula[25] to the "continuous probe" Landauer-Büttiker formula in order to take into account continuously distributed inelastic scatterings,[31] Equation 12 can be extended to

$$\bar{P}(\bar{r};\ \omega) = \int_{-\infty}^{\infty} d\bar{r}' \left\{ T_p(\bar{r}',\bar{r};\ \omega)\bar{I}(\bar{r};\ \omega) - T_p(\bar{r},\bar{r}';\ \omega)\bar{I}(\bar{r}';\ \omega) \right\}. \qquad (27)$$

Imposing the power conservation at all positions $\bar{r}$, i.e., $\bar{P}(\bar{r};\ \omega) = 0$, we have

$$\bar{I}(\bar{r};\ \omega) = \int_{-\infty}^{\infty} d\bar{r}' T_p(\bar{r},\bar{r}';\ \omega)\bar{I}(\bar{r}';\ \omega) \qquad (28)$$

where $\int_{-\infty}^{\infty} d\bar{r}' T_p(\bar{r}',\bar{r};\ \omega) = 1$ was used (see Appendix A). $\bar{I}(\bar{r};\ \omega)$ is the fluctuating force added from an ER at $\bar{r}$, which is related via the fluctuation-dissipation theorem to the field damping as follows:[16]

$$I(\bar{r};\ \omega) = \left| F(\bar{r}) \right|_\omega^2 = \frac{u_E(\bar{r};\ \omega)}{\tau(\bar{r};\ \omega)} \qquad (29)$$

where for simplicity, we have assumed that the fluctuations are both temporally and spatially uncorrelated, i.e., $\langle F(\bar{r},\ t)\ F^+(\bar{r}',\ t') \rangle_\omega \sim \delta(t - t')\ \delta(\bar{r} - \bar{r}')$, u_E is the electromagnetic energy density, and τ is the dephasing time of VFs. Since the fluctuation field is assumed to be added from a pointlike source in this model, the phase and amplitude of the field is equal at any two points located at equal distances along a line through the point source. This is more realistic in a thin ER with dimensions much less than a wavelength like a QW layer.[2] With the use of Equation 29, Equation 28 can be written by

$$\frac{u_E(\bar{r};\ \omega)}{\tau(\bar{r};\ \omega)} = \int_{-\infty}^{\infty} d\bar{r}' T_p(\bar{r},\ \bar{r}';\ \omega)\frac{u_E(\bar{r}';\ \omega)}{\tau(\bar{r}';\ \omega)}. \qquad (30)$$

The LHS of Equation 30 expresses the fluctuating force added at $\bar{r}$ and the RHS expresses the dissipation (at $\bar{r}$) of the fluctuating force added at $\bar{r}'$ (see Figure 8). $T_p(\bar{r},\ \bar{r}';\ \omega)$ is defined by (see Appendix A for the derivation)

$$T_p(\bar{r},\ \bar{r}';\ \omega) = \frac{\omega\mu U_E(\bar{r},\ \bar{r}';\ \omega)}{\text{Im}\left[G_E(\bar{r}',\ \bar{r}';\ \omega) \right]\tau(\bar{r};\ \omega)} \qquad (31)$$

where $U_E(\bar{r},\ \bar{r}';\ \omega) = \varepsilon(\bar{r};\ \omega)\left| G_E(\bar{r},\ \bar{r}';\ \omega) \right|^2$, $\varepsilon(\bar{r};\ \omega)$ is the dielectric constant, μ is the permeability and $G_E(\bar{r},\ \bar{r}';\ \omega)$ obeys

$$\left[\frac{d^2}{d\bar{r}^2} + k_0^2 N^2(\bar{r};\ \omega) \right] G_E(\bar{r},\ \bar{r}';\ \omega) = \delta(\bar{r} - \bar{r}') \qquad (32)$$

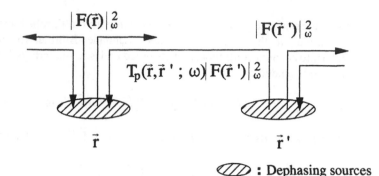

: Dephasing sources

FIGURE 8   Illustration for dephasing of vacuum fields due to electron reservoirs in model II.

where $k_0^2 N^2(\bar{r}; \omega) = [k_0^2 n^2(\bar{r}; \omega) - \alpha^2(\bar{r}; \omega)/4] - j k_0 n (\bar{r}; \omega) \alpha(\bar{r}; \omega)$, k_0 is the wave number in vacuum, $n(\bar{r}; \omega)$ is the refractive index, and $\alpha(\bar{r}; \omega)$ is related to the dephasing time, τ, by $\tau(\bar{r}; \omega)^{-1} = \alpha(\bar{r}; \omega) c(\bar{r}; \omega) (c(\bar{r}; \omega)$ is the speed of light). The physical meaning of $T_p(\bar{r}, \bar{r}'; \omega)$ is the probability that a field, introduced at position $\bar{r}'$, propagates to $\bar{r}$ coherently, i.e., without losing its phase memory. Elastic scatterings due to boundary roughnesses would also destroy phase coherence of VFs in a microcavity. This effect can also be taken into account in this model if the corresponding dephasing time can be determined.

From Equation 30, we finally arrive at the central result of model II, which is the expression for the field strength of the VFs at any positions, given by (see Appendix A for the derivation)

$$\left|E_{vac}\left(\bar{r}; \omega\right)\right|^2 = \frac{\omega^3}{\pi \varepsilon_0 c_0^2} \int_{-\infty}^{\infty} d\bar{r}' \left|G_E\left(\bar{r}, \bar{r}'; \omega\right)\right|^2 \frac{\alpha(\bar{r}'; \omega)}{c(\bar{r}'; \omega)} \qquad (33)$$

where ε_0 and c_0 are the dielectric constant and the speed of light in vacuum. At first sight, the field strength, $|E_{vac}|^2$ seems to vanish at all positions in the case of $\alpha = 0$. However, $|E_{vac}|^2$ does not vanish because of the α-dependence of $|G_E(\bar{r}, \bar{r}'; \omega)|^2$. This is shown in the subsequent section, in which the SE rate in a uniform medium obtained by Equation 33 is independent of α.

B.   SPONTANEOUS EMISSION RATE IN A BULK DIELECTRIC

In order to justify model II, let us calculate the SE rate in a bulk dielectric using model II. The standard expression for the SE rate is given by Fermi's golden rule:

$$R_{sp} = \frac{2\pi}{\hbar} \sum_{v=1}^{2} \left| \vec{d} \cdot \vec{e}_v \right|^2 \left| E_{vac}(\vec{r};\, \omega) \right|^2 \tag{34}$$

where $\vec{d}$ is the transition dipole moment given by $q\langle ex|\vec{r}|g\rangle$ located at $\vec{r}$ (ex and g denote the excited and ground states, respectively and q is the magnitude of the charge of an electron) and $\vec{e}_v$, with $v = 1,2$, are unit polarization vectors. In a homogeneous system, $G_E(\vec{r}, \vec{r}';\, \omega)$ is analytically obtained by

$$G_E(\vec{r}, \vec{r}';\, \omega) = \frac{1}{4\pi} \frac{1}{|\vec{r} - \vec{r}'|} e^{jk_0 n|\vec{r} - \vec{r}'|} e^{-\frac{\alpha}{2}|\vec{r} - \vec{r}'|} \tag{35}$$

where $\alpha/k_0 n \ll 1$ was assumed. Inserting Equation 35 into Equation 33 and using Equation 34, we obtain the familiar expression for the SE rate:[5]

$$R_{sp}^{free} = \frac{\omega^3 |\vec{d}|^2}{3\pi c_0^3 \hbar \varepsilon_0} n \tag{36}$$

which differs from the SE rate in vacuum by a factor equal to the real part of the refractive index (n) and is not affected by the imaginary part ($\propto \alpha$).[18]

C.   NUMERICAL EXAMPLES

We have applied model II to a one-dimensional dielectric microcavity in order to study the effect of dephasing of VFs on the spontaneous emission (SE) in a more practical microcavity structure. A detailed description of the way of calculations lies in Appendix B. The structure used for calculations shown in Figure 9(a) is a half-wavelength ($\lambda/2$) AlAs cavity sandwiched by two distributed Bragg reflectors (DBRs) each of which is made by alternating layers of $Al_{0.2}Ga_{0.8}As$ and AlAs. Each of the DBR reflectivities, R, is set to be 0.999. A single planar GaAs QW is embedded in the middle of the cavity as a SE source. All dipole moments are assumed to be along the QW sheet.[30] In the case where the dipole-moment relaxation time, T_2, typically $\sim 10^{-13}$ s at room temperature,[29] is much shorter than the photon lifetime, τ_{ph}, in a cavity, the parameter α, which is related to the dephasing time of VFs, is approximately equal to an absorption coefficient. Since $\tau_{ph} \sim 10^{-11}$ s for R = 0.999, we have employed this approximation. The absorption coefficients used here are $\alpha_{QW} = 1 \times 10^4$ cm^{-1} [32] for excitonic optical transitions in the GaAs QW sheet and $\alpha_f = 3 \times 10^{-18} n + 7 \times 10^{-18} p$ (cm^{-1})[33] for free carrier absorption in the DBR regions (n (p) is the free electron (hole) concentration). Figure 9(b) shows the SE rate for an S-polarized wave as a function of emission angle relative to the cavity normal. The corresponding dephasing parameter, κ, for the QW region is $\kappa = 0.01$. For comparison,

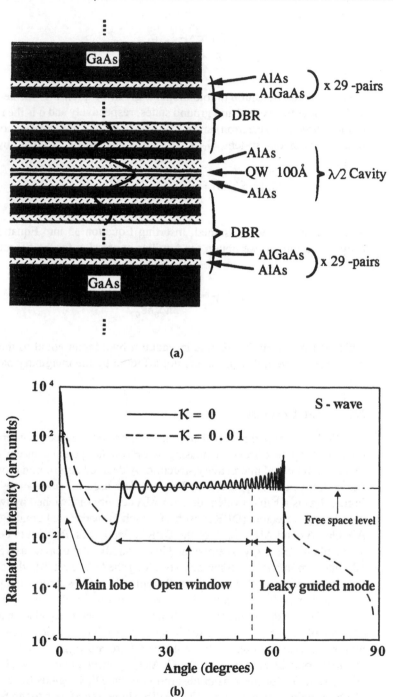

FIGURE 9   (a) One-dimensional DBR microcavity structure used for calculations. (b) The corresponding spontaneous radiation patterns. For comparison, the result for no dephasing (due to the electron reservoir) is shown by a solid line.

the result for the no dephasing case ($\kappa = 0$) is shown by a solid line. As expected, the dephasing effect causes the main lobe to significantly lower and broaden. However, the spurious modes, including open window modes and leaky guided modes, are almost unchanged. This is because the self-interference of the VFs does not play an important role in the spurious modes. The inhibited modes, emitted at angles over ~62°, tend to be enhanced and approach their free space levels.

IV. SUMMARY

The alteration of the spontaneous emission (SE) in a microcavity can be viewed as a manifestation of the self-interference of the quantized fields, which are vacuum fields (VFs).[24] Therefore, the destruction of the self-interference would affect the alteration of the SE in a microcavity. Since such an effect has not been taken into account so far in cavity QED, we have presented two simple theoretical models to incorporate the effect of dephasing of VFs on the SE in a microcavity, caused not only by interacting with external phonon reservoirs through output coupling mirrors but also by interacting with internal electron reservoirs composed of electron systems surrounded by photon heat baths. They are based on the reservoir approach[20,21] in which fluctuation forces are added to preserve the quantum mechanical consistency in the commutation relation for field operators. For simplicity, we have restricted ourselves to the case in which the dephasing effect is uncorrelated both temporally and spatially. Although this simplifying assumption might not be always realistic, we believe that the model can be used to describe the essential physics of the effect of dephasing of VFs on the SE in a microcavity. Also, we believe that the calculations implemented here provide some ideas about how the dephasing of VFs affects the SE in a microcavity.

V. ACKNOWLEDGMENTS

The author is grateful for valuable discussions with Prof. M. Yamanishi, Hiroshima University, and Prof. Y. Yamamoto, Stanford University. He also thanks Mr. M. Inoue, Hiroshima University, for the numerical calculations presented here and thanks Prof. K. Ujihara, The University of Electro-Communications, and Dr. G. Björk, Stanford University, for critical comments for this manuscript.

VI. APPENDIX A

The objective of this appendix is to derive the central result of model II, i.e., Equation 33.

Let us consider a plane electromagnetic wave traveling in the z-direction with an electric field: $\vec{E} = (E_x, 0, 0)$, a magnetic field: $\vec{H} = (0, H_y, 0)$ and an angular frequency of ω. The time-averaged power flow for the electromagnetic field, i.e., the Poynting vector: $\vec{P}_z$ is defined by

$$\vec{P}_z = \frac{1}{2} \text{Re}\left(\vec{E} \times \vec{H}^*\right) \tag{37}$$

where Re denotes a real part. By eliminating the magnetic field, $\vec{H}$, with use of the relation between $\vec{E}$ and $\vec{H}$ obtained by Maxwell's equations and replacing $\vec{E}$ by the corresponding retarded Green's function, $G_E(\vec{z}, \vec{z}'; \omega)$, Equation 37 can be rewritten as

$$\vec{P}_z = -\frac{j}{2\omega\mu}\left[G_E(\vec{z}, \vec{z}'; \omega)\frac{\partial}{\partial z}G_E^*(\vec{z}, \vec{z}'; \omega) \right.$$

$$\left. -G_E^*(\vec{z}, \vec{z}'; \omega)\frac{\partial}{\partial z}G_E(\vec{z}, \vec{z}'; \omega) \right] \tag{38}$$

where μ is the permeability of a medium of interest and $G_E(\vec{z}, \vec{z}'; \omega)$ obeys the following Maxwell's equation with a pointlike source:

$$\left[\vec{\nabla}^2 + k_0^2 N^2(\vec{z}; \omega)\right]G_E(\vec{z}, \vec{z}'; \omega) = \delta(\vec{z} - \vec{z}') \tag{39}$$

where $k_0^2 N^2(\vec{z}; \omega) = [k_0^2 n^2(\vec{z}; \omega) - \alpha^2(\vec{z}; \omega)/4] - j k_0 n (\vec{z}; \omega) \alpha(\vec{z}; \omega)$, k_0 is the wave number, $n(\vec{z}; \omega)$ is the refractive index at $\vec{z}$ and $\alpha(\vec{z}; \omega)$ is related to the dephasing time of a field, $\tau(\vec{z}; \omega)^{-1} = c_0 \alpha(\vec{z}; \omega)/n(\vec{z}; \omega)$. Differentiating Equation 38 with respect to $\vec{z}$ and using Equation 39, we have

$$\vec{\nabla} \cdot \vec{P}_z + \frac{U_E(\vec{z}, \vec{z}'; \omega)}{\tau(\vec{z}; \omega)} = \frac{-j}{2\omega\mu}\left[G_E(\vec{z}, \vec{z}'; \omega) - G_E^*(\vec{z}, \vec{z}'; \omega)\right]$$

$$\times \delta(\vec{z} - \vec{z}') \tag{40}$$

where $U_E(\vec{z}, \vec{z}'; \omega)$ is defined by

$$U_E(\vec{z}, \vec{z}'; \omega) \equiv \varepsilon(\vec{z}; \omega)\left|G_E(\vec{z}, \vec{z}'; \omega)\right|^2 \tag{41}$$

and $\varepsilon(\vec{z}; \omega)$ is the dielectric constant at $\vec{z}$ and c_0 the speed of light in vacuum. By integrating Equation 40 over all volume and using the divergence theorem under the assumption that the boundaries are far away so that no electromagnetic power flows out of the surface, i.e., $\int \vec{\nabla} \cdot \vec{P}_z dV = \int \vec{P}_z \cdot d\vec{S} = 0$, we have

$$\int_{-\infty}^{\infty} d\vec{z} T_p(\vec{z}, \vec{z}'; \omega) = 1 \tag{42}$$

where

$$T_p(\bar{z}, \bar{z}'; \omega) \equiv \frac{\omega\mu U_E(\bar{z}, \bar{z}'; \omega)}{Im[G_E(\bar{z}', \bar{z}'; \omega)]\tau(\bar{z}; \omega)} \tag{43}$$

(Im denotes an imaginary part). The physical meaning of the $T_p(\bar{z}, \bar{z}'; \omega)$ is the probability that an electromagnetic field propagates from $\bar{z}'$ to $\bar{z}$ without losing its phase coherence.

The derivation of Equation 33 is as follows. For convenience, we repeat Equation 30:

$$\frac{u_E(\bar{z}; \omega)}{\tau(\bar{z}; \omega)} = \int_{-\infty}^{\infty} d\bar{z}' T_p(\bar{z}, \bar{z}'; \omega) \frac{u_E(\bar{z}'; \omega)}{\tau(\bar{z}'; \omega)}. \tag{44}$$

Then we express $u_E(\bar{z}; \omega)$ and $u_E(\bar{z}'; \omega)$ by

$$u_E(\bar{z}; \omega) = \varepsilon(\bar{z}; \omega)|E_{vac}(\bar{z}; \omega)|^2$$

$$u_E(\bar{z}'; \omega) = \frac{1}{2}\hbar\omega \times \frac{1}{\pi} Im[G_E(\bar{z}', \bar{z}'; \omega)] \times \left(\frac{2\omega}{\hbar c(\bar{z}'; \omega)^2} \right) \tag{45}$$

where $E_{vac}(\bar{z}; \omega)$ is the electric field of a vacuum field, $\hbar\omega/2$ is the energy of a vacuum field, Im $[G_E(\bar{z}', \bar{z}', \omega)]/\pi \times (2\omega/\hbar c(\bar{z}'; \omega)^2)$ is the photon density of states per unit energy and volume[34] and $c(\bar{z}'; \omega)$ is the speed of light at $\bar{z}'$. Substituting Equation 45 into Equation 44 and carrying out a simple calculation with use of $\tau(\bar{z}; \omega)^{-1} = c_0\alpha(\bar{z}; \omega)/n(\bar{z}; \omega)$ and $\mu = \varepsilon_0^{-1}c_0^{-2}$, we finally arrive at the key result in the model II:

$$|E_{vac}(\bar{z}; \omega)|^2 = \frac{\omega^3}{\pi\varepsilon_0 c_0^2} \int_{-\infty}^{\infty} d\bar{z}' |G_E(\bar{z}, \bar{z}'; \omega)|^2 \frac{\alpha(\bar{z}'; \omega)}{c(\bar{z}'; \omega)} \tag{46}$$

where ε_0 is the dielectric constant in vacuum. Since we can arbitrarily choose the $\bar{z}$ direction in space, Equations 43 and 46 become

$$T_p(\bar{r}, \bar{r}'; \omega) = \frac{\omega\mu U_E(\bar{r}, \bar{r}'; \omega)}{Im[G_E(\bar{r}', \bar{r}'; \omega)]\tau(\bar{r}; \omega)} \tag{47}$$

$$|E_{vac}(\bar{r}; \omega)|^2 = \frac{\omega^3}{\pi\varepsilon_0 c_0^2} \int_{-\infty}^{\infty} d\bar{r}' |G_E(\bar{r}, \bar{r}'; \omega)|^2 \frac{\alpha(\bar{r}'; \omega)}{c(\bar{r}'; \omega)} \tag{48}$$

by replacing $\bar{z}$ by $\bar{r}$.

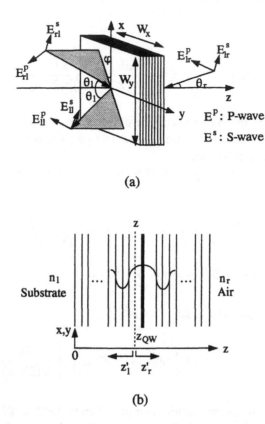

(a)

(b)

FIGURE 10 (a) Spherical coordinates used for the calculations; (b) cross-sectional view of the microcavity.

VII. APPENDIX B

In this appendix, our objective is to describe the way of calculating the SE radiation pattern and the SE lifetime in a one-dimensional DBR microcavity with model II. Let us consider a system shown in Figure 10. Assuming that the system is uniform along the x- and y-directions and their dimensions, W_x and W_y, are much larger than a wavelength of the SE under consideration, the problem can be tackled as a one-dimensional problem. Averaging $|E_{vac}(\bar{r}; \omega)|^2$ in Equation 33 over the x-y plane gives

$$\langle |E_{vac}(z; \omega)|^2 \rangle = \frac{1}{W_x W_y} \int dx\, dy |E_{vac}(\bar{r}; \omega)|^2$$

$$= \frac{\omega^3}{\pi \varepsilon_0 c_0^2} \frac{1}{W_x W_y} \int dx\, dy \int_{-\infty}^{\infty} d\bar{r}' |G_E(\bar{r}, \bar{r}'; \omega)|^2 \frac{\alpha(\bar{r}'; \omega)}{c(\bar{r}'; \omega)}. \tag{49}$$

In general, one can expand Green's function, $G_E(\bar{r}, \bar{r}'; \omega)$, in terms of a complete orthonormal set of eigenfunctions, $\{\phi_i\}$, (subject to the same boundary conditions) and the corresponding set of eigenvalues, $\{k_i^2\}$, as follows[34]

$$G_E(\bar{r}, \bar{r}'; \omega) = \sum_{l,m,n} \frac{\phi_l(x)\phi_m(y)\phi_n(z)\phi_l^*(x')\phi_m^*(y')\phi_n^*(z')}{k_0^2 N(z; \omega)^2 - (k_l^2 + k_m^2 + k_n^2)} \tag{50}$$

where k_0 is the wave number in vacuum and $N(z; \omega)$ is the complex refractive index given by $N(z; \omega) = n(z; \omega) - j\,\alpha(z; \omega)/2k_0$. By using Equation 50 and assuming that $\phi_l(x)$ and $\phi_m(y)$ are plane-wave modes given by $\phi_l(x) = e^{jk_l x}/\sqrt{W_x}$ and $\phi_m(y) = e^{jk_m y}/\sqrt{W_y}$, respectively, we have

$$\frac{1}{W_x W_y} \iint dx\,dy \iint dx'dy' |G_E(\bar{r}, \bar{r}'; \omega)|^2 \frac{\alpha(\bar{r}'; \omega)}{c(\bar{r}'; \omega)} = \frac{1}{W_x W_y}$$

$$\sum_{l,m} \left[\sum_n \frac{\phi_n(z)\phi_n^*(z')}{k_0^2 N^2(z; \omega) - k_l^2 - k_m^2 - k_n^2} \times \sum_{n'} \frac{\phi_{n'}^*(z)\phi_{n'}(z')}{k_0^2 N^2(z; \omega) - k_l^2 - k_m^2 - k_{n'}^2} \right]$$

$$\frac{\alpha(z'; \omega)}{c(z'; \omega)} = \frac{1}{W_x W_y} \sum_{l,m} |G_E^{1-D}(z, z'; k_z^2)|^2 \frac{\alpha(z'; \omega)}{c(z'; \omega)} \tag{51}$$

where $k_z^2 = k_0^2 N^2(z; \omega) - k_l^2 - k_m^2$ and $G_E^{1-D}(z, z'; k_z^2)$ is a solution for

$$\left[\frac{d^2}{dz^2} + k_z^2(z; \omega) \right] G_E^{1-D}(z, z'; k_z^2) = \delta(z - z'). \tag{52}$$

Furthermore, by replacing the summation over modes l's and m's by $W_x W_y/(2\pi)^2 \int_0^{2\pi} d\varphi \int_0^\infty k_t dk_t$, integrating over k_t satisfying $k_t^2 = k_0^2 N^2(z; \omega) - k_z^2$ (because the integration over k_t given by $k_t^2 = k_0^2 N^2(z; \omega) - k_z^2$ is dominant) and transforming k_t into $k_z = k_0 N(z; \omega)\cos\theta$, Equation 51 becomes

$$\frac{1}{W_x W_y} \iint dx\,dy \iint dx'dy' |G_E(\bar{r}, \bar{r}'; \omega)|^2 \frac{\alpha(\bar{r}'; \omega)}{c(\bar{r}'; \omega)}$$

$$= \frac{1}{2\pi^2} k_0^2 n^2(z; \omega) \int_0^{2\pi} d\varphi \int_0^{\pi/2} d\theta\, \sin\theta\, \cos\theta |G_E^{1-D}(z, z'; k_z^2)|^2 \frac{\alpha(z', \omega)}{c(z', \omega)} \tag{53}$$

where $N(z; \omega) \sim n(z; \omega)$ was used for the coefficient in front of the integration by assuming $\alpha/k_0 \ll 1$. With use of the above result, we obtain

$$\left\langle \left| E_{vac}(z;\ \omega) \right|^2 \right\rangle = \frac{\omega^3 k_0^2 n^2(z;\ \omega)}{2\pi^3 \varepsilon_0 c_0^2} \int_0^{2\pi} d\varphi \int_0^{\pi/2} d\theta\ \sin\theta\ \cos\theta$$

$$\times \int_{-\infty}^{\infty} dz' \left| G_E^{1-D}\left(z,\ z';\ k_z^2\right) \right|^2 \frac{\alpha(z';\ \omega)}{c(z';\ \omega)}$$

$$= \int_0^{2\pi} d\varphi \int_0^{\pi/2} d\theta\ \sin\theta\ F(z,\theta) \tag{54}$$

where

$$F(z,\theta) \equiv \frac{\omega^3 k_0^2 n^2(z;\ \omega)}{2\pi^3 \varepsilon_0 c_0^2} \int_{-\infty}^{\infty} dz' \left| G_E^{1-D}\left(z,\ z';\ k_z^2\right) \right|^2 \frac{\alpha(z';\ \omega)}{c(z';\ \omega)}\ \cos\theta. \tag{55}$$

For a vacuum field incident from the left-hand side (substrate side) at an angle of θ_l,

$$F_l^{s(p)}\left(z,\ \theta_l\right) \equiv \frac{\omega^3 k_0^2 n^2(z;\ \omega)}{2\pi^2 \varepsilon_0 c_0^2} \int_{-\infty}^z dz_l' \left| G_E^{1-D,s(p)}\left(z,\ z_l';\ k_z^2\right) \right|^2 \frac{\alpha(z_l';\ \omega)}{c(z_l';\ \omega)}\ \cos\theta_l \tag{56a}$$

and for a vacuum field incident from the right-hand side (air side) at an angle of θ_r,

$$F_r^{s(p)}\left(z,\ \theta_r\right) \equiv \frac{\omega^3 k_0^2 n^2(z;\ \omega)}{2\pi^3 \varepsilon_0 c_0^2} \int_z^{\infty} dz_r' \left| G_E^{1-D,s(p)}\left(z,\ z_r';\ k_z^2\right) \right|^2 \frac{\alpha(z_r';\ \omega)}{c(z_r';\ \omega)}\ Re(\cos\theta_r) \tag{56b}$$

where s and p denote S- and P-polarized waves, respectively, and $Re(\cos\theta_r)$ in Equation 56b indicates total internal reflections (This factor does not appear in Equation 56a because the refractive index of the substrate is assumed to be highest). With the help of Equations 56a and 56b, the SE rates in a quantum well located at $z = z_{QW}$ as a function of emission angle, $\gamma(\theta,\varphi)$, are expressed by

$$\gamma_l^s(\theta_l;\ \varphi) = \frac{2\pi}{\hbar} F_l^s\left(z_{QW};\ \theta_l\right)\left[d_x^2 \sin^2\varphi + d_y^2 \cos^2\varphi\right] \tag{57a}$$

$$\gamma_l^p(\theta_l;\ \varphi) = \frac{2\pi}{\hbar} F_l^p\left(z_{QW};\ \theta_l\right)\left\{\left[d_x^2 \cos^2\varphi + d_y^2 \sin^2\varphi\right]\left|\cos\theta_{QW}\right|^2 + d_z^2 \left|\sin\theta_{QW}\right|^2\right\} \tag{57b}$$

$$\gamma_r^s(\theta_r;\ \varphi) = \frac{2\pi}{\hbar} F_r^s\left(z_{QW};\ \theta_r\right)\left[d_x^2 \sin^2\varphi + d_y^2 \cos^2\varphi\right] \tag{57c}$$

$$\gamma_r^p(\theta_r;\ \varphi) = \frac{2\pi}{\hbar} F_r^p\left(z_{QW};\ \theta_r\right)\left\{\left[d_x^2 \cos^2\varphi + d_y^2 \sin^2\varphi\right]\left|\cos\theta_{QW}\right|^2 + d_z^2 \left|\sin\theta_{QW}\right|^2\right\} \tag{57d}$$

where γ_i and γ_r denote the SE rates on the substrate side and air side, respectively, and d_x, d_y, and d_z are the x-, y-, and z-components of an atomic dipole moment. The SE lifetime, τ_{sp}, can now be calculated simply by the inverse of the SE rate integrated over all solid angles:

$$\frac{1}{\tau_{sp}} = \int_\Omega d\Omega \left(\gamma_1^s(\theta_1;\ \varphi) + \gamma_1^p(\theta_1;\ \varphi) + \gamma_r^s(\theta_r;\ \varphi) + \gamma_1^p(\theta_r;\ \varphi) \right)$$

$$= \int_0^{2\pi} d\varphi \int_0^{\pi/2} d\theta_1 \sin\theta_1 \left(\gamma_1^s(\theta_1;\ \varphi) + \gamma_1^p(\theta_1;\ \varphi) + \gamma_r^s(\theta_1;\ \varphi) + \gamma_r^p(\theta_1;\ \varphi) \right) \quad (58)$$

where Snell's law, $\theta_r = \sin^{-1}\left[\dfrac{n_1 \sin(\theta_1)}{n_r} \right]$, was used. Finally, we would like to

note that the present model is applicable to only the case in which no guided modes exist.

VIII. APPENDIX C

The objective of this appendix is to describe the numerical calculations of the retarded Green's function $G_E^{1-D}(z, z';\ k_z^2)$ introduced in the previous appendix.[35]

Let us consider a system with two perfect absorbers at both ends shown in Figure 11 where T_n expresses a transfer matrix for the n-th region, which is either a free space matrix or a boundary matrix, as was shown at the end of this chapter. These absorbers allow us to neglect the reflections of electromagnetic waves after they escape from a cavity. As shown in Appendix B, $G_E^{1-D}(z, z';\ k_z^2)$ is the solution for

$$\left[\frac{d^2}{dz^2} + k_z^2(z;\ \omega) \right] G_E^{1-D}\left(z, z';\ k_z^2(z;\ \omega) \right) = \delta(z - z') \quad (59)$$

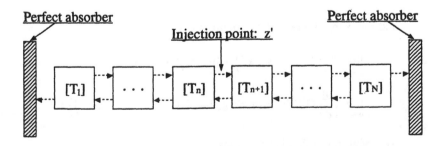

FIGURE 11 The system used for the calculation of $G_E^{1-D}(z, z';\ k_z^2(z))$, which is connected to two perfect absorbers at both ends.

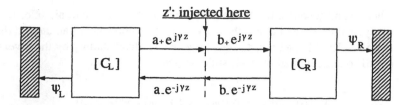

FIGURE 12   Illustration for the first step of the calculation of $G_E^{1-D}(z, z'; k_z^2(z))$ around $z = z'$. The matrices C_L and C_R are the overall matrices on the LHS and on the RHS with respect to z', respectively.

where $k_z^2(z; \omega) = k_{z0}^2 N^2(z; \omega) = [k_{z0}^2 n^2(z; \omega) - \alpha^2(z; \omega)/4] - j k_{z0} n(z; \omega) \alpha(z; \omega)$, and k_{z0} is the z-component of a wave number in vacuum.

We begin by the situation described in Figure 12. The field amplitudes in Figure 12 are related by

$$
\begin{bmatrix} 0 \\ \Psi_L \end{bmatrix} = \begin{bmatrix} t_{11}^L & t_{12}^L \\ t_{21}^L & t_{22}^L \end{bmatrix} \begin{bmatrix} a_+ e^{j\gamma z'} \\ a_- e^{-j\gamma z'} \end{bmatrix} \tag{60}
$$

$$
\begin{bmatrix} b_+ e^{j\gamma z'} \\ b_- e^{-j\gamma z'} \end{bmatrix} = \begin{bmatrix} t_{11}^R & r_{12}^R \\ t_{21}^R & t_{22}^R \end{bmatrix} \begin{bmatrix} \Psi_R \\ 0 \end{bmatrix} \tag{61}
$$

where Ψ_L and Ψ_R are the field amplitudes absorbed by the perfect absorbers and $\gamma \equiv \left[k_{z0}^2 N^2(z; \omega) \right]^{\frac{1}{2}}$. From the above relations, we have

$$
a_+ = -\left(\frac{t_{12}^L}{t_{11}^L} \right) a_- e^{-2j\gamma z'}, \quad b_- = \left(\frac{t_{21}^R}{t_{11}^R} \right) b_+ e^{2j\gamma z'}. \tag{62}
$$

Next, by assuming a solution for $G_E^{1-D}(z, z'; k_z^2)$ around z' takes the form:

$$
\begin{aligned}
G_E^{1-D}(z, z'; k_z^2) &= a_+ e^{j\gamma z} + a_- e^{-j\gamma z} \quad (z \le z') \\
G_E^{1-D}(z, z'; k_z^2) &= b_+ e^{j\gamma z} + b_- e^{-j\gamma z} \quad (z > z')
\end{aligned} \tag{63}
$$

and imposing the boundary conditions given by

$$
\lim_{\delta \to 0} G_E^{1-D}(z' - \delta, z'; k_z^2) = \lim_{\delta \to 0} G_E^{1-D}(z' + \delta, z'; k_z^2)
$$

$$
\lim_{\delta \to 0} \left\{ \frac{d}{dz} G_E^{1-D}(z' + \delta, z'; k_z^2) - \frac{d}{dz} G_E^{1-D}(z' - \delta, z'; k_z^2) \right\} = 1, \tag{64}
$$

we obtain the following relations

$$a_+ = b_+ - \frac{e^{-j\gamma z'}}{2j\gamma}, \quad a_- = b_- + \frac{e^{j\gamma z'}}{2j\gamma}.$$

(65)

Then, a combination of Equation 62 and Equation 65 gives the amplitudes: a_+, a_-, b_+, and b_- as

$$a_+ = -\frac{C_L(1+C_R)e^{-j\gamma z'}}{2j\gamma(1+C_L C_R|)}, \quad a_- = \frac{(1+C_R)e^{j\gamma z'}}{2j\gamma(1+C_L C_R)}$$

$$b_+ = \frac{(1-C_L)e^{-j\gamma z'}}{2j\gamma(1+C_L C_R)}, \quad b_- = \frac{C_R(1-C_L)e^{j\gamma z'}}{2j\gamma(1+C_L C_R)}$$

(66)

where $C_L = (t_{12}^t/t_{11}^t)$, $C_R = (t_{21}^R/t_{11}^R)$. Thus $G_E^{L-D}(z, z'; k_z^2)$ around z' have been determined. The remaining task is to determine $G_E^{L-D}(z, z'; k_z^2)$ for the other regions. To do this, we derive the general expressions for field amplitudes of the positively and negatively traveling waves in an arbitrary region. By summing all possible paths involving multiple reflections in the region under consideration (the typical paths are shown in Figure 13(a) and (b)), ψ_n^+ and ψ_n^- are obtained by

$$\Psi_n^+ = r'_{n-1}\left[1 - r_n r'_{n-1}\right]^{-1} t'_n \cdot \Psi_{r \to l}$$

$$\Psi_n^- = \left[1 - r_n r'_{n-1}\right]^{-1} t'_n \cdot \Psi_{r \to l}$$

(67a)

and similarly,

$$\Psi_{n+2}^+ = \left[1 - r_{n+2} r'_{n+1}\right]^{-1} t_{n+1} \cdot \Psi_{l \to r}$$

$$\Psi_{n+2}^- = \left[1 - r_{n+2} r'_{n+1}\right]^{-1} t_{n+1} r_{n+2} \cdot \Psi_{l \to r}$$

(67b)

where t_i (t_i') and r_i (r_i') are the transmission and reflection coefficients for field amplitudes shown in Figure 13(a) and (b). The matrices C_{n-1} and C_{n+2} are the overall matrices for the region 1 to $n-1$ and for the region $n+2$ to N, respectively. They are obtained simply by matrix multiplications. Starting from the results obtained in Equation 66 and using Equations 67a and 67b, one can determine $G_E^{L-D}(z, z'; k_z^2)$ in each region by iteration and finally obtain $G_E^{L-D}(z, z'; k_z^2)$ for the entire structure (see Figure 14).

To close this appendix, we show the expressions for the transfer matrices of a free space region and a boundary region, and also show how to obtain t_i (t_i') and r_i (r_i') used in Equations 67a and 67b from the elements of a transfer matrix. Suppose a system as depicted in Figure 15. For a homogeneous planar dielectric, the corresponding transfer matrix equation is given by

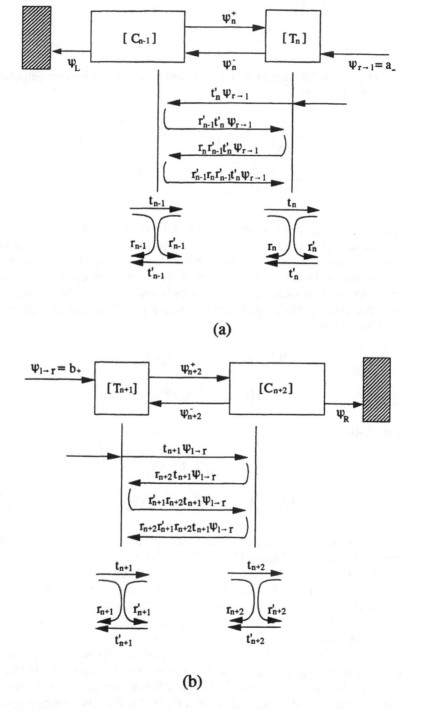

FIGURE 13 Illustration for the calculations of $G_E^{L-D}(z, z'; k_z^2(z))$ for the remaining positions; (a) $z < z'$, (b) $z > z'$. The matrices C_{n-1} and C_{n+2} are the overall matrices for the region 1 to $n - 1$ and for the region $n + 2$ to N, respectively.

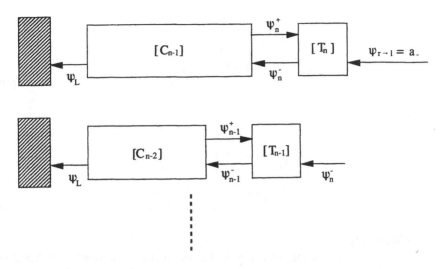

FIGURE 14   Illustration for the iterative calculation of $G_E^{I-D}(z, z'; k_z^2(z))$.

$$\begin{bmatrix} \Psi_0^+ \\ \Psi_0^- \end{bmatrix} = \begin{bmatrix} \exp(j\gamma\Delta z) & 0 \\ 0 & \exp(-j\gamma\Delta z) \end{bmatrix} \begin{bmatrix} \Psi_1^+ \\ \Psi_1^- \end{bmatrix} \tag{68}$$

for both S- and P-polarized waves[2] where $\gamma = \left[k_{z0}^2 N^2(z; \omega)\right]^{\frac{1}{2}}$, $N(z;\omega)$ is the complex refractive index defined by $N(z;\omega) = n(z;\omega) - j\alpha(z;\omega)/2k_{z0}$ and Δz is the distance of a free space. For a planar interface between two different dielectrics, the corresponding transfer matrix equation for S- and P-polarized waves are given by

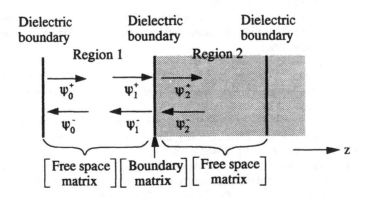

FIGURE 15   Transfer matrix representation of an inhomogeneous dielectric. The transfer matrix for the whole structure is obtained by cascading the free space matrices and boundary matrices.

$$\begin{bmatrix} \Psi_1^+ \\ \Psi_1^- \end{bmatrix} = \frac{1}{2} \begin{bmatrix} 1 + \dfrac{\gamma_2}{\gamma_1} & 1 - \dfrac{\gamma_2}{\gamma_1} \\ 1 - \dfrac{\gamma_2}{\gamma_1} & 1 + \dfrac{\gamma_2}{\gamma_1} \end{bmatrix} \begin{bmatrix} \Psi_2^+ \\ \Psi_2^- \end{bmatrix}$$

(69)

and

$$\begin{bmatrix} \Psi_1^+ \\ \Psi_1^- \end{bmatrix} = \frac{N_2}{2N_1} \begin{bmatrix} 1 + \dfrac{N_1^2 \gamma_2}{N_2^2 \gamma_1} & 1 - \dfrac{N_1^2 \gamma_2}{N_2^2 \gamma_1} \\ 1 - \dfrac{N_1^2 \gamma_2}{N_2^2 \gamma_1} & 1 + \dfrac{N_1^2 \gamma_2}{N_2^2 \gamma_1} \end{bmatrix} \begin{bmatrix} \Psi_2^+ \\ \Psi_2^- \end{bmatrix},$$

(70)

respectively.[2]

Once an overall transfer matrix for the region of interest is obtained by using the above matrices, one can obtain t, t', r, and r' used in Equations 67a and 67b by

$$t = \frac{1}{t_{11}}, \qquad t' = t_{22} - \frac{t_{12}t_{21}}{t_{11}}$$

$$r = \frac{t_{21}}{t_{11}}, \qquad r' = -\frac{t_{12}}{t_{11}}$$

(71)

where the $t_{i,j}$'s are the (i,j) elements of the transfer matrix.

REFERENCES

1. Haroche, S. and Raimond, J.-M., Cavity quantum electrodynamics, *Adv. At. Mol. Phys.*, 20, 350, 1985.
2. Björk, G., Machida, S., Yamamoto, Y., and Igeta, K., Modification of spontaneous emission rate in planar dielectric microcavity structures, *Phys. Rev. A*, 44, 669, 1991.
3. Loudon, R. and Knight, P. L., Squeezed light, *J. Mod. Opt.*, 34, 709, 1987.
4. Dailbard, J., Dupont-Roc, J., and Cohen-Tannoudji, C., Vacuum fluctuations and radiative reaction: identification of their respective contributions, *J. Phys.*, 43, 1617, 1982.
5. Yablonovich, E., Gmitter, T. J., and Bhat, R., Inhibited and enhanced spontaneous emission from optically thin AlGaAs/GaAs double heterostructures, *Phys. Rev. Lett.*, 61, 2546, 1988.
6. Yokoyama, H., Nishi, K., Anan, T., Yamada, H., Brorson, S. D., and Ippen, E. P., Enhanced spontaneous emission from GaAs quantum wells in monolithic microcavities, *Appl. Phys. Lett.*, 57, 2814, 1990.
7. Roger, T. J., Deppe, D. G., and Streeman, B. G., Effect of an AlAs/GaAs mirror on the spontaneous emission of an InGaAs-GaAs quantum well, *Appl. Phys. Lett.*, 57, 1858, 1990.

8. Yamamoto, Y., Machida, S., Horikoshi, Y., Igeta, K., and Björk, G., Enhanced and inhibited spontaneous emission of free excitons in GaAs quantum wells in a microcavity, *Opt. Commun.*, 80, 337, 1991.
9. Ochi, N., Shiotani, T., Yamanishi, M., Honda, Y., and Suemune, I., Controllable enhancement of excitonic spontaneous emission by quantum confined Stark effect in GaAs quantum wells embedded in quantum microcavities, *Appl. Phys. Lett.*, 58, 2735, 1991.
10. Yamauchi, T., Arakawa, Y., and Nishioka, M., Enhanced and inhibited spontaneous emission in GaAs/AlGaAs vertical microcavity lasers with two kinds of quantum wells, *Appl. Phys. Lett.*, 58, 2339, 1991.
11. Tezuka, T., Nunoue, S., Yoshida, H., and Noda, T., Spontaneous emission enhancement in pillar-type microcavities, *Jpn. J. Appl. Phys.*, 32, L54, 1993.
12. Vredenberg, A. M., Hunt, N. E. J., Schubert, E. F., Jacobson, D. C., Poate, J. M., and Zydzik, G. J., Controlled atomic spontaneous emission from E_r^{3+} in a transparent S_i/S_iO_2 microcavity, *Phys. Rev. Lett.*, 71, 517, 1993.
13. Yokoyama, H. and Brorson, S. D., Rate equation analysis of microcavity lasers, *J. Appl. Phys.*, 66, 4801, 1989.
14. Yamamoto, Y., Machida, S., and Björk, G., Microcavity semiconductor laser with enhanced spontaneous emission, *Phys. Rev. A*, 44, 657, 1991.
15. See, for instance, Yariv, A., *Quantum Electronics*, 3rd ed., John Wiley & Sons, Singapore, 1988, chap. 5.
16. Haken, H., *Light*, North-Holland, Amsterdam, 1981, chap. 9.
17. Huttner, B., Baumberg, J. J., and Barnett, S. M., Canonical quantization of light in a linear dielectric, *Europhys. Lett.*, 16, 177, 1991.
18. Barnett, S. M., Huttner, B., and Loudon, R., Spontaneous emission in absorbing dielectric media, *Phys. Rev. Lett.*, 68, 3698, 1992.
19. Johnson, F. S., Physical cause of group velocity in normally dispersive, nondissipative media, *Am. J. Phys.*, 58, 1044, 1990.
20. Lax, M., Quantum noise X: Density matrix treatment of fields and population difference fluctuations, *Phys. Rev.*, 157, 213, 1967.
21. Haken, H., *Laser Theory*, Encyclopedia of Physics XXV/2C, Springer-Verlag, New York, 1970, chap. 4.
22. Lee, Y., Yamanishi, M., Yamamoto, Y., Interaction of vacuum field fluctuations with electron system in a microcavity, in *Proceedings of the Satellite Workshop, Quantum Control and Measurement*, ed. Ezawa, H. and Murayama, Y., North-Holland, Amsterdam, 1993, 161.
23. Marcuse, D., *Engineering Quantum Electrodynamics*, Harcourt, Brace & World, Inc., San Francisco, 1970, chap. 6.
24. Ueda, M. and Imoto, N., Anomalous commutation relation and modified spontaneous emission inside a microcavity, *Phys. Rev. A*, 50, 89, 1994.
25. Büttiker, M., Role of quantum coherence in series resistors, *Phys. Rev. B*, 33, 3020, 1986.
26. von Hippel, A. R., *Dielectrics and Waves*, John Wiley & Sons, New York, 1954, chap. 2.
27. Yamamoto, Y. and Imoto, N., Internal and external field fluctuations of a laser oscillator: part I — quantum mechanical Langevin treatment, *IEEE. J. Quantum Electron.*, QE-22, 2032, 1986.
28. Sergent III, M., Scully, M. O., and Lamb, Jr, W. E., *Laser Physics*, Addison-Wesley, Cambridge, MA, 1974, chap. 20.

29. Yamanishi, M. and Lee, Y., Phase dampings of optical dipole moments and gain spectra in semiconductor lasers, *IEEE. J. Quantum Electron.*, QE-23, 367, 1987.
30. Yamanishi, Y. and Suemune, I., Comment on polarization dependent momentum matrix elements in quantum well lasers, *Jpn. J. Appl. Phys.*, 23, L35, 1984.
31. Datta, S., Steady-state quantum kinetic equation, *Phys. Rev. B*, 40, 5830, 1989.
32. Masumoto, Y., Matsuura, M., Tarucha, S., and Okamoto, H., Direct experimental observation of two-dimensional shrinkage of the exciton wavefunction in quantum wells, *Phys. Rev. B*, 32, 4275, 1985.
33. Casey, Jr., H. C. and Panish, M. B., *Heterostructure Lasers*, Academic Press, London, 1978, chap. 3.
34. Economou, E. N., *Green's Functions in Quantum Physics*, Springer-Verlag, Berlin, 1990, chap. 3.
35. Datta, S. and McLennan, M. J., Technical Report TR-EE 88-42, Purdue University, Lafayette, IN, 1988.

3 Effects of Atomic Broadening on Spontaneous Emission in an Optical Microcavity

Kikuo Ujihara

TABLE OF CONTENTS

0-8493-3786-0/95/$0.00+$.50
© 1995 by CRC Press, Inc.

I.   INTRODUCTION

Spontaneous emission in an optical microcavity is studied for spectrally broadened atoms. Effects of homogeneous broadening and inhomogeneous broadening are analyzed by a master equation approach. Under the Markov approximation the expressions for the emission rate and the emission line profile are obtained. Numerical examples are given.

A.   HISTORICAL BACKGROUND

The mechanism and characteristics of spontaneous emission in a cavity have been studied in a number of papers.[1-5] When the coupling ratio of spontaneous emission to a particular cavity mode is close to unity, it is possible to construct a laser of very low threshold.[6-8] For this purpose a microcavity in the optical region can be used. In this context it is important to understand the characteristics of spontaneous emission in such an optical microcavity. Most theoretical or experimental analyses of spontaneous emission in a microcavity have been made assuming two-level atoms.[9,10] On the other hand, attempts to construct a very-low-threshold laser have been made using homogeneously and inhomogeneously broadened atoms or media, dyes[8,11] and excitons in semiconductors.[12,13] At room temperature dyes are broadened by vibrational-rotational energy levels and by frequent collisions to have a width of, typically, 2×10^{14} rad/s or 30 nm at 600 nm. The transverse relaxation time of a dye is[14] of the order of 10^{-12} s, which amounts to a homogeneous width of about 2×10^{12} rad/s or 0.3 nm. The emission width at 77K of free excitonic transition between the lowest conduction band and the lowest heavy hole band in a GaAs quantum well is 0.5 to 0.7 nm or 1.4 to 2.0×10^{12} rad/s, which includes both homogeneous and inhomogeneous broadening.[15] Therefore, we need a theory to calculate the spontaneous emission rate and emission spectrum for an atom in a microcavity that has, simultaneously, homogeneous broadening, inhomogeneous broadening, and energy-level continuum. Several authors have intuitively introduced atomic broadening in their theory to treat spontaneous emission rate and the coupling ratio in an optical microcavity.[16-18]

In this chapter we develop a consistent theory of spontaneous emission in an optical microcavity from a broadened atom. We derive the atomic broadening from first principles by assuming the presence of reservoirs coupled to the atoms. Solving the master equation we find expressions for the spontaneous emission rate and the emission spectrum. We do not consider superradiant[19,20] or collective behavior of broadened atoms.

A useful, very small optical cavity has only one (or limited number of) resonant mode(s), which leads to a large coupling ratio of spontaneous emission to a particular quasi mode of interest. However, it is important to note that the cavity must have finite output coupling to be of practical use. The cavity structures actually employed to date[6-8] mostly belong to the family of very short planar Fabry-Pérot cavity with mirrors of finite transmission. These features will be taken into account in constructing a model cavity.

In the remainder of this section physical concepts and mathematical techniques that appear in this chapter are explained. Especially, the master equation for an atom interacting with a reservoir is derived in Section I.C. In Section II the spontaneous emission in an optical microcavity is analyzed. In Section II.A the cavity model and the expression for the radiation field as well as the model of atom-field coupling are described. In Section II.B the density matrix equations are derived from the master equation. In Section II.C the density matrix equations are solved and the expressions for the spontaneous emission rate and for the emission spectrum for homogeneously broadened atoms are given. Section II.D deals with inhomogeneous broadening. In Section III the present calculation is discussed and the conclusions are given.

B.   NATURE OF THE PROBLEM

We treat here the phenomenon of spontaneous emission. Spontaneous emission is "induced" by the vacuum-field fluctuation, and the latter is properly described only by quantum theory of the radiation field. Spontaneous emission is an irreversible process and is necessarily related to an infinite number of degrees of freedom of the radiation field. Before the spontaneous emission event, ideally, every degree of freedom is at vacuum state initially, i.e., there is no photon in any mode of the field. Spontaneous emission is a process whereby an excited atom releases a photon to a mode of the field as it relaxes to a lower energy state. In free space, the photon leaves the atom and never comes back.

In a perfect cavity, however, an excited atom can emit a photon, but the photon will eventually come back to the atom and can be absorbed by the atom. Thus the process is reversible. In a realistic cavity, the cavity has a finite loss and the energy will eventually be dissipated from the atom-cavity system to the loss mechanism. In this case, the field energy follows an oscillatory damped evolution tending to zero. In this case, if the loss rate is very large, (more specifically, if the loss rate is larger than the so-called Rabi frequency) the oscillation is suppressed and a monotonic decay of the field energy will occur. The last situation is what will be treated in this chapter.

Thus, in our problem there is a cavity mode to which the excited atom emits a photon. However, this field mode is strongly damped. The damping may come from absorption at the cavity walls and in the laser medium or from transmission through the cavity wall (mirror) to outside free space. From an engineering point of view, the former damping mechanism is undesirable but the latter mechanism is inevitable if the emitted photon is to be used outside the cavity. In this chapter, we will treat only the latter damping mechanism. The effect of former mechanism on spontaneous emission in a cavity is treated in Chapter 2 by Yong Lee in this volume.

The loss of the cavity due to transmission to the outside space cannot be described properly by the cavity mode only. Since in this case the cavity mode is associated with a damping mechanism, the quantization of the cavity mode can be performed only with special care; the quantum mechanical description of the damping will be mentioned shortly.

In this chapter we use a method to treat the transmission loss which comprises writing down the whole radiation field, i.e., the field inside and outside the cavity as well as within the mirror layers. This requires continuously distributed modes in the frequency and propagation directions. We call these modes the "spatial modes". The spatial modes are described by a set of mode functions that satisfy the boundary conditions of the cavity and a cyclic boundary condition in a very large box including the cavity. The mode functions are orthonormal. Their amplitudes outside the cavity are the same for every function, but the amplitude inside the cavity differs for different spatial modes. The difference in amplitude, when quantized, gives the measure of the strength of the vacuum field for various spatial modes. Thus an excited atom in a cavity sees a differing strength of the vacuum field for different frequencies and for different directions. This is the origin of the so-called controlled spontaneous emission in a cavity. By the way, the cavity mode of interest which is lossy is sometimes called a "quasi mode" because such modes have no rigorous orthogonal property.

This method of describing the radiation field over the whole space by the "spatial modes" has a strong advantage over other methods of describing the damping mechanism in that the spatial distribution of the spontaneously emitted photons can be obtained straightforwardly because the form of the spatial mode function outside the cavity and its relation to that inside the cavity is known.

Since spontaneous emission is a quantum mechanical process, the emitting atom should also be described in a proper quantum mechanical manner. The simplest model of the atom is the so-called two-level atom, which has only two energy levels, the excited and the ground level. However, realistic atoms are surrounded by media that affect the atoms, giving them homogeneous or inhomogeneous broadening. Physically, these phenomena occur by random action on the atoms of the surrounding media, sometimes called the reservoir, causing damping of the population in each level and damping of the atomic polarization as well as shifting the transition frequency.

Damping of a system, field mode, or atom requires a special treatment in quantum mechanics in order not to lose the uncertainty relation for the system. There are three known ways of treating the damping effect of a reservoir:[21] the Langevin force method, the density matrix method, and the distribution function method. In this chapter we use the density matrix method, which is alternatively called the master equation method.

The density matrix is the matrix representation of the density operator, which comprises the "outer products" of the state vectors and satisfies an equation similar to the Heisenberg equation of motion. The density matrix describes an ensemble of quantum systems. If the density matrix is known, the

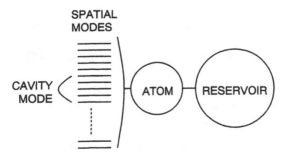

FIGURE 1   Diagram depicting interaction of an atom with the spatial modes and with an atomic reservoir. The cavity quasi mode is composed of a group of spatial modes.

double average, the statistical average of the quantum mechanical expectation value of any observable variable over the ensemble of the system, can be calculated. For details see Reference 22. The motion of the density operator for an atom coupled to a reservoir will be discussed in the next subsection, which leads to the density matrix equation or the master equation.

It is advantageous to describe the atom in the second quantized form when it interacts with other systems. In this scheme an atomic state is described by its creation and annihilation operators, and the transition from the excited to the ground state with emission of a photon can be written down in terms of the annihilation operator of the excited state and the creation operator of the ground state as well as the creation operator of the field mode. Likewise, the interaction of the atom with its reservoir can be described by the atomic operators and the operators for the reservoir.

Figure 1 summarizes the whole system we are to treat in this chapter. The atom interacts with the continuous spectrum of field modes (spatial modes) describing the cavity and the outside space, while the atom interacts with its reservoir. The cavity mode of interest (quasi mode) is composed of a portion of the continuous field mode spectrum.

C.   DERIVATION OF THE MASTER EQUATION

The interaction of the atom with its reservoir makes the knowledge of the state of the atom incomplete. To treat this situation the density operator is introduced, which is written as

$$\rho = \sum_{\Psi} |\Psi\rangle P_{\Psi} \langle\Psi|$$ (1.1)

where $|\Psi\rangle$ is a state of the system, an atom in this case, and P_{Ψ} is the probability that the system is in the state $|\Psi\rangle$. As stated above, if the density operator is known, we can calculate, on an observable A, the statistical average of the quantum mechanical expectation value by

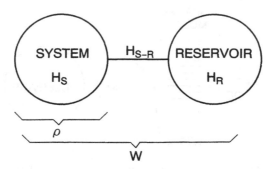

FIGURE 2   Diagram depicting interaction of an atom with its reservoir.

$$\langle A \rangle = \text{Tr}(\rho A) = \sum_{\Psi} P_{\Psi} \langle \Psi | A | \Psi \rangle, \tag{1.2}$$

as can be easily verified assuming $\langle \Psi | \Psi \rangle = 1$.

In this subsection, following Reference 21, we give the outline of the derivation of the equation of motion for the density operator, the master equation, for a system interacting with a reservoir. Our system and reservoir are depicted in Figure 2, where H_S and H_R are the Hamiltonian of the system (atom) and the reservoir (environment), respectively. H_{S-R} is the interaction Hamiltonian, ρ is the density operator for the system, and W is the density operator for the whole system. The system is described by $H_S = \sum_i \hbar\omega_i a_i^\dagger a_i$ where a_i and $a_i^\dagger$ are the annihilation and creation operators, respectively, for the i-th state of the system. $H_{S-R} = \hbar \sum_k V_k B_k$ where $V_k = a_i^\dagger a_j$ and $B_k = B_{ij}$. The operator V_k describes the change in the state of the system from j to i by the action of the reservoir, while B_{ij} is the operator for the element of the reservoir which causes the change from j to i in the system.

Using the Schrödinger equation for a wave function of the total system (system + reservoir) it is easy to show that W obeys the equation

$$(d/dt)W = -(i/\hbar)(H_T W - W H_T) \equiv -(i/\hbar)[H_T, W], \tag{1.3}$$

where $H_T = H_S + H_R + H_{S-R}$, the Hamiltonian of the total system: We start with $W(t=0) = W_0$ and solve Equation 1.3 in the interaction picture where the "free motion" by $H_0 = H_S + H_R$ is truncated from the wave equation.[22] We write

$$\tilde{W} = \exp(iH_0 t/\hbar) \, W \exp(-iH_0 t/\hbar), \tag{1.4}$$

and

$$\tilde{H}_{S-R} = \exp(iH_0 t/\hbar) \, H_{S-R} \exp(-iH_0 t/\hbar)$$

$$= \hbar \sum_k V_k(t) B_k(t) = \hbar \sum_{i,j} a_i^\dagger a_j \exp(i\omega_{ij} t) B_{ij}(t). \tag{1.5}$$

We write

$$a_i^\dagger a_j \exp\left(i\omega_{ij}t\right) = V_k \exp\left(i\Delta_k t\right). \tag{1.6}$$

The new density operator obeys the following equation:

$$(d/dt)\tilde{W} = -(i/\hbar)\left[\tilde{H}_{S-R},\ \tilde{W}\right]. \tag{1.7}$$

Starting from the initial value $\tilde{W}(t=0) = W_0$, we solve $\tilde{W}$ perturbatively to the second order in $\tilde{H}_{S-R}$ to obtain

$$\tilde{W}(t) = W_0 - \frac{i}{\hbar}\int_0^t\left[\tilde{H}_{S-R},\ W_0\right]dt' + \left(-\frac{i}{\hbar}\right)^2\int_0^t dt'\int_0^{t'}dt''\{\cdots\}, \tag{1.8}$$

where

$$\{\cdots\} = \tilde{H}_{S-R}(t')\tilde{H}_{S-R}(t'')W_0 - \tilde{H}_{S-R}(t')W_0\tilde{H}_{S-R}(t'')$$

$$- \tilde{H}_{S-R}(t'')W_0\tilde{H}_{S-R}(t') + W_0\tilde{H}_{S-R}(t'')\tilde{H}_{S-R}(t'). \tag{1.9}$$

As we are interested in the motion of ρ rather than that of W, we take the trace of $\tilde{W}$ over reservoir states to obtain $\tilde{\rho} = \text{Tr}_R\,\tilde{W}$ where the subscript R is for the reservoir. Then Equation 1.8 becomes

$$\tilde{\rho}(t) = \rho(0) - \int_0^t dt'\int_0^{t'}dt''\{\ldots\}, \tag{1.10a}$$

where

$$\{\ldots\} = \sum_{k,k'}\left\{V_k(t')V_{k'}(t'')\rho(0)\text{Tr}_R\left[B_k(t')B_{k'}(t'')\rho_R(0)\right]\right.$$

$$-V_k(t')\rho(0)V_{k'}(t'')\text{Tr}_R\left[B_k(t')\rho_R(0)B_{k'}(t'')\right]$$

$$-V_k(t'')\rho(0)V_{k'}(t')\text{Tr}_R\left[B_k(t'')\rho_R(0)B_{k'}(t')\right]$$

$$\left.+\rho(0)V_k(t'')V_{k'}(t')\text{Tr}_R\left[\rho_R(0)B_k(t'')B_{k'}(t')\right]\right\}. \tag{1.10b}$$

In deriving Equation 1.10, it was assumed that the initial density operator at the $t = 0$ factorizes as $W_0 = \rho(0)\,\rho_R(0)$ and that the terms that are linear in the reservoir variable vanish on taking the trace: $\text{Tr}_R(B_k\rho_R) = 0$. The latter assumption is not essential but means that the expectation value of the (random) reservoir variable B_k is zero. The second term in Equation 1.8 was thus eliminated.

The next step is to obtain a coarse-grained time derivative of the density operator ρ. We assume that the time scale is now larger than the coherence time τ_0 of the reservoir and that ρ changes by a little amount during time t. Because of the short memory in the reservoir, $\mathrm{Tr_R}[B_k(t')B_k'(t'')\rho_R(0)]$, etc. are nonzero only for $|\tau| = |t' - t''| < \tau_0$. This is essentially the Markov approximation. Then the double time integral in Equation 1.10 can be shown to be proportional to t.

To show this we set $t'' = t' - \tau$ and change the second integral to that over τ. The integrand $\mathrm{Tr_R}[B_k(t')B_k'(t'')\rho_R(0)]$, etc. becomes $\mathrm{Tr_R}[B_k(\tau)B_k'(0)\rho_R(0)]$ etc. by assuming the reservoir is stationary. Then, by Equations 1.5 and 1.6, the integral over t' will have an integrand of the form $\exp[i(\Delta_k + \Delta_{k'})t']$. The integral of this quantity will vanish unless $\Delta_k + \Delta_{k'} = 0$. For the combination of k and k' that satisfy this condition, the first integral over t' from 0 to t yields t. The second integral over τ from t' to 0 will contribute a constant because the integrand is finite only for $|\tau| = |t' - t''| < \tau_0$ by assumption. The factor $V_k V_{k'}\rho(0)$, etc. remains as a multiplying factor.

Thus Equation 1.10 reduces to the form $\tilde{\rho}(t) = \rho(0) - t\sum_{k,k'}[V_k V_{k'}\,\rho(0)A_{kk'} - \cdots]$, where $-A_{kk'}$ is the constant resulting from the second integral. The time t is long compared with the coherence time τ_0 of the reservoir but short enough that the system changed only by a little amount. Thus we can regard $[\tilde{\rho}(t) - \rho(0)]/t = -\sum_{k,k'}[V_k V_{k'}\rho(0)A_{kk'}) - \cdots]$ as the coarse-grained time derivative of $\tilde{\rho}$ at t = 0.

Finally we repeat the above calculation for $t < t' < 2t$, $2t < t' < 3t$, and so on and obtain the derivative at a general time t.

$$\frac{d\rho(t)}{dt} = -\sum_{k,k'}\left[V_k V_{k'}\tilde{\rho}(t)A_{kk'} - V_k\tilde{\rho}(t)V_{k'}A'_{kk'}\right.$$

$$\left. -V_{k'}\tilde{\rho}(t)V_k A_{kk'} + \tilde{\rho}(t)V_{k'}V_k A'_{kk'}\right],$$

(1.11)

where $A'_{kk'}$ is another constant. That the integrals in Equation 1.11 result in two constants, rather than four different constants, comes from the cyclic property of the traces. Recalling that $\Delta_k + \Delta_{k'} = 0$ (which suggests a detailed balance) and $V_k = a_i^{\dagger}a_j$, we have k = (i,j) and k' = (j,i) and therefore $V_{k'} = a_j^{\dagger}a_i$. This condition also suggests that $B_{k'}^{\dagger} = B_k$ or $B_{j1}^{\dagger} = B_{ij}$. Using these relations, these two constants are shown to be related by $A'_{kk'} = A^*_{kk'}$ through the argument that the interaction Hamiltonian H_{S-R} is Hermitian.

On going back to the Schrödinger picture these replacement results in the form of the equation:

$$\frac{d\rho}{dt} = -\frac{i}{\hbar}[H_s,\,\rho] + \left(\frac{\partial\rho}{\partial t}\right)_{\text{incoh, atom}},$$

(1.12)

where

$$\left(\frac{\partial\rho}{\partial t}\right)_{incoh,\ atom} = \sum_i \sum_j \left\{\left[\left(a_i^\dagger a_j\right)\rho\left(a_j^\dagger a_i\right) - \left(a_j^\dagger a_j\right)\rho\right]A_{jiij}\right.$$

$$\left. + \left[\left(a_i^\dagger a_j\right)\rho\left(a_j^\dagger a_i\right) - \rho\left(a_j^\dagger a_j\right)\right]A_{jiij}^*\right\}.$$

(1.13)

The ratio of the constants $A_{kk'}$ and $A_{k'k}$ is given by a factor that depends on the temperature of the reservoir. The expression for $A_{kk'}$ in terms of the variables (B's) can be obtained. However, such a calculation does not give us any practical information unless we know the precise structure of the reservoir. To apply the above density operator equation, one rather determines the meaning of the constants (A's) by comparing the density matrix equations that the above equation yields and the empirical equation containing phenomenological relaxation constants. Thus the model reservoir acts as an intermediary to let one write down a quantum mechanically consistent equation for the system with damping in terms of the phenomenological relaxation constants.

D. RULES OF CALCULATION

Here we summarize the rules of operator action on the state vectors and the operator algebra for the atom. For the j-th field mode we define the annihilation and creation operators a_j and $a_j^\dagger$, respectively, that act on the n-th number state $|n\rangle_j$. We have

$$a_j^\dagger|n\rangle_j = \sqrt{n+1}|n+1\rangle_j, \quad (n \geq 0)$$

$$a_j|n\rangle_j = \sqrt{n}|n-1\rangle_j, \quad (n \geq 1)$$

$$a_j|0\rangle_j = 0,$$

$$a_j^\dagger a_j|n\rangle_j = n|n\rangle_j \quad (n \geq 0)$$

(1.14)

For the m-th two-level atom, which has upper state $|2_m\rangle$ and lower state $|1_m\rangle$, we define the creation and annihilation operators a_{mi} and $a_{mi}^\dagger$, respectively, for the upper (i = 2) and lower (i = 1) states. We have[21,23]

$$a_{mi}^\dagger a_{mj}|k_m\rangle = \delta_{jk}|i_m\rangle, \quad (i, j, k = 1, 2)$$

$$a_{mi}^\dagger a_{mj}a_{mk}^\dagger a_{ml} = \delta_{jk}a_{mi}^\dagger a_{ml},$$

(1.15)

which are valid in one-electron space, where there is always one electron although its state can change.

Adjoint relations to Equations 1.14 and 1.15, e.g.,

$$_j\langle n|a_j = \sqrt{(n+1)}\,_j\langle n+1|, \quad (n \geq 0) \tag{1.16}$$

hold. All the state vectors are normalized:

$$_j\langle n|n\rangle_j = 1, \quad (n \geq 0)$$

$$\langle i_m|i_m\rangle = 1. \quad (i = 1, 2) \tag{1.17}$$

II. ANALYSIS OF SPONTANEOUS EMISSION

A. THE MODEL CAVITY AND INTERACTION OF THE ATOM WITH THE FIELD

1. The Characteristics of the Cavity

Let us consider an optical microcavity which has a finite coupling loss. For example, we may take a cavity of planar Fabry-Pérot type of length of the order of an optical wavelength.[6-8,10] Such a short cavity can be considered to have only one longitudinal cavity quasi mode within the spectral region of the emitting atom even though the atomic bandwidth (in free space) is as wide as, e.g., a tenth of the atomic transition frequency. The mode diameter and the angular emission pattern corresponding to the mode are determined by the mirror reflectivity and the cavity length.[10,24,25] The characteristics of the cavity in the frequency domain representing the resonance and the finite loss can then be given approximately by a Lorentzian as in the case of a microwave cavity[26] with a resonance frequency ω_c and a characteristic constant γ_c, the cavity half width at half maximum of the resonance. This approximation is valid on a time scale much larger than the cavity round-trip time, which is of the order of the reciprocal optical frequency in our case: we take a time scale which is large enough for any wave transit time in the cavity[27] to be neglected in calculating the atomic decaying process associated with spontaneous emission. Consequently, the characteristic resonance spectrum of the cavity takes the form:

$$U(\omega) = \frac{\gamma_c}{\omega - \omega_c + i\gamma_c}. \tag{2.1}$$

The cavity mode function will be expanded in terms of orthonormal radiation modes of the large quantization box of volume V which includes the cavity and is much larger than the cavity.[10] These modes distribute continuously and uniformly and their amplitudes *in the cavity* trace the Lorentzian character around the cavity resonance frequency as given in Equation 2.1. However, the amplitudes of these modes are uniform outside the cavity.

Note that this is complementary to the view that we have uniform amplitudes inside the cavity but with nonuniform mode density in the frequency domain as the cause of the alteration of spontaneous emission rate.[1,16,28] The difference stems from the fact that our model cavity has finite output coupling through the mirrors while a model based on a hollow metallic cavity is usually assumed to be completely or partially closed. In a completely closed cavity the mode density is calculated in terms of cavity quality factor,[1] while in partially closed cavities (or wave guides) the density is calculated using the dispersion equation of the allowed field mode.[28]

2.   Interaction Between Atom and Field

We consider two-level atoms in the cavity. They interact with continuously distributed radiation modes of which the amplitudes in the cavity are characterized by Equation 2.1. We assume that each atom interacts with its reservoir, which causes phase fluctuation of the atomic dipole, longitudinal relaxation, and frequency shift. The radiation modes of the large space of volume V (space modes) are lossless and are described by a set of orthonormal mode functions. The damping of the cavity quasi mode due to the output coupling, which is described by Equation 2.1, is implicitly included in the normalization factor of the mode functions that apply inside the cavity.[10] We assume that the cavity is otherwise lossless and therefore we need no reservoir for the field modes. We also assume that the presence of the cavity does not affect the properties of the atomic reservoirs.

We consider each atom separately, assuming that the emission processes of different atoms are independent of each other. The interaction of an atom with the field modes and with the atomic reservoir is depicted in Figure 1. We do not consider collective or cooperative effects in this chapter.

Taking the mth atom, the total Hamiltonian H_s of the atom-field system describing the coherent interaction reads

$$H_s = \sum_j \hbar\omega_j a_j^\dagger a_j + \hbar\omega_m a_{m2}^\dagger a_{m2} + \hbar \sum_j \left(g_{12mj} a_j^\dagger a_{m1}^\dagger a_{m2} + g_{21mj} a_{m2}^\dagger a_{m1} a_j \right), \quad (2.2)$$

under the rotating wave approximation, where a_j and $a_j^\dagger$ are the annihilation and the creation operator, respectively, of the jth field mode and ω_j is the eigen frequency of the jth mode. The mth atom is described by the annihilation and creation operators $a_{m1,2}$ and $a_{m1,2}^\dagger$, respectively, of the lower state $|1_m\rangle$ and the upper state $|2_m\rangle$. The mth atomic transition frequency is ω_{ma}.

Under the dipole approximation the coupling constants read

$$g_{12mj} = i\left(\omega_j/2\hbar\right)^{1/2} U_j(r_m) \cdot d_m, \quad (2.3)$$

and

$$g_{21mj} = g_{12mj}^*, \quad (2.4)$$

where $\mathbf{U}_j(\mathbf{r})$ is the jth normalized mode function and $\mathbf{r}$ the position vector. The mth atom is located at $\mathbf{r} = \mathbf{r}_m$ and has a dipole matrix element $\mathbf{d}_m$.

B.   The Master Equation for the System

Now we take into account the reservoir for the atom. The master equation for the above described system plus reservoir for the atom reads[21]

$$\frac{d\rho}{dt} = -\frac{i}{\hbar}[H_s, \rho] + \left(\frac{\partial \rho}{\partial t}\right)_{\text{incoh, atom}},$$

(2.5)

where

$$\left(\frac{\partial \rho}{\partial t}\right)_{\text{incoh, atom}} = \sum_i \sum_j \left\{ \left[\left(a_{mi}^\dagger a_{mj}\right)\rho\left(a_{mj}^\dagger a_{mi}\right) - \left(a_{mj}^\dagger a_{mj}\right)\rho\right]A_{jiijm} \right.$$

$$\left. + \left[\left(a_{mi}^\dagger a_{mj}\right)\rho\left(a_{mj}^\dagger a_{mi}\right) - \rho\left(a_{mj}^\dagger a_{mj}\right)\right]A_{jiijm}^* \right\}.$$

(2.6)

Here, ρ is the density operator of the field-atom system. The coefficient A_{jiijm} represents the interaction of the mth atom with its reservoir, and their physical meanings are to be determined later.

Under the influence of the reservoir the atom can go up and down between the two levels. However, we are interested in the decay of the atom which is initially prepared in its excited state, the decay being associated with spontaneous emission of a photon. We assume that the field is initially in the vacuum state. We follow the atom's motion as far as its first emission of a photon. Therefore, our system can be described as a superposition of the product states $|2_m\rangle|\{0\}\rangle$ and $|1_m\rangle|1_j\rangle$ where $|2_m\rangle$ or $|1_m\rangle$ denotes the atomic upper or lower state, respectively. The field state $|\{0\}\rangle$ denotes the vacuum state of the whole radiation field while $|1_j\rangle$ denotes the state of the field in which one photon exists in the jth mode and no photon exists in other modes. For simplicity, we write

$$|2_m\rangle|\{0\}\rangle = |0\rangle,$$

(2.7)

$$|1_m\rangle|1_j\rangle = |mj\rangle.$$

(2.8)

We derive, from Equation 2.5, the equations of motion for the density matrix elements

$$\rho_{pq} = \langle p|\rho|q\rangle.$$

(2.9)

The equation of motion for ρ_{pq} is obtained by sandwiching the quantities on both sides of Equation 2.5 by $\langle p|$ and $|q\rangle$ and applying the rules in Equations

1.15 to 1.17. We are especially interested in the element $\rho_{00} = \langle 0|\rho|0\rangle$, of which the equation of motion is found to contain the elements $\rho_{0mj} = \langle 0|\rho|mj\rangle$ and $\rho_{mj0} = \langle mj|\rho|0\rangle$:

$$\frac{d\rho_{00}}{dt} = i\left\{\sum_{j}\left(g_{12mj}\rho_{0mj} - g_{21mj}\rho_{mj0}\right) - \rho_{00}\left(A_{2112m} + A_{2112m}^{*}\right)\right.$$

$$\left. + \rho_{0'm0'm}\left(A_{1221m} + A_{1221m}^{*}\right)\right\}, \tag{2.10}$$

$$\frac{d\rho_{0mj}}{dt} = -i\left(\omega_{ma} - \omega_{j}\right)\rho_{0mj} + ig_{21mj}\rho_{00}$$

$$- \rho_{0mj}\left(A_{1111m}^{*} + A_{2222m} + A_{1221m}^{*} + A_{2112m}\right), \tag{2.11}$$

$$\frac{d\rho_{mj0}}{dt} = \left\{\frac{d\rho_{0mj}}{dt}\right\}^{*}, \tag{2.12}$$

$$\frac{d\rho_{0'm0'm}}{dt} = 2\,\text{Re}\,A_{2112m}\rho_{00} - 2\,\text{Re}\,A_{1221m}\rho_{0'm0'm}. \tag{2.13}$$

Here, $\rho_{0'm0'm}$ is the matrix element between $|0'm\rangle = |1_{m}\rangle|\{0\}\rangle$, the state of a de-excited atom with a vacuum field and its adjoint $\langle 0'm|$.

We can identify the factor $(A_{2112m} + A_{2112m}^{*})$ as the damping rate, $(A_{1221m} + A_{1221m}^{*})$ as the upward relaxation rate, $\text{Re}(A_{1111m}^{*} + A_{2222m} + A_{1221m}^{*} + A_{2112m})$ as the dephasing rate, and $\text{Im}(A_{1111m}^{*} + A_{2222m} + A_{1221m}^{*} + A_{2112m})$ as the frequency shift, all due to the reservoir of the mth atom, which we write

$$\left(A_{2112m} + A_{2112m}^{*}\right) = W_{21m}, \tag{2.14}$$

$$\left(A_{1221m} + A_{1221m}^{*}\right) = W_{12m}, \tag{2.15}$$

$$\text{Re}\left(A_{1111m}^{*} + A_{2222m} + A_{1221m}^{*} + A_{2112m}\right) = \gamma_{m}, \tag{2.16}$$

$$\text{Im}\left(A_{1111m}^{*} + A_{2222m} + A_{1221m}^{*} + A_{2112m}\right) = \Delta\Omega_{m}. \tag{2.17}$$

We are interested in the situation where the radiative decay rate is larger than or comparable to the damping rate. The upward relaxation rate W_{12m} can usually be ignored compared to the downward rate W_{21m} when an optical transition is relevant.

$$W_{12m} \ll W_{21m}. \tag{2.18}$$

Under these circumstances the density matrix equations simplify to

$$\frac{d\rho_{00}}{dt} = i \sum_j \left(g_{12mj} \rho_{0mj} - g_{21mj} \rho_{mj0} \right) - W_{21m} \rho_{00}, \qquad (2.19)$$

and

$$\frac{d\rho_{0mj}}{dt} = \left(i\Delta_{mj} - \gamma_m \right) \rho_{0mj} + i g_{21mj} \rho_{00},$$

$$\frac{d\rho_{mj0}}{dt} = -\left(i\Delta_{mj} + \gamma_m \right) \rho_{mj0} - i g_{12mj} \rho_{00}, \qquad (2.20)$$

where

$$\Delta_{mj} = \omega_j - \omega_{ma} - \Delta\Omega_m. \qquad (2.21)$$

Equations 2.19 to 2.21 are the starting equations for the subsequent analysis.

C. THE CASE OF HOMOGENEOUS BROADENING

1. Solving the Master Equation

First we examine the case where all the reservoirs are of the same property giving a homogeneous broadening of the atoms. We then write, dropping the subscript m, $g_{12mj} = g_{12j}$, $\omega_{ma} = \omega_a$, $\Delta\Omega_m = \Delta\Omega$, $\gamma_m = \gamma$, $W_{12m} = W_{12}$, and $W_{21m} = W_{21}$. In this case $\Delta_{mj} = \omega_j - \omega_{ma} - \Delta\Omega_m = \omega_j - \omega_a - \Delta\Omega = \Delta_j$. Here, we are assuming that all the atoms are located at equivalent points of the field mode distribution and are oriented in the same direction. From Equations 2.19 to 2.20 the decay of ρ_{00}, the density matrix element corresponding to the initial state, is described by the equation of the form:

$$\frac{d\rho_{00}}{dt} = -\sum_j |g_{12j}|^2 \int_0^t \exp\left[\left(i\Delta_j - \gamma \right)(t - t') \right] \rho_{00}(t') dt'$$

$$- \sum_j |g_{12j}|^2 \int_0^t \exp\left[-\left(i\Delta_j + \gamma \right)(t - t') \right] \rho_{00}(t') dt' - W_{21} \rho_{00}. \qquad (2.22)$$

If necessary, inclusion of the atomic distribution in space and in orientation is straightforward and can be done, to a first-order approximation, by replacing $|g_{12j}|^2$ by its average.

When the Markov approximation is valid, i.e., when the factor $\exp[-\gamma(t - t')]$ decays rapidly for $t' < t$ as compared to $\rho_{00}(t)$, the ρ_{00} in the integrands in the evolution Equation 2.22 can be set outside the integral signs and we have a simpler equation:

$$\frac{d\rho_{00}}{dt} = -\left\{ \sum_j 2\pi |g_{12j}|^2 L(\Delta_j, \gamma) + W_{21} \right\} \rho_{00}, \tag{2.23}$$

where

$$L(\Delta_j, \gamma) = \frac{\gamma/\pi}{\Delta_j^2 + \gamma^2}, \tag{2.24}$$

is a normalized Lorentzian. The first term on the right-hand side of Equation 2.23 gives the radiative decay rate of the atom while the second term gives the damping rate due to the reservoir. We denote the total decay rate by A_t. Then we have

$$A_t = \sum_j 2\pi |g_{12j}|^2 L(\Delta_j, \gamma) + W_{21}. \tag{2.25}$$

If the atomic reservoir is absent, the longitudinal and the transverse relaxation rates as well as the frequency shift vanish and it is easy to see that the right-hand side then reduces to an expression of the usual Fermi golden rule where the function L goes to a δ-function.

2.　Spontaneous Emission Rate for a Homogeneously Broadened Atom

We will concentrate on the spontaneous emission rate A given by the first term on the right-hand side of Equation 2.25. Then, by Equation 2.3,

$$A = \sum_j \frac{\pi \omega_j}{\hbar} \alpha_c |U_j|^2 |d|^2 L(\Delta_j, \gamma). \tag{2.26}$$

Here we have written

$$|U_j(r_m) \cdot d_m|^2 = \alpha_j |U_j|^2 |d|^2$$

$$\sim \alpha_c |U_j|^2 |d|^2, \tag{2.27}$$

where $|d|$ is the magnitude of the atomic dipole matrix element and $|U_j|$ is the average mode modulus of the jth mode in the cavity. The latter quantity may be large or small compared to the amplitude in the region outside the cavity, according as the jth mode is within or outside the bandwidth of the cavity quasi mode of interest. The constant α_j is the measure of the alignment of the electric field of the jth mode and the atomic dipole. It also measures the degree of coincidence of atomic location and the nodes or antinodes of the jth mode. In the second line we have replaced α_j by α_c, the value at $\omega_j = \omega_c$, assuming that the continuously distributed field modes that constitute a particular cavity quasi mode of interest do not differ much in shape in the cavity.

We assume that the average mode modulus *in the cavity* has the form, in accordance with Equation 2.1,

$$\left|U_j\right|^2 = \left[F/(3\alpha_c)\right]\left(\epsilon_0 V\right)^{-1} \pi\gamma_c L\left(\omega_j - \omega_c, \gamma_c\right), \tag{2.28}$$

where ϵ_0 is the dielectric constant of the vacuum, ω_c and γ_c are the resonance frequency and the half width at half maximum of the cavity, respectively, and V is the volume of the large quantization box that includes the cavity. The factor F is the enhancement factor of spontaneous emission measured at the cavity resonance frequency. This factor is defined in the absence of the atom's reservoir as the ratio of the spontaneous emission rate in the cavity to that in a free vacuum. This factor depends on the atomic orientation and atomic location and is proportional to α_c. The ratio $F' = F/\alpha_c$ is the spontaneous-emission-enhancement parameter characteristic of the cavity quasi mode, which is independent of atomic location or atomic dipolar orientation.

Here we briefly check the validity of the expression for $|U_j|^2$ given in Equation 2.28. In the absence of the reservoir, Equation 2.25 takes the form:

$$A = \sum_j 2\pi\left|g_{12j}\right|^2 \delta\left(\Delta_j\right). \tag{2.29}$$

Substituting Equations 2.3 and 2.28 into Equation 2.29, using Equation 2.27 with $\alpha_j \rightarrow \alpha_c$, and noting that the density of spatial modes $D(\omega_j)d\omega_j = \omega_j^2 V/(\pi^2 c^3)d\omega_j$, we have

$$A = FA_0 \frac{\gamma_c^2}{\left(\omega_c - \omega_a\right)^2 + \gamma_c^2}, \tag{2.30}$$

in the absence of the reservoir. Here A_0 is the spontaneous emission rate in a free vacuum and is given by Equation 2.32 below. Equation 2.30 is consistent with the definition of F: the quantity F was defined as the enhancement factor of spontaneous emission measured in the absence of the reservoir at the cavity resonance frequency ($\omega_a = \omega_c$). This proves the validity of the expression for $|U_j|^2$ in Equation 2.28.

The average mode modulus in the cavity, $|U_j|$ in Equation 2.28, should not depend on the atomic orientation or location. Therefore, as was mentioned above, the enhancement factor F should be proportional to α_c, and the ratio $F' = F/\alpha_c$ expresses the effectiveness of the cavity structure for enhancing the vacuum field amplitude of a particular cavity quasi mode.

Substituting Equation 2.28 into Equation 2.26 we have

$$A = \int_0^\infty d\omega_j D(\omega_j) \frac{\pi^2 \omega_j |d|^2 \gamma_c}{3\hbar\varepsilon_0 V} FL(\omega_j - \omega_c, \gamma_c) L(\Delta_j, \gamma),$$

$$\sim FA_0 \pi\gamma_c \int_0^\infty d\omega_j L(\omega_j - \omega_c, \gamma_c) L(\omega_j - \omega_a, \gamma),$$

(2.31)

where $D(\omega_j) = \omega_j^2 V/(\pi^2 c^3)$ is the density of spatial modes, and

$$A_0 = \frac{\omega_a^3 |d|^2}{3\pi\varepsilon_0 \hbar c^3},$$

(2.32)

is the spontaneous emission rate in a free vacuum and c is the light velocity in a vacuum. We have replaced $\omega_a + \Delta\Omega$ by ω_a for simplicity.

Equation 2.31 is the first of the two main results of this chapter. It gives the spontaneous emission rate for an atom in a cavity for the case where the cavity has a Lorentzian resonance and the atom has another homogeneous Lorentzian line shape due to its reservoir. The enhancement factor F for an atom without a reservoir is assumed to be known for the atomic frequency at resonance with the cavity ($\omega_a = \omega_c$).

Note that the factor F appears as a simple multiplying factor as long as Equation 2.28 is valid. This point may be proved by arguing that the sponta-neous-emission-enhancing character of a microcavity quasi mode (at reso-nance with the atom) is solely determined by the spatial and spectral distribu-tion of the vacuum field in the cavity and is irrelevant to the broadening of the atoms. This is an important result because it allows us to analyze the enhance-ment factors due to the cavity structure and the effect of atomic broadening separately. This holds also for the case of inhomogeneous broadening or energy-level continuum to be examined in the next subsection.

The expression for the factor F has been obtained for several cavity structures.[1,10,16,28] In Reference 10 we have given the expression for the factor F for a two-level atom in a very short optical cavity with plane parallel dielectric mirrors.

Two limiting cases of Equation 2.31 can be examined immediately. One is the case of small reservoir effect ($\gamma \ll \gamma_c$) when we have $A \sim FA_0 \pi\gamma_c L(\omega_a - \omega_c, \gamma_c)$. In this case, $A = FA_0$ at $\omega_c = \omega_a$ and A decreases with detuning and falls to $FA_0/2$ at $\omega_c = \omega_a \pm \gamma_c$. The other is the case of a large reservoir effect ($\gamma_c \ll \gamma$) when we have

$$A = FA_0 \frac{\gamma_c \gamma}{(\omega_c - \omega_a)^2 + \gamma^2},$$

(2.33)

which gives $A \ll FA_0$. The reservoir causes the spontaneous emission rate to decrease roughly by a factor of γ_c/γ for small atom-cavity detunings.

By the way, we note that, if the cavity structure disappears, $F \to 1$ and $\pi \gamma_c L(\omega_j - \omega_c, \gamma_c) \to 1$ as $\gamma_c \to \infty$. Therefore, Equation 2.31 yields $A = A_0$. That is, under our approximation, the transverse relaxation of the atom does not affect the spontaneous emission rate in an electromagnetically free space.

The emission spectrum under the Markov approximation may be obtained from the integrand in Equation 2.31. We see that the spectral width is roughly determined by the narrower of the two Lorentzians.

Here we show that the emission spectrum which is obtained after a time long enough for the atom to decay is indeed the same as that obtained from the integrand in Equation 2.31. For this purpose we assume that the total decay rate A_t applies for times up to $t \gg A_t^{-1}$ and integrate Equation 2.22 to $t \to \infty$ with $\rho_{00}(t')$ on the right-hand side replaced by $\exp(-A_t t')$ where A_t is given by Equation 2.25. We obtain

$$\rho_{00}(\infty) = 1 - \left[\sum_j |g_{12j}|^2 \frac{1}{\gamma - A_t - i\Delta_j} \left\{ \frac{1}{A_t} - \frac{1}{\gamma - i\Delta_j} \right\} + \text{c.c.} \right] - \frac{W_{21}}{A_t}$$

$$= 1 - \sum_j 2\pi |g_{12j}|^2 \frac{L(\Delta_j, \gamma)}{A_t} - \frac{W_{21}}{A_t}, \qquad (2.34)$$

where the summation over the radiation mode j's may be interpreted as displaying the distribution over j of the probability that the photon has been emitted into mode j. This spectrum is the same as that given by the first term in Equation 2.25 or by Equation 2.31. A further discussion on the approximations used will be given in Section III.A. Incidentally, the value of $\rho_{00}(\infty)$ is zero by virtue of Equation 2.25, as is expected.

In Figures 3(a) and (b) the emission rate A and the emission full width at half maximum $\Delta \omega$ are plotted, respectively, as functions of the homogeneous half width at half maximum γ. The emission rate is normalized by the emission rate in a free vacuum A_0, while both the emission full width and the homogeneous half width are normalized by the cavity half width at half maximum γ_c. In these figures the parameters used are $\omega_c/\gamma_c = 200$, $F = 10$, and $\omega_c - \omega_a = 0$ (zero detuning). It can be shown that the results are virtually independent of the value of ω_c/γ_c and of the cavity quality factor $Q[= \omega_c/(2\gamma_c)]$. In reality, even if the emission rate is very small, to get a large inhibition, or a very small overall decay rate, may be difficult because of possible nonradiative damping.

D. THE CASE OF INHOMOGENEOUS BROADENING

1. General Formula

Let us consider N atoms with transition frequency ω_0 each of which is in contact with its own reservoir. If the characteristics of the reservoirs affecting

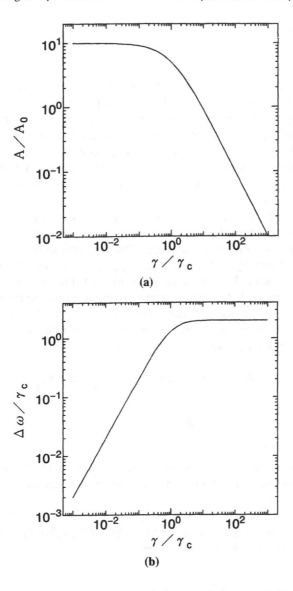

FIGURE 3   (a) Scaled spontaneous emission rate A/A_0 vs. scaled homogeneous half width at half maximum γ/γ_c. (b) Emission full width at half maximum normalized by the cavity half width at half maximum $\Delta\omega/\gamma_c$ vs. homogeneous half width at half maximum normalized by the cavity half width at half maximum γ/γ_c. The parameters are $\omega_o/\gamma_c = 200$, $F = 10$, and $\omega_a - \omega_c = 0$.

the atoms vary from one to another, they give the atoms inhomogeneous frequency shifts, which results in an inhomogeneous broadening. Then, the evolution equation (2.23) for the density matrix element ρ_{00} should be rewritten for an average atom as

$$\frac{d\rho_{00}}{dt} = -N^{-1}\left\{\sum_m\left[\sum_j 2\pi|g_{12mj}|^2 L\left(\Delta_{mj}, \gamma_m\right)\right]\rho_{00} + \sum_m W_{21m}\rho_{00}\right\}, \qquad (2.35)$$

where $\Delta_{mj} = \omega_j - \omega_0 - \Delta\Omega_m$. We concentrate again on the spontaneous emission, i.e., the first term on the right-hand side of Equation 2.35. We assume for simplicity that the transverse relaxation rate γ_m is the same for all the atoms, while the frequency shift $\Delta\Omega_m$ differs for different atoms. The above assumptions apply to a fairly broad class of fluorescing substances.

In this case, among the inhomogeneously broadened atoms, atoms with almost the same frequency shift $\Delta\Omega_p$ within a bandwidth $d(\Delta\Omega_p)$ constitute a subgroup p of the atoms. The atoms in subgroup p may have different orientations or locations. Therefore, we may need to take the sum over the atoms within the subgroup p before we sum over the atoms of different frequency shifts, i.e., over different subgroups. Thus the sum over m in Equation 2.35 may be replaced by a double sum $\sum_m = \sum_p \sum_{np}$, where np denotes the nth atom in the subgroup p. Here we assume that the distribution of the atomic orientation and location in any one subgroup is similar to that in the whole group of atoms. We retrace, for each subgroup p, the procedure to obtain the emission rate of a homogeneously broadened atom [from Equation 2.23 to Equation 2.31]. Then the emission rate is given by

$$A = N^{-1}\sum_p\sum_{np}\left[\sum_j 2\pi|g_{12npj}|^2 L\left(\omega_j - \omega_p, \gamma\right)\right]$$

$$= \sum_p \frac{N_p}{N} A\left(\omega_p\right), \qquad (2.36)$$

where $\omega_p = (\omega_0 + \Delta\Omega_p)$ and N_p is the number of atoms belonging to the subgroup p. Here $A(\omega_p)$ is the emission rate A in Equation 2.31, with ω_a being replaced by ω_p.

The probability of the atom being at ω_p within $d\omega_p$ [= $d(\Delta\Omega_p)$] is N_p/N. We define the probability density function $g(\omega_p)$ by $(N_p/N)d\omega_p = g(\omega_p)d\omega_p$. Thus the spontaneous emission rate becomes

$$A = \int_0^\infty d\omega_p g\left(\omega_p\right) A\left(\omega_p\right)$$

$$= FA_0\pi\gamma_c \int_0^\infty\int_0^\infty d\omega_j d\omega_p L\left(\omega_j - \omega_c, \gamma_c\right) L\left(\omega_j - \omega_p, \gamma\right) g\left(\omega_p\right), \qquad (2.37)$$

where Equation 2.31 has been used in the second line. This is the second of the two main results of this chapter. It gives the spontaneous emission rate for an atom in a cavity for the case where the cavity has a Lorentzian resonance and the atom has another homogeneous Lorentzian line shape as well as inhomogeneous broadening of an arbitrary shape. Note that the enhancement factor F appears as a multiplying factor. The emission spectrum (obtained after the atom has decayed) may again be given by the integrand in Equation 2.37. This may be proved by a procedure similar to that employed to derive Equation 2.34.

When we have an energy-level continuum in both the upper and the lower levels, as in a large molecule, we have to sum the previously calculated quantities over relevant state pairs $|2_m\rangle$ and $|1_m\rangle$. Then, Equation 2.35 may be rewritten in the form:

$$\frac{d\rho_{00}}{dt} = -N^{-1} \sum_m \sum_{2m} \sum_{1m} C_{2m} \left\{ \left[\sum_j 2\pi |g_{12mj}|^2 L\left(\Delta_{21mj}, \gamma_{21m}\right) \right] \rho_{00} + W_{21m}\rho_{00} \right\}. \quad (2.38)$$

Here C_{2m} is the probability that the mth molecule is prepared in a particular state $|2m\rangle$ in the upper continuum. Now the probability C_2 can depend on the atom number m. Further, the frequency shift Δ, the transverse relaxation constant γ, and the damping constant due to the reservoir w_{21} can depend on the state pair as well as on the atom number m. Equation 2.38 is fairly general but is too complicated to be of practical use.

If we limit our consideration to the case where we have an energy-level continuum but no inhomogeneous broadening, and if we also assume that the probability C_2 is independent of m and that transverse relaxation constant is the same for all the level pairs, then, as can be shown easily, Equation 2.38 will yield an emission rate in the form of Equation 2.37 with redefinition of $g(\omega_p)$ as a normalized function describing the number of effective level pairs at ω_p.

2. A Few Limiting Cases

Here we examine Equation 2.37 in some limiting parameter regions. In the following we assume a Gaussian profile for the probability density function $g(\omega_p)$ with center frequency ω_0 and half width at half maximum γ_i, i.e.,

$$g(\omega_p) = \frac{1}{\sqrt{\pi}\,\gamma_i} \exp\left[-\left(\frac{\omega_p - \omega_0}{\gamma_i} \right)^2 \right]. \quad (2.39)$$

The constant γ_i denotes the inhomogeneous width or the spread of the energy-level continuum. For simplicity, let us call it from now on the inhomogeneous width.

Now we can give some qualitative discussions on the emission rate in Equation 2.37. First, if the inhomogeneous width γ_i is much smaller than both the homogeneous width γ and the cavity width γ_c ($\gamma_i \ll \gamma, \gamma_c$), the results are almost the same as in the case of homogeneous broadening and are expressed by Equation 2.31. Second, if the homogeneous width is small, i.e., $\gamma \ll \gamma_i, \gamma_c$, then we have

$$A = FA_0 \pi \gamma_c \int_0^\infty d\omega_p L\left(\omega_p - \omega_c, \gamma_c\right) g(\omega_p), \quad (2.40)$$

which describes the effect of inhomogeneous broadening only. Further, if $\gamma \ll \gamma_i \ll \gamma_c$, then we have

$$A = A_0 F\pi\gamma_c L(\omega_c - \omega_0)$$

$$= A_0 F \frac{1}{\left[(\omega_c - \omega_0)/\gamma_c\right]^2 + 1}.$$

(2.41)

These two observations and the discussion made above Equation 2.33 suggest that when both homogeneous and inhomogeneous widths are smaller than the cavity width; i.e., $\gamma,\gamma_i \ll \gamma_c$, the emission rate is approximately $A_0 F$. Also, as an inspection of Equation 2.37 shows, the spectral profile of the atoms, homogeneous or inhomogeneous, will be reflected in the emission spectrum if the total atomic width is small compared with the cavity width.

Third, if the inhomogeneous width is much greater than both the homogeneous width and the cavity width, $\gamma,\gamma_c \ll \gamma_i$, then we have

$$A = A_0 F \frac{\sqrt{\pi}\,\gamma_c}{\gamma_i}\, \exp\left\{-\left[\frac{(\omega_c - \omega_0)}{\gamma_i}\right]^2\right\},$$

(2.42)

which reduces, if $|\omega_c - \omega_0| < \gamma_i$, to $A \sim \sqrt{\pi}\, A_0 F \gamma_c/\gamma_i$, which is much smaller than $A_0 F$. Finally, for $\gamma_c \ll \gamma_i \ll \gamma$ we have a situation similar to that expressed by Equation 2.33 for homogeneous broadening (with $\gamma_c \ll \gamma$) and we have $A \ll A_0 F$.

The last two observations show that, if the cavity width is smaller than either the homogeneous or inhomogeneous width, $\gamma_c \ll \gamma$ or $\gamma_c \ll \gamma_i$, the emission rate is smaller than $A_0 F$. In these cases the emission spectral profile will reflect the Lorentzian line shape of the cavity.

In Sections II.C and II.D we have considered mainly the spontaneous emission rate A rather than the total decay rate A_t, putting aside the longitudinal relaxation rate W_{21}. The decay rate A_{obs}, which we observe experimentally by detection of spontaneously emitted light under pulsed excitation is in general different from A and A_t. The interpretation of A_{obs} depends on the relative magnitudes of γ_c, γ, and γ_i. For example, when $\gamma \ll \gamma_c \ll \gamma_i$, which is the case for a typical dye-planar microcavity system, we have $A \ll FA_0$. In this case the decay of the atoms outside the cavity bandwidth is in general governed by the nonradiative relaxation W_{21}. (In the case of a planar microcavity, emission by atoms outside the cavity bandwidth is not necessarily suppressed because of resonance in oblique directions.) On the other hand, inside the cavity bandwidth the atoms decay radiatively and nonradiatively by the rate approximately given by $A_{obs} = FA_0 + W_{21}$. Thus, in this case, if W_{21} is known, we can determine the enhancement factor F by experimental observation.

3.   Numerical Calculations on Inhomogeneous Broadening

We performed numerical calculations using Equation 2.37 with the parameter values to cover those of a dye solution (Rhodamine 6G in ethanol) in a planar microcavity half a wavelength long with mirror reflectivity of about 99%. Then we have, typically, $\omega_0 \sim \omega_c \sim 3 \times 10^{15}$, $\gamma \sim 10^{12}$ (Reference 14), $\gamma_c \sim 6 \times 10^{12}$, and $\gamma_i \sim 10^{14}$, all in radians per second.

We show in Figures 4(a) and (b) the scaled spontaneous emission rate A/A_0 and the scaled spectral full width ($\Delta\omega/\gamma_c$) vs. scaled inhomogeneous width (γ_i/γ_c), respectively. The curves are calculated for $\gamma/\gamma_i = 10^{-1}$. Here the ratio $\omega_c/\gamma_c = 200$ [the cavity quality factor $Q = \omega_c/(2\gamma_c) = 100$], the enhancement factor $F = 10$, and $\omega_c - \omega_0 = 0$ (zero detuning).

If we examine the results in Figures 3(a) and 4(a) carefully, we see that the emission rate is in general higher in the case of inhomogeneous broadening than in homogeneous broadening. This is because, for the same half width, the assumed Gaussian atomic spread function is more tightly covered by the cavity Lorentzian function than a Lorentzian atomic spread function is, yielding a larger integrand. It can be shown that there is virtually no difference between the results for zero homogeneous width $\gamma = 0$ and for small values of γ/γ_c. This result depends on the assumption of the Gaussian inhomogeneous profile, which may be justified in many cases.

The emission spectrum for small γ_i/γ_c is a Gaussian which displays the narrow, assumed atomic spread function, and the corresponding total normalized emission rate A/A_0 is close to the F value. The spectrum for $\gamma_i/\gamma_c \gg 1$ is a Lorentzian showing the cavity resonance characteristics, and the corresponding total normalized emission rate is much smaller than F. The spectrum for $\gamma_i/\gamma_c \sim 1$ is of a mixed type, and the corresponding total normalized emission rate is a little smaller than F.

III.   DISCUSSION AND CONCLUSION

A.   Discussion

In this chapter the cavity is described in terms of continuously distributed spatial mode functions. This is in contrast to theories assuming a well-defined single mode of the cavity.[1,3] Our formulation allows, in principle, to examine the spatial distribution of vacuum field, if the k-vector of the function U_j is explicitly taken into account. This in turn allows the determination of spatial distribution of the spontaneously emitted field[10] as well as its spectrum.

In Equations 2.25 and 2.34 we have obtained the same emission spectrum by the Markov approximation and by an approximation similar to that due to Wigner and Weisskopf. The essential assumption implied in these equations can be seen more clearly if we solve Equation 2.22 by Laplace transform. By

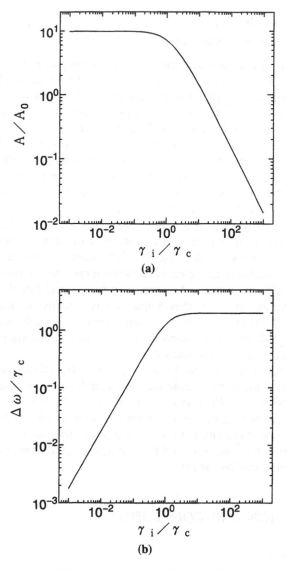

FIGURE 4   (a) Scaled spontaneous emission rate A/A_0 vs. scaled inhomogeneous width (γ_i/γ_c). (b) The scaled spectral width ($\Delta\omega/\gamma_c$) vs. the scaled inhomogeneous width (γ_i/γ_c). The parameters are $\gamma/\gamma_i = 10^{-1}$, $\omega_c/\gamma_c = 200$, $F = 10$, and $\omega_0 - \omega_c = 0$.

this procedure we can show that our results in Sections II.C and II.D, including the emission spectrum, are subject to the condition that $A_t \ll \gamma$. In our dye-microcavity system $\gamma \sim 10^{12}$ rad · s^{-1} and the measured $A_{obs} \sim 5 \times 10^8$ s^{-1} (Reference 11). Also, in an excitonic transition in GaAs quantum well γ is of the order of 10^{12} rad · s^{-1} (Reference 15) and A_{obs} is of the order of 10^8 s^{-1} (Reference 29). As was mentioned in the last paragraph of subsection II.D.3, it is expected that $A_{obs} > A_t$ in these systems. Therefore, the criterion, $A_t \ll \gamma$, is satisfied both in the dye and the GaAs systems.

The emission spectrum as given by the second term in Equation 2.34 contains a Lorentzian factor which comes from the homogeneous broadening of the atoms. It is noted that a similar factor appears, as to the emission spectrum, when the atom has no reservoir in the sense of Figure 1 but decays at a finite rate.[30] This is because a decaying atomic dipole has a Lorentzian Fourier spectrum.

The important parameter of coupling ratio of spontaneous emission to a desired quasi mode cannot be exactly treated by the present mathematical model of the cavity, although it is easily seen qualitatively that the ratio will decrease with increasing atomic broadening.[17,18] An improved treatment of this problem will be published elsewhere.[31]

In this chapter we have assumed that each atom emits independently and ignored the possible interaction among atoms during spontaneous emission. It is known that N-atom system can decay at a rate NA.[32] For a typical dye solution in a half-wavelength cavity with effective area of the Fabry-Pérot,[10,33] we estimate that $N \sim 10^8$ and $A \sim A_0 \sim 10^8$ s^{-1} with $F \sim 1$.[10] Thus $NA \sim 10^{16}$ s^{-1}. To observe this fast decay (the decay time shorter than one optical period) will be impossible and to excite the atoms to their upper states faster than this rate will also be very difficult. Therefore, for the dye-microcavity system, this collective system decay is of only theoretical interest.

B.   CONCLUSION

We have developed a theory on the effects of atomic broadening on spontaneous emission in an optical microcavity where spontaneous emission is strongly affected by the cavity structure. We have treated homogeneous broadening and inhomogeneous broadening. The atoms are assumed to be in contact with their respective reservoirs which cause broadening of the atoms. The radiation field is expanded in terms of continuously distributed modes in a large quantization box that includes the microcavity. Inside the cavity, the field has Lorentzian resonance characteristics and the cavity enhancement factor for the spontaneous emission rate in the absence of atomic reservoirs is assumed to be known. Solving the master equation for the density matrix elements, we have derived formulas for spontaneous emission rate and emission spectrum for various situations concerning the relative widths of homogeneous or inhomogeneous broadening (or energy-level continuum) of the atom and of cavity resonance. The broadening of the atom is important when its width is greater than the cavity width. Although numerical calculations have been done with a dye-microcavity system in mind, we stress that the present analytical results are also applicable to the excitonic emission in microcavities made of semiconductors.

In conclusion, experimental evidence for the present theory concerning the emission spectrum, for the case of small cavity width compared to the dye fluorescence spectrum, has been obtained for a rhodamine 6G–planar microcavity system, which will be described elsewhere.[31]

REFERENCES

1. Purcell, E. M., Spontaneous emission probabilities at radio frequencies, *Phys. Rev.* **69,** 681, 1964.
2. Stehle, P., Atomic radiation in a cavity, *Phys. Rev.* **A2,** 102, 1970.
3. Sanchez-Mondragon, J. J., Narozhny, N. B., and Eberly, J. H., Theory of spontaneous-emission line shape in an ideal cavity, *Phys. Rev. Lett.* **51,** 550, 1983.
4. Goy, P., Raimond, J. M., Gross, M., and Haroche, S., Observation of cavity-enhanced single-atom spontaneous emission, *Phys. Rev. Lett.* **50,** 1903, 1983.
5. Meschede, D. and Walther, H., One-atom maser, *Phys. Rev. Lett.* **54,** 551, 1985.
6. Yokoyama, H. and Brorson, S. D., Rate equation analysis of microcavity lasers, *J. Appl. Phys.* **66,** 4801, 1989.
7. Yamamoto, Y., Machida, S., Igeta, K., and Björk, G., Controlled spontaneous emission in microcavity semiconductor lasers, *Coherence, Amplification, and Quantum Effects in Semiconductor Lasers,* Y. Yamamoto, Ed., John Wiley & Sons, New York, 1991, p. 561.
8. DeMartini, F. and Jacobovitz, G. R., Anomalous spontaneous-stimulated-decay phase transition and zero-threshold laser action in a microscopic cavity, *Phys. Rev. Lett.* **60,** 1711, 1988.
9. Heinzen, D. J., Childs, J. J., Thomas, J. E., and Feld, M. S., Enhanced and inhibited visible spontaneous emission by atoms in a confocal resonator, *Phys. Rev. Lett.* **58,** 1320, 1987.
10. Ujihara, K., Nakamura, A., Manba, O., and Feng, X. P., Spontaneous emission in a very short optical cavity with plane-parallel dielectric mirrors, *Jpn. J. Appl. Phys.* **30,** 3388, 1991.
11. Yokoyama, H., Suzuki, M., and Nambu, Y., Spontaneous emission and laser oscillation properties of microcavities containing a dye solution, *Appl. Phys. Lett.* **58,** 2598, 1991.
12. Yokoyama, H., Nishi, K., Anan, T., Yamada, H., Brorson, S. D., and Ippen, E. P., Enhanced spontaneous emission from GaAs quantum wells in monolithic microcavities, *Appl. Phys. Lett.* **57,** 2814, 1990.
13. Björk, G., Machida, S., Yamamoto, Y., and Igeta, K., Modification of spontaneous emission rate in a planar dielectric microcavity structures, *Phys. Rev.* **A44,** 669, 1991.
14. Drexhage, K. H., Interaction of light with monomolecular dye layers, *Progress in Optics,* vol. XII, E. Wolf, Ed., North-Holland, Amsterdam, 1974, p. 163.
15. Yamamoto, Y., Machida, S., and Björk, G., Micro-cavity semiconductor lasers with controlled spontaneous emission, *Opt. Quantum Electron.* **24,** S215, 1992.
16. Brorson, S. D., Yokoyama, H., and Ippen, E. P., Spontaneous emission rate alteration in optical waveguide structures, *IEEE J. Quantum Electron.* **26,** 1492, 1990.
17. Yamamoto, Y., Machida, S., and Björk, G., Microcavity semiconductor laser with enhanced spontaneous emission, *Phys. Rev.* **A44,** 657, 1991.
18. Baba, T., Hamano, H., Koyama, F., and Iga, K., Spontaneous emission factor of a microcavity DBR surface-emitting laser, *IEEE J. Quantum Electron.* **27,** 1347, 1991.

19. Agarwal, G. S., Master equation approach to spontaneous emission. III. Many-body aspects of emission from two-level atoms and the effect of inhomogeneous broadening, *Phys. Rev.* **A4**, 1791, 1971.
20. Jodin, R. and Mandel, L., Superradiance in an inhomogeneously broadened atomic system, *Phys. Rev.* **A9**, 873, 1974.
21. Haken, H., *Handbuch der Physik,* L. Genzel, Ed., Springer-Verlag, Berlin, 1970, vol. XXV/2c, p.51.
22. Louisell, W. H., *Radiation and Noise in Quantum Electronics,* Robert E. Krieger Publishing Co., Huntington, New York, 1977, p. 222.
23. Loudon, R., *The Quantum Theory of Light,* Clarendon, Oxford, 1982, Ch. 8.
24. Ujihara, K., A simple relation between directivity and linewidth of a planar microcavity laser, *Jpn. J. Appl. Phys.,* **33**, 1059, 1994.
25. Björk, G., Heitmann, H., and Yamamoto, Y., Spontaneous-emission coupling factor and mode characteristics of planar dielectric microcavity lasers, *Phys. Rev.* **47**, 4451, 1993.
26. Meystre, P., Cavity QED, *Nonlinear Optics in Solids,* O. Keller, Ed., Springer-Verlag, Berlin, 1990, p. 26.
27. Feng, X. P. and Ujihara, K., Quantum theory of spontaneous emission in a one-dimensional optical cavity with two-side output coupling, *Phys. Rev.* **A41**, 2668, 1990.
28. Kleppner, D., Inhibited spontaneous emission, *Phys. Rev. Lett.* **47**, 233, 1981.
29. Yokoyama, H., Nishi, K., Anan, T., Nambu, Y., Brorson, S. D., Ippen, E. P., and Suzuki, M., Controlling spontaneous emission and threshold-less laser oscillation with optical microcavities, *Opt. Quantum Electron.* **24**, S245, 1992.
30. Ujihara, K., Decay rate dependence of spontaneous emission pattern from an atom in an optical microcavity, *Opt. Commun.* **101**, 179, 1993.
31. Osuge, M. and Ujihara, K., *Journ. Appl. Phys.,* **76**, 2588, 1994.
32. Haroche, S. and Raimond, J. M., Radiative properties of Rydberg states in resonant cavities, *Advances in Atomic and Molecular Physics,* Vol. 20, Academic Press, New York, 1985, p. 347.
33. DeMartini, F., Marrocco, M., and Murra, D., Transverse quantum correlations in the active microscopic cavity, *Phys. Rev. Lett.* **65**, 1853, 1990.

4 Microcavities And Semiconductors: The Strong-Coupling Regime

Claude Weisbuch, Romuald Houdré, and Ross P. Stanley

TABLE OF CONTENTS

I. INTRODUCTION

The interaction of light with semiconductors has always attracted great attention due to its importance in the fundamental understanding of semiconductors and in the many applications which depend on it. On the former aspect, we can cite the photoconductivity property, light emission capability, and various electro-optic and nonlinear effects. Applications are numerous and of major importance in many growth markets: light detectors, LEDs and lasers, photovoltaic generators, displays, etc.

0-8493-3786-0/95/$0.00+$.50

The field has therefore been a very active one over the years, and much attention has been devoted to the fundamentals, i.e., measurement and understanding of bandgaps, impurity levels, photocurrents, absorption mechanisms, excitons, high density effects (bandgap renormalization), optical gain formation, minority carrier injection, etc.

The control and improvement of light-matter interaction is also obviously of great interest, for the same fundamental and applications reasons. A major breakthrough in this respect was brought along in the late 1970s and 1980s by the advent of quantum wells (QWs).[1,2] The improved light-matter interaction relies at the first-order level on the freezing of one degree of freedom in such structures, yielding more spectrally concentrated optical features such as absorption and gain. Additional, unexpected improvements also came from the increased exciton binding energy, allowing exciton effects at room temperature and therefore more efficient electro-optic and nonlinear effects.[3] Quantum wells thus exhibit outstanding optical features when compared to other solids, and this is due to their improved light-matter interaction.

In view of the successes of QWs, great effort has been devoted since the mid-1980s to the fabrication and evaluation of lower dimensionality systems such as quantum wires (QWWs) and quantum boxes (QBs), in an attempt to freeze more degrees of freedom, and therefore reach sharper optical features in the spectrum of such structured solids as determined by the density of states[1,2] (Figure 1). It must be remembered that in bulk solids, absorption bands extend over the whole conduction and valence bands, whereas the emission linewidth in spontaneous emission and the gain linewidth in lasers are usually of the order of kT (25 meV at room temperature) due to thermal population of the 3-D continuum of states. Therefore, there has been a major thrust of study to develop solids which would have sharper optical features. Besides the quantum-confinement scheme of low dimensional structures already mentioned, another widely explored avenue was that of localized defects in solids: due to their well-defined, atom-like energy levels, they indeed exhibit sharp lines. However, the most successful system is that of rare earths in insulating matrices, such as Nd ions in YAG or glasses. Therefore, they can only be excited in an optical pumping scheme.

The various other schemes involving optically active centers in semiconductor matrices, such as rare earths in semiconductors, crystalline or amorphous, have a major intrinsic drawback: in order to yield sharp atom-like levels, they only weakly interact with band levels. Thus, the energy transfer between delocalized semiconductor excitations and ions is usually poor, and the direct electrical excitation of these systems is always a difficult point. Also, due to their very localized nature, rare-earth ions have rather weak oscillator strengths (also due to the forbidden nature of many transitions) and therefore do not supply the strong light-matter coupling which is required for many applications. Increasing the density of the ions is no solution, as ion-ion interactions lead to level broadening and nonradiative recombination increases faster than the light-matter interaction. Quantum boxes, with their discrete

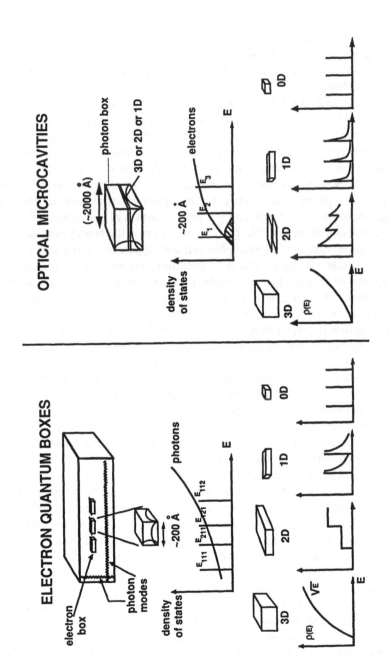

FIGURE 1 Schematics of the densities of states for low dimensional electronic systems (quantum wells, wires, or dots) or photonic systems (planar, waveguide, or closed microcavities).

energy levels, might also suffer from intrinsic relaxation limitations,[4] in addition to the difficulty of fabricating such minute structures with small enough size fluctuations.[5] It is therefore still a matter of exploratory research in the fields of solid-state electronics and chemistry to find the right solution of QBs or ions with sharp, strong optical transitions interacting "just enough" with a continuum of band levels to yield efficient energy transfer without "too much" level broadening.

Only recently was it recognized that another way exists to improve the light-matter interaction in solids, although it was recognized long ago for atoms and widely studied in the 1980s: it is that based on the control of photon modes. It comes from the understanding that actually two continua of states come into play to yield broad optical features in solids: one only obtains broad features because for each allowed electron-hole pair transition there exist an energy- and momentum-matched photon state. Consider a small piece of active material placed in a large environment, as described by a large photon confined volume. Whatever the photon quantizing condition is (dielectric or metallic boundaries), due to the large photon "box", the allowed photon modes appear as a continuum (with the usual 3-D spectral density $8\pi\nu^2/c^3$). Therefore, any probe light will contain a continuum of modes which will probe the density of allowed transitions of the active material. Conversely, if we now consider that the active material is enclosed in an optical cavity of very small size, then the electronic continuum of states can only interact with discrete photon modes, if energy and momentum are to be conserved in optical transitions (Figure 1). Instead of being emitted in all directions in all the photon modes matching the populated electron-hole pair states, photons can only be emitted by those electron-hole pairs which match the photon modes, in well-determined directions and polarizations. The fate of all other electron-hole pairs not matching a photon cavity mode is either to recombine nonradiatively or, in good enough samples (the usual case), to reach the photon-matched electronic quantum states through energy and momentum relaxation, which usually occur at a much faster pace ($10^{-10} - 10^{-12}$ s) than radiative and nonradiative recombination ($\geq 10^{-9}$ s).

The control of photon modes therefore appears as a very promising field which should allow new control of the emission process, but also of all other phenomena involving light-matter interaction. Actually, two possibilities are opened for photon mode control: one is the microcavity approach, similar to the more usual microwave resonator, where multiple reflections on mirrors select those modes which have in-phase multiple reflections and reject all other electromagnetic modes. The detailed analysis[6-9] leads to an increased E-field intensity of an allowed mode by the quality factor Q, an allowed mode width of $\Delta\nu/\nu \approx Q^{-1}$, and rejection of the amplitude of modes at other frequencies by a factor Q^{-1}. The other approach is to use a spatially periodically modulated dielectric medium, which at the optical frequencies determined by the Bragg condition will hinder wave propagation, therefore opening frequency gaps in the allowed frequencies for propagating waves in the medium, very similar to the energy gaps of solids determined by the periodic ionic potential. This approach of photonic bandgap (PBG) materials, pioneered by Yablonovitch, is quite similar in its results to the microcavity approach, as will be discussed in Section II.B. It relies on multiple reflections from distributed scatterers, instead of the multiple reflections from localized mirrors of microcavities. Of course, placing inside a PBG structure an active material with energies within the photonic bandgap gives too much suppression of light-matter interaction: no photon mode can interact with the active material. In order to have a single mode interaction, one has to artificially create such a mode, which is done by introducing an optical-impurity mode. It suffices to suppress the exact translational invariance of the PBG structure by locally changing the index or phase, i.e., by adding or suppressing some dielectric material. Then, a localized photon state can build up, similar to the localized electronic states of chemical impurities or defects in otherwise perfect solids. If the frequency of the optical impurity is well chosen to interact with the active medium, there again the active material will be coupled to that single "impurity" photon mode.

From the preceeding analysis we see the emergence of the various concepts which are being put to good use to control the light-matter interaction by photon mode control: one or a few photon modes are singled-out and increased by multiple, in-phase reflections. All other modes are suppressed by multiple destructive interferences. Overall, the light-matter interaction is hardly changed when integrated over all interacting photon modes, but this interaction is now concentrated in one (or a few) photon modes.

This selection of photon modes has many effects, which have already been put to good use. The multiple reflections lead to the buildup of light emission in a given direction, thus leading to directionality. This effect alone could have dramatic impact in LEDs and displays in that it solves the major difficulty in light emission in solids. Lambertian sources have an external efficiency in the percent range in useful outside angles due to the refraction factor at the solid-air interface ($n_2/n_1 \approx 3.5$). In contrast, the multiple reflections in a microcavity select an escape cone which is much smaller than the 2π angle of a Lambertian source (of the order of π (1–R), where R is the mirror reflectivity). This leads to a small outside emission angle, even when taking into account the beam widening due to the refractive index change. Another useful effect is that of spectral narrowing: due to the spectral width of allowed photon modes, light within a photon mode can be much narrower than the usual thermally broadened emission. This has been shown to increase by a factor of three the transmission capacity of optical fiber systems where chromatic dispersion is the limiting factor.[14]

So far, we have only assumed that the interaction of electromagnetic modes and active material is weak and can be described in a perturbative manner. All phenomena can then be described by evaluating the modification of the electromagnetic spectral energy density brought along by microcavity or photonic bandgap effects.

Many of these aspects are being discussed in detail in the other chapters of this book. We will concentrate in this chapter on the new phenomenon which arises when the optically active material placed inside a cavity is in strong coupling with the cavity electromagnetic modes. Then, one has to take into account the deterministic, coherent evolution of the coupled system made of the interacting electromagnetic modes and electronic excitations of the active material. New descriptions of the system are required, either quantum or classical, which can explain and predict the effects that can be expected from such a novel optical material system, very unique due to its strong light-matter coupling, which can persist up to room temperature. This chapter will be organized along the following lines: Section II will discuss the modeling of semiconductor microcavities based on distributed Bragg reflector (DBR) mirrors. In Section III we will describe the various single models of strong light-matter interaction, i.e., the usual strong-field Rabi oscillation of atoms and its relation to perturbation theory, the vacuum-field Rabi oscillation, the dressed-atom description, and the classical dispersion model based on Lorentz oscillators. Section IV will describe how the elementary excitations in semiconductor quantum wells, the excitons, can be modeled as Lorentz oscillators. With that background, we will describe in Section V the various experimental results obtained in GaAs microcavities.

II. THE FABRY-PÉROT RESONATOR: A PLANAR MICROCAVITY

The structure which has mainly been used up to now is the Fabry-Pérot resonator (FP). As it lends itself to a simple complete description, we give here a detailed analysis both of the classical localized-mirror-based FP resonator and of the semiconductor based distributed Bragg mirror (DBR) resonator.

A. THE FP RESONATOR

The detailed usual multiple pass interference description of the FP resonator can be found in many textbooks.[15] The physics of a FP mode is the following: An incident beam undergoes a series of partial reflections and transmissions at each mirror. At resonance the phase shift for a round trip inside the cavity is an integer number times 2π. All the transmitted beams are in phase, so they add up. The secondary reflected beams are in phase, and their sum just cancels the primarily reflected beam. An alternative view is that, at resonance, the built-in field inside the cavity is intense enough that it just balances the low transmission of the second mirror, making the total transmission unity. The resonant photon mode, for which all reflected waves in the cavity are in phase, has an intensity, at the antinode position, increased by a quality factor Q:

$$Q = \frac{(1+r)^2}{(1-r^2)} \approx \frac{4}{(1-R)}$$

(1)

where R (r) is the reflection coefficient for the optical (electric) field intensity (amplitude) ($R = r^2$). Conversely, the nonresonant photon modes are suppressed by a factor Q':

$$Q' = \frac{(1+r)(1-r^2)}{(1+r^2)} \approx 1 - R$$

(2)

The cavity acts as if it concentrates the electric field in the cavity from all nonresonant modes into the resonant one. It follows that the FP cavity can be regarded as a photon mode selector with a selectivity S:

$$S \approx \frac{4}{(1-R)^2}$$

(3)

S grows as the square of the finesse F:

$$F = \frac{\pi r}{(1-r^2)}$$

(4)

Note that these considerations are also valid for vacuum-field fluctuations. In the microwave domain, finesses up to 10^{10} can be achieved, and an object located inside the cavity can only couple to a single electromagnetic mode.

For planar FP's, simple geometrical calculations show that the FP mode eigenenergy E observed at a finite angle of incidence θ varies as:

$$E(\theta_{internal}) = \frac{E_\perp}{\cos(\theta_{internal})}$$

(5)

Alternatively, because of the invariance under in-plane translation, it is possible to define an in-plane momentum $k_{//}$. Rewriting the previous relation as a function of $k_{//}$ gives the dispersion curves of the FP mode (note that there is no $k_\perp$ anymore)

$$E(k_{//}) = \sqrt{E_0^2 + \frac{c^2 \hbar^2}{n^2} k_{//}^2}$$

(6)

B. THE DBR RESONATOR

1. One-Dimensional Photonic Bandgap and DBR Fabry-Pérot

The same calculation can be made for a distributed Bragg mirror (DBR), which consists of a series of alternating high and low index quarter wavelength

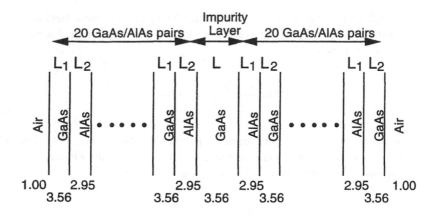

FIGURE 2 Schematic representation of Bragg stack used in the model. The middle layer has a variable width and acts as an impurity.

layers.[16] There is an instructive and alternative one-dimensional model that unifies impurities mode in a photonic bandgap and FP mode.[17] We start with a well-known physical system, the multilayer quarter wave stack DBR, as it can be a perfect 1-D photonic bandgap material when all the layers are equal or a PBG with an impurity mode when one layer has a different thickness. This analogy between DBR's and photonic bandgap systems has already been mentioned by Yablonovitch.[18]

The modeled structure is based on classic Bragg mirror structures using alternate $\lambda/4$ layers of GaAs and AlAs, consisting of 20 pairs of GaAs/AlAs— a GaAs "impurity" layer—20 pairs of AlAs/GaAs (Figure 2). The front and back surfaces are air, and we choose the dimensions of the layers such that $hc/\lambda = 1$ eV. The standard method of matrices can be used to calculate the reflectivity and transmission of the stack as well as the field intensity throughout the structure.[15] The matrices are based on a scalar plane-wave solution to Maxwell's equations that relates the incident and outgoing electric field when at an interface.

The transmission of the structure with a $\lambda/4$ "impurity" layer, for which the complete structure acts as a Bragg mirror is shown in Figure 3(b) (the curve denoted $\lambda/4$). The transmission has a broad minimum or stop band from 0.94 to 1.06 eV. On either side of the stop band is a series of side lobes, with maxima reaching unity transmission. For a $\lambda/2$ "impurity" layer we satisfy the FP condition, and a transmission maximum occurs in the center of the stop band (Figure 3(a), curve $2\lambda/4$). This structure is a $\lambda/2$ cavity between two DBR mirrors. For intermediate "impurity" layer widths, the structure is neither a perfect mirror nor a perfect Fabry-Pérot. Instead the first high energy transmission peak (edge mode) of the perfect mirror moves into the stop band and becomes a "FP-like" maximum. The variation of all the transmission maxima as a function of layer width and energy is shown in Figure 3(a). This plot shows that with increasing layer width, the high energy band edge transmission peak splits away from the others and moves into the stop band. The peak moves across the stop band and merges with the band of

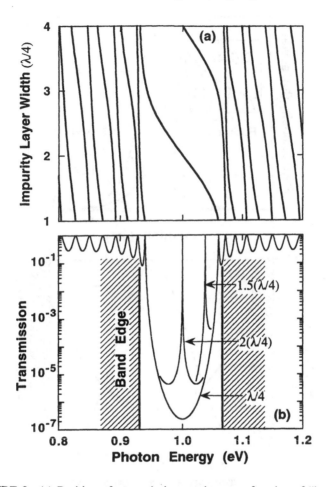

FIGURE 3 (a) Position of transmission maxima as a function of "impurity" layer width. (b) Transmission of λ/4 DBR stack as a function of photon energy. The transmission peak on the high energy side of the stop band moves into the stop band as shown for 1.5 (λ/4) and 2 (λ/4) "impurity" layer widths. The horizontal scale (photon energy) is the same for both (a) and (b).

transmission maxima at low energies. This pattern repeats itself as a function of layer width every λ/2.

To evaluate different impurity modes it is important to know the maximum field intensity inside the structure and how the field is distributed spatially. Figure 4(a) shows the envelope of the field intensity across the length of the test structure for various conditions. A λ/4 layer width generates a DBR with edge modes. Increasing the layer width to λ/2 produces a FP-impurity mode in the stop band, and its maximum field intensity is two orders magnitude larger than for the band edge mode. The strong peak of the field at the "impurity" layer and the exponential decay of the field away from the impurity illustrates the localization of light at an impurity center. By varying the "impurity" layer thickness we can examine the variation of maximum field intensity of the

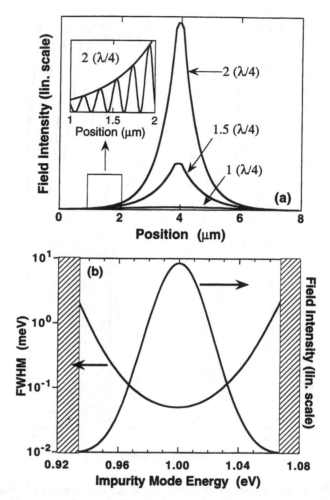

FIGURE 4 (a) Envelope field intensity inside the DBR stack for different "impurity" layer widths. Magnifications of five and twenty are used to make the broad envelope visible for the 1.5 (λ/4) and λ/4 "impurity" layer widths, respectively. Inset shows the oscillating electric field for the 2(λ/4) case. (b) Shows the variation of the maximum field intensity as a function of impurity mode energy (right axis) and the spectral width (full width half maximum) in meV (left axis).

impurity mode as it traverses the PBG. This is shown in Figure 4(b). The largest field occurs when the impurity mode is exactly at the mid-gap energy, and it decays rapidly when it shifts toward the band edges where the localization of light is also weak. In addition to the decrease in the spatial extent of the field intensity of the impurity mode, there is also a decrease in its spectral width. The variation in linewidth is shown in Figure 4(b). The decrease of the spectral width follows the increase of the effective FP finesse. This model shows that, in one dimension, edge modes, impurity modes, and FP microcavities are general features of PBG's. The perfect FP is identical to a mid-gap impurity mode, and this mode is the best for localizing light and maximizing the field intensity.

2. Experimental Results on a DBR-FP

The reflectivity spectrum of a cavity with a bottom mirror consisting of 27 pairs of 675-Å Al_1Ga_9As and 764-Å AlAs, a 2670-Å GaAs cavity and a 20-pair top mirror is shown in Figure 5. Because of the energy dependence of the phase shift of the reflectivity of a DBR several definitions of the finesse can be used:

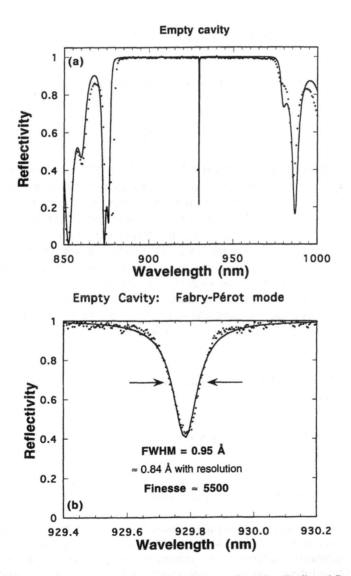

FIGURE 5 (a) Reflectivity spectrum of a 27 × 20 pairs distributed Bragg mirror Fabry-Pérot microcavity. Measured curve (dots) and calculated curve (solid line). (b) Detail of Fabry-Pérot mode. The solid line is a Lorentzian with a full width half maximum of 0.95Å.

$$(i) \; F = \frac{\text{Free spectral range}}{\text{FWHM}} = \frac{\lambda}{m\Delta\lambda} \tag{7}$$

where m is the fringe order $\left(m = \frac{2 \times \text{cavity length}}{\lambda} \right)$. Finesse can also be defined according to a local free spectral range or an effective cavity length. Both are strictly equivalent.

$$(ii) \; F_{eff} = \frac{\text{Local free spectral range}}{\text{FWHM}} = \frac{\lambda}{m_{eff}\Delta\lambda} \tag{8}$$

with $m_{eff} = m + m_0$. In the high reflectivity limit.

$$m_0 = \frac{n_1}{\left(n_h - n_1 \right)} \tag{9}$$

For a GaAs/Al$_{.1}$Ga$_{.9}$As DBR-FP, $n_h = 3.45$, $n_l = 2.95$ and $m_0 = 7.5$. A λ sized cavity is a second-order cavity (m = 2), so with $\lambda = 9300$ Å and $\Delta\lambda = 0.84$ Å this gives a finesse of 5530. The measured Fabry-Pérot mode has a linewidth of 0.84 Å at 930 nm. This implies a finesse in excess of 5500 and an effective (mirror corrected) finesse greater than 1450 in comparison to values of 160 by Jewell et al.[19] who calculated the local FSR, and to values of 700 by Oudar et al.[20] who measured the local FSR. Moreover, comparison with theoretical calculations[21] for such a structure shows that (i) the growth rates are stable to 0.25% over 14 hours and (ii) the internal losses are less than 1 cm^{-1}. Although such finesses are much lower than for microwave cavities, they are more than adequate to reach the strong coupling regime in semiconductors at optical frequencies.

III. MODELS OF STRONG LIGHT-MATTER COUPLING

A. FROM REVERSIBILITY TO IRREVERSIBILITY

. . . AND BACK

The reasons that make a quantum system undergoes a reversible or irreversible evolution are essential and somewhat subtle.

1. Let us start with a two-level system that will be called (A). It is well known that if the system is coupled to a single oscillator level of the same energy, the true eigenstates of the system are a symmetric and antisymmetric combination of the uncoupled eigenstates; if the system is prepared at time t = 0 in an eigenstate of (A) (i.e., if energy is put in (A)) the energy will oscillate in time between (A) and the outside oscillator. This behavior is completely reversible, and energy never tends to leave (A) when t goes to infinity (Figures 6(a) and 7(a)).

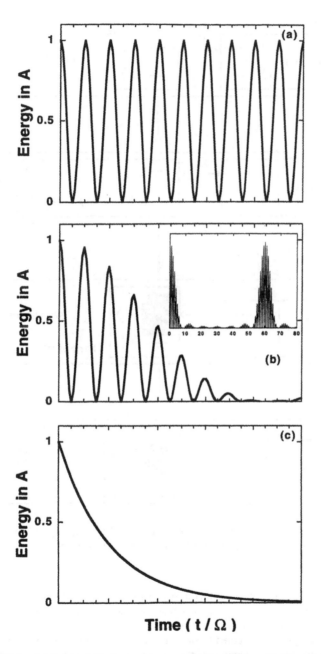

FIGURE 6 (a) Rabi oscillations of two coupled oscillators. (b) Damping and echoes in a finite number of coupled oscillators. (c) Irreversible decay of a discrete level coupled to a continuum.

2. If now a set of a finite number of undamped oscillators is coupled to (A) (Figure 7(b), the energy in (A) oscillates as shown in Figure 6(b), but the oscillations are damped. This damping is only an initial trend toward an irreversible decay:

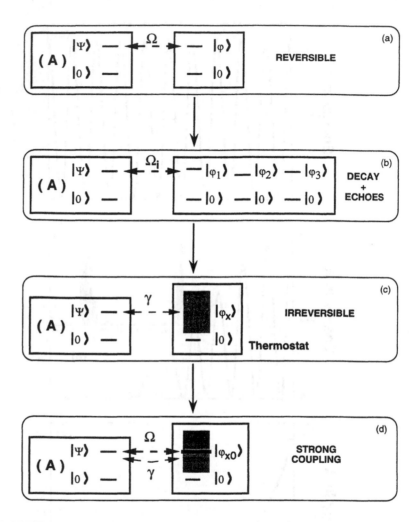

FIGURE 7 (a) Coupling to one oscillator. (b) Coupling to several oscillators. (c) Coupling to a continuum. (d) Vacuum-field Rabi splitting.

because of the finite number of oscillators, there is always a time for which all the oscillators are in phase again and the initial state is restored. The inset of Figure 6(b) shows such echoes.

3. If the system is coupled to an infinite number of oscillators (i.e., a continuum, Figure 7(c)), the damping of the oscillations becomes the predominant behavior and an irreversible decay is observed. In other words, as the number of oscillators increases the separation of the echoes tends to infinity; hence, the damped oscillators are irreversible. Energy tends to leave (A) into the continuum, which acts as a thermostat, at a rate which is always slower than the oscillation frequency defined by the interaction energy with each oscillator. A finite number of damped oscillators would give the same effect.

The time evolution of a single level coupled to a continuum of states can be calculated using perturbation theory, Fermi's well-known golden rule follows.[22] The probability density W of the transition from the initial discrete level $|i\rangle$ into a final state $|f,E_f\rangle$ is:

$$W_{i \to f} = \frac{2\pi}{\hbar} \left| \langle f, E_f = E_i |W|i\rangle \right|^2 \rho(E_f = E_i) \tag{10}$$

where $\rho(E)$ is the density of final states. An example of interest in our case is the irreversible decay of an excited atomic state $|e\rangle$. In the atom-field space the $|e\rangle|n = 0$ photon$\rangle$ state is coupled to the continuum formed by the $|n = 1, E, \bar{k}_2$ states, which has a continuous spectrum in energy E and in polarization and wavevector $\bar{k}$ (Figure 8(a)). Application of Fermi's golden rule leads to the emission rate:

$$\frac{d}{dt} P|e\rangle \to |f\rangle = \frac{2\pi}{\hbar} \left(1 + n_{\bar{k},E}\right) \frac{E_{e \to f}}{2\varepsilon_0 L^3} \left| \langle e_i |H_{int}|f\rangle \right|^2 \delta\left(E_{e \to f} - E_{\bar{k}}\right) \tag{11}$$

where L is the size of an arbitrarily large box and n is the number of photons in mode $\bar{k}$, E. It includes both spontaneous and stimulated emission rates. Usually one then makes L$\to\infty$ and sums over states to a density of states. This gives the classical radiative emission rates and the Einstein relations.

The concept of strong light-matter coupling in microcavities arises from the central question: having a system whose excited states exhibits an irreversible decay because of the coupling to a thermostat, is it possible to recover a reversible behavior? From the previous consideration it follows that a recipe is to select one state of the continuum and very strongly enhance the coupling with this given state up to the point at which coupling with the other states of the continuum can be neglected (Figure 7(d)). There are two ways to achieve this in quantum optics: The first one is the intense-field Rabi oscillations, where an intense driving field of a single photon mode makes negligible the vacuum properties. The second one directly deals with a modification of the vacuum properties as compared to free space. These two points will be discussed in the following sections.

B.   THE INTENSE FIELD RABI OSCILLATIONS

Let us consider a two-level system (defined by its Hamiltonian $\mathcal{H}$ and the two states $|\varphi_1\rangle$, $|\varphi_2\rangle$ with $\mathcal{H}|\varphi_i\rangle = Ei|\varphi_i\rangle$ $E_1 < E_2$ $\hbar\omega_0 = E_2 - E_1$) on which is applied a sinusoidal external perturbation W that couples the $|\varphi_1\rangle$ and $|\varphi_2\rangle$ states. In our case this perturbation will be the electric dipole of the light-matter interaction with an intense incident driving field. Intense means that the field is strong enough so that the interaction with this given photon mode is much stronger that coupling with any other vacuum photon modes, as well as any other scattering mechanisms (i.e., lifetime, dephasing time, etc.). W can be written as:

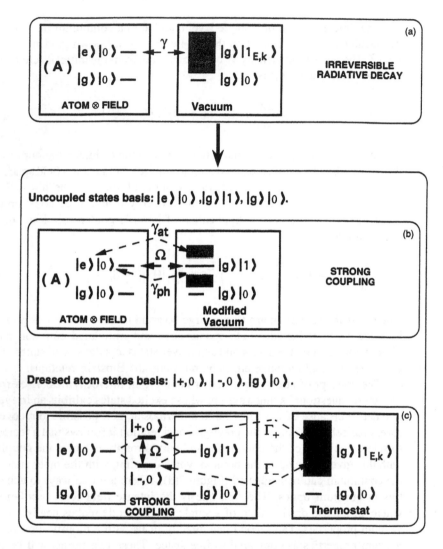

FIGURE 8   (a) Irreversible radiative decay. (b) Vacuum-field Rabi oscillations in the uncoupled states basis. (c) Vacuum-field Rabi splitting in the dressed atom states basis.

$$W = \sum_{i,j} W_{i,j} \sin \omega t \left| \varphi_i \right\rangle\!\left\langle \varphi_j \right|$$

$$= \sum_{i,j} W_{i,j} \frac{e^{i\omega t} - e^{-i\omega t}}{2i} \left| \varphi_i \right\rangle\!\left\langle \varphi_j \right| \tag{12}$$

For short times ($t \ll |W_{1,2}|/h$) it is well known[22] that a perturbative calculation will describe a transition from the $|\varphi_1\rangle$ to the $|\varphi_2\rangle$ state (absorption) or from the $|\varphi_2\rangle$ to the $|\varphi_1\rangle$ state (stimulated emission). We are interested here in the evolution of the system for long times. At a time t the states of the system can be written along the $|\varphi_1\rangle$, $|\varphi_2\rangle$ basis states:

$$|\Psi(t)\rangle = c_1(t)|\varphi_1\rangle + c_2(t)|\varphi_2\rangle \tag{13}$$

Without interaction the $c_i(t)$ have the usual form:

$$c_i(t) = a_i e^{-\frac{iE_i t}{\hbar}}, \quad a_i \in C \tag{14}$$

With the interaction we will look for solutions of the form:

$$c_i(t) = a_i(t)e^{-\frac{iE_i t}{\hbar}} \tag{15}$$

This leads to the equations:

$$\begin{cases} i\hbar\dot{a}_1(t) = \dfrac{1}{2i}\left[\left(e^{i\omega t} - e^{-i\omega t}\right)W_{11}a_1(t) + \left(e^{i(\omega-\omega_0)t} - e^{-i(\omega+\omega_0)t}\right)W_{12}a_2(t)\right] \\[3mm] i\hbar\dot{a}_2(t) = \dfrac{1}{2i}\left[\left(e^{i(\omega+\omega_0)t} - e^{-i(\omega-\omega_0)t}\right)W_{21}a_1(t) + \left(e^{i\omega t} - e^{-i\omega t}\right)W_{22}a_2(t)\right] \end{cases} \tag{16}$$

Neglecting the terms in ω, ω_0, and $\omega + \omega_0$, (usually called a secular approximation or rotating wave approximation in quantum optics) leads to the probability P_2 of the system being in the $|\varphi_2\rangle$ state:

$$P_2 = \frac{|W_{12}|^2}{|W_{12}|^2 + \hbar^2(\omega-\omega_0)^2}\sin^2\sqrt{\frac{|W_{12}|^2}{\hbar^2} + (\omega-\omega_0)^2}\,\frac{\tau}{2} \tag{17}$$

The function is shown in Figure 9 for different values of the detuning $\delta = \omega - \omega_0$. At $\delta = 0$, the system will oscillate at the Rabi frequency $\Omega = |W_{1,2}|/\hbar$. The long-term evolution is not an irreversible transition from one state to another but a periodic oscillation between both states according to the following scheme:

$$|\varphi_1\rangle \rightarrow \text{absorption} \rightarrow |\varphi_2\rangle \rightarrow \text{stimulated emission} \rightarrow |\varphi_1\rangle \rightarrow \cdots \tag{18}$$

C.   CAVITY QUANTUM ELECTRODYNAMICS MODEL

In the field of atomic physics the influence of photon quantization due to optical cavity resonances has long been a matter of interest.[23-34] This was

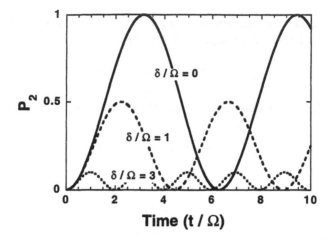

FIGURE 9   Intense-field Rabi oscillations for different values of the detuning $\delta = \omega-\omega_0$.

recently revived by the observation of the deterministic, coherent behavior of single atoms interacting with single photon modes, leading to so-called vacuum-field Rabi oscillations. The area has steadily developed over the years because of the information it has provided on many fundamental aspects of quantum electrodynamics (QED), i.e., the interaction of atoms with electromagnetic fields. It is currently called cavity quantum electrodynamics (CQED), and has implications for fundamental quantum measurement theories ("Schrödinger cats") and photon squeezed states.

Vacuum-field Rabi splitting has been extremely well described previously,[25,30-32,50] we only intend here to give a summary of the model and of its physical consequences. The point we are interested in is, what happens when an atomic two level system is resonantly coupled to a single cavity photon mode? The atomic oscillator can be described by a 1/2-spin-type Hamiltonian with a ground state $|g\rangle$ of energy $-\dfrac{\hbar\omega_{at}}{2}$ and an excited state $|e\rangle$ of energy $+\dfrac{\hbar\omega_{at}}{2}$. The Hamiltonian can be written as:

$$\mathcal{H}_{at} = \frac{1}{2}\hbar\omega_{at}\left(c^+c - cc^+\right) \tag{19}$$

with c and c^+ representing the annihilation and creation operators, respectively.

$$c = |g\rangle\langle e|, \; c^+ = |e\rangle\langle g|, \; \text{and} \left[c, \, c^+\right]_+ = 1. \tag{20}$$

The single-photon mode can be represented by an harmonic oscillator with an infinite series of $|n\rangle$ states with n = 0, 1, . . . photons. Its Hamiltonian is:

$$\mathcal{H}_{ph} = \frac{1}{2}\hbar\omega_{ph}\left(a^+a + aa^+\right) \tag{21}$$

with a and a^+ representing the annihilation and creation operators, respectively, and $a^+|n\rangle = \sqrt{n+1}\,|n+1\rangle$, $a|n\rangle = \sqrt{n}\,|n-1\rangle$, and $[a, a^+] = 1$. The total Hamiltonian of the atom + photon system is:

$$\mathcal{H}_T = \mathcal{H}_{at} + \mathcal{H}_{ph} + \mathcal{H}_{inter} \qquad (22)$$

with $\mathcal{H}_{inter}$ an interaction Hamiltonian that mixes photon and atom operators[25]

$$\mathcal{H}_{inter} = -\hbar g\left(ac^+ + a^+c\right) \text{ with } \hbar g = dE_0 = d\sqrt{\frac{\hbar\omega_{ph}}{\varepsilon_0 V_{cav}}}. \qquad (23)$$

E_0 is called the electric field per photon mode and d is the electric dipole matrix element. States are written as $|e$ or $g\rangle|$in photons$\rangle$. Close to the resonance condition $\delta = \omega_{ph}-\omega_{at} \approx 0$. If $g = 0$ there is a ladder of pairs of degenerated states (Figure 10). The $|e\rangle|n\rangle$ is degenerated with the $|g\rangle|n+1\rangle$ state; these two states are also the states that $\mathcal{H}_{inter}$ connects. Turning on the light-matter interaction ($g \neq 0$) lifts the degeneracy (Figure 10). A straightforward calculation shows that the energy of both eigenstates is:

$$E_\pm = (n+1)\hbar\omega_{ph} \pm \frac{1}{2}\sqrt{4g^2(1+n)+\delta^2} \qquad (24)$$

At exact resonance the eigenstates are a symmetric and antisymmetric combination of the uncoupled states:

$$|\mp, 0\rangle_{\delta=0} = \frac{1}{\sqrt{2}}\left(|e\rangle|n\rangle \pm |g\rangle|n+1\rangle\right) \qquad (25)$$

This series of states is called the dressed atom states ladder.[31] At resonance the energy separation of the states is:

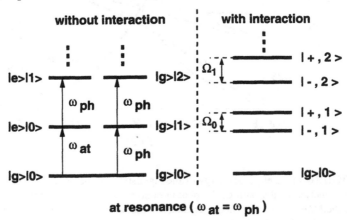

at resonance ($\omega_{at} = \omega_{ph}$)

FIGURE 10   The dressed atom states ladder.

$$\Delta_n = \hbar\Omega_{Rabi} = 2d\sqrt{\frac{\hbar\omega_{ph}}{\epsilon_0 V_{cav}}}\sqrt{1+n}$$

(26)

The very remarkable feature is that even for no photon initially present in the cavity ($n = 0$) a splitting still occurs. This effect is related to the textbook case of intense-field Rabi splitting[35] which, in this case, is induced by the zero-point field fluctuations in the cavity. It can be regarded as coupling between the atomic oscillator and the vacuum field of the cavity. It was first called vacuum-field Rabi splitting by J.J. Sanchez-Mondragon et al.[35] If several atomic oscillators are present, it can be shown[36] that the splitting increases as the square root of the number N_{at} of atoms, $\Omega_{Rabi}^{n=0}(N_{at}) = \Omega_{Rabi}^{n=0}\sqrt{N_{at}}$. If the system is prepared in the pure atomic oscillator or photon oscillator state, it will oscillate between these two states at the Rabi frequency Ω_{Rabi}. In a classical oscillator description, the overall system exhibits an anticrossing behavior when both oscillators are resonant, with two split modes corresponding to the normal modes of the system (Figure 11). The relevant physical parameters that are used as experimental evidence of vacuum-field Rabi splitting are the emitted spectrum,[35] absorption (i.e., the dielectric susceptibility)[36,37] time resolved oscillations[26,38] and photon statistics.[29] As will be discussed, neither reflectivity nor transmission are *a priori* relevant parameters.

Extension of the model to imperfect oscillators (i.e., including damping) has been treated by Haroche[25] using a density matrix approach formalism. There are two interesting basis states that can be considered: (1) the uncoupled

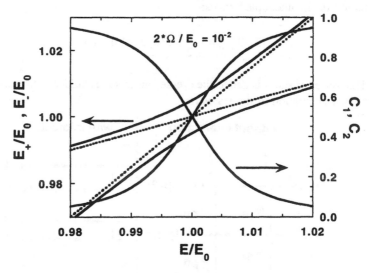

FIGURE 11 Left axis: energy of the eigenstates of the dressed atom states ladder as the function of the ratio of the photon energy over the atomic transition energy. The dashed lines are for the uncoupled states. Resonance occurs at $E/E_0 = 1$. The interaction energy is $0.01 \cdot E/E_0$. Right axis: projection of the eigenstates wavefunction onto the uncoupled states.

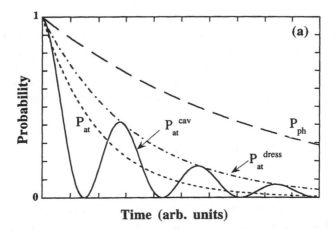

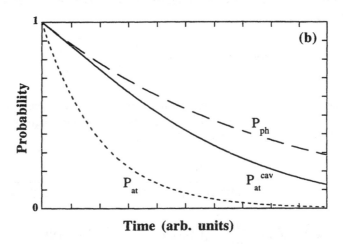

FIGURE 12 Time evolution of the probability of finding an atom in the excited state (a) in the strong coupling regime and (b) in the weak coupling regime. The dashed line shows the time evolution for the uncoupled photon mode and the "natural" decay of the atom. The dotted dashed line in (a) shows the decay of a dressed atom basis state in the case of subnatural radiative linewdith ($\gamma_{cav} < \gamma_{at}$).

states basis (Figure 8), which gives rise to damped Rabi oscillations in the time domain (Figure 12) and (2) the dressed atom states basis (Figure 8), where eigenstates are coupled to a "usual" continuum, so an irreversible decay is expected. Nevertheless, interesting configurations can be achieved: as the decay rate is on the average of the decay rates of both oscillators,[27] subnatural linewidth can be observed when the photon decay rate is smaller than the natural atomic decay rate[27] (Figure 12). It follows that in order to observe vacuum field Rabi splitting, several conditions must be fulfilled: Rabi oscillations must occur at a faster rate than both atomic excited decay rate $1/\tau$ and the photon life time τ_c in the cavity, i.e., $\Omega\tau_c > 1$ and $\Omega\tau > 1$. If these conditions are satisfied the system is in strong coupling regime, if not the cavity acts as a reservoir and the excited state decays at a rate[30] $1/\tau_{at} = \Omega^2\tau_c$. This regime is the usual weak coupling regime where energy levels are weakly affected and perturbative approaches such as Fermi's golden rule are valid.

D.   A CLASSICAL DESCRIPTION: THE LINEAR DISPERSION MODEL

In a letter, Zhu et al.[39] demonstrated that the linear dispersion theory can be used to described the so-called vacuum field Rabi splitting. The cavity is modeled by the standard Airy description of a Fabry-Pérot (FP) oscillator (mirror reflectivity, R; cavity length, L_c; FP mode frequency, n_{FP}; mode order, m; free spectral range, FSR; cavity finesse, F; and FP linewidth, d_c) and the two-level atomic system of thickness L by a Lorentz oscillator dispersive dielectric constant:

$$\varepsilon(v) = n(v)^2 = \varepsilon_\infty + \frac{Nf\,e^2}{m\varepsilon_0}\frac{1}{v_0^2 - v^2 - i\delta_H v} \tag{27}$$

where f is the oscillator strength per atom, e (m) the charge (mass) of the electron, N the oscillator density, n_0 the resonance frequency and δ_H the oscillator linewidth. It is approximated by:

$$n(v) = n_\infty - \alpha_0 \frac{c}{2\pi v}\frac{(v-v_0)\delta_H}{4(v-v_0)^2 + \delta_H^2}, \quad \alpha = \alpha_0\frac{\delta_H^2}{4(v-v_0)^2 + \delta_H^2}$$

$$\text{with } \alpha_0\delta_H = \frac{2Nf\,e^2}{mnc}\,\text{(CGS)}, \quad \alpha_0\delta_H = \frac{Nf\,e^2}{2\pi\varepsilon_0 mnc}\,\text{(SI)} \tag{28}$$

The transmission ($T[v]$), reflectivity ($R[v]$) and absorption ($A[v]$) of the (cavity + atom) system is then:

$$T(v) = \frac{(1-R)^2 e^{-\alpha L}}{(1-Re^{-\alpha L})^2 + 4\,Re^{-\alpha L}\sin^2\left(\frac{\varphi}{2}\right)} \quad \text{and}$$

$$R(v) = \frac{(1-e^{-\alpha L})^2 R + 4\,Re^{-\alpha L}\sin^2\left(\frac{\varphi}{2}\right)}{(1-Re^{-\alpha L})^2 + 4\,Re^{-\alpha L}\sin^2\left(\frac{\varphi}{2}\right)} \tag{29}$$

$$A(v) = 1 - T(v) - R(v) = \frac{(1-R)(1-e^{-\alpha L})(1+Re^{-\alpha L})}{(1-Re^{-\alpha L})^2 + 4\ Re^{-\alpha L}\sin^2\left(\frac{\varphi}{2}\right)}$$

$$\varphi = \frac{2\pi(v-v_{FP})}{FSR} + \frac{4\pi(n-n_\infty)Lv}{c} \quad \text{and} \quad FSR = F\delta_c = \frac{v_{FP}}{m} \tag{30}$$

where ε is the round-trip phase shift. This model is valid for low single pass absorption ($\alpha_0 L \ll 1$) and thick oscillator layers ($L \leq L_c$) as it neglects the multiple reflections at the oscillator / cavity interface and the location of the oscillators with respect of the node of the optical field in the cavity. As it will be mentioned later it can easily be extended to the case of semiconductor systems. It can lead to discrepancy of up to a factor of 2. The absorption line splitting is not correct in Reference 39:

$$\Omega = \sqrt{\Omega_{max}^2 - \frac{\delta_c^2 + \delta_H^2}{2}} \quad \text{with} \quad \Omega_{max} = \sqrt{(\alpha_0 L\delta_H)\frac{c}{2\pi L_c}} \propto \sqrt{\frac{Nf}{L_c}} \tag{31}$$

where $\alpha L\delta_H$ is proportional to the oscillator strength or to the matrix dipole element in a quantum model. The maximum splitting is only a function of total oscillator strength and the cavity size, but not of the finesse, which is the exact analog of the full quantum mechanical model.[30]

The simple physical picture is that in order to form a FP resonance the round-trip phase shift has to be an integer multiple of 2π. Due to the form of the real part of the refractive index of the Lorentz oscillator the round-trip phase shift vs. photon energy is changed according to Figure 13, up to the point where the phase shift conditions are fulfilled three times. This gives rise to the doublet structure because the central solution, which also corresponds to a maximum of absorption, does not create a FP resonance.

Figure 14 demonstrates that neither reflectivity nor transmission are *a priori* relevant parameters. The existence of two split levels in the cavity + atom system is evidenced by transitions that can occur at these energies. Existence of a structure in reflectivity may have a different origin, because it cannot distinguish between true absorption transitions and a change in the reflectivity/transmission balance. A reinterpretation of Figure 2 of Reference 39 clearly demonstrates that the observation of a splitting in reflectivity or transmission is not an unambiguous signature of vacuum-field Rabi splitting, as a splitting can be observed on reflectivity/transmission curves while the absorption (1-R-T) is not split. (Figure 14(c)). Moreover, high Q cavities are not even necessary to generate this situation, which can be regarded as a variation in the reflectivity/transmission balance induced by the phase shift created by the atoms or an absorption dip in a broad FP mode.

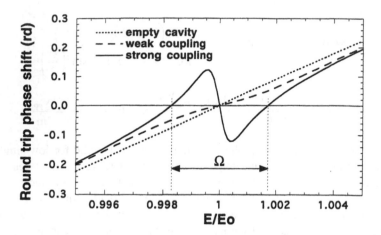

FIGURE 13   Round-trip phase shift vs. photon energy in a linear dispersion model for an empty cavity (dotted line), cavity in the weak coupling regime (dashed line) and in the strong coupling regime (continuous line). Fabry-Pérot modes develops for a zero phase shift.

IV.   OPTICS OF SEMICONDUCTORS

It is remarkable that the optical properties of solids can be described with concepts similar to those used for atoms, i.e., oscillator strengths.[40,41] The Lorentz oscillator description of the dielectric susceptibility of atoms gives under an exciting wave at frequency ω:[41-43]

$$\chi^{(1)}(\omega) = \frac{Ne^2 f \varepsilon_0 m_0}{\omega_0^2 - \omega^2 + \gamma \omega} \qquad (32)$$

where N is the atom density, m_0 the force electron mass, ω_0 the dominant atomic transition frequency, and γ the level broadening. In atoms, f is given by:

$$f = \frac{2 m_0 \omega}{\hbar} |\langle 1|\vec{r}|2 \rangle|^2 = \frac{2}{m_0 \hbar \omega} |\langle 1|\vec{p}|2 \rangle|^2 \qquad (33)$$

between two states $|1\rangle$ and $|2\rangle$.

In solids, Equation 32 is still valid when using Bloch states for electron-hole pairs, and summing over all states in the Brillouin zone,[40-43] taking into account the occupancy of levels by the Fermi-Dirac distribution function. Upward (in energy) transitions must be counted positively, downward ones negatively. Therefore, one obtains with the same formalism the expression of gain in semiconductor lasers.[43]

For electron-hole pairs in a crystal of any dimensionality (bulk, quantum well, wire, or dot), momentum conservation implies that only one Bloch state

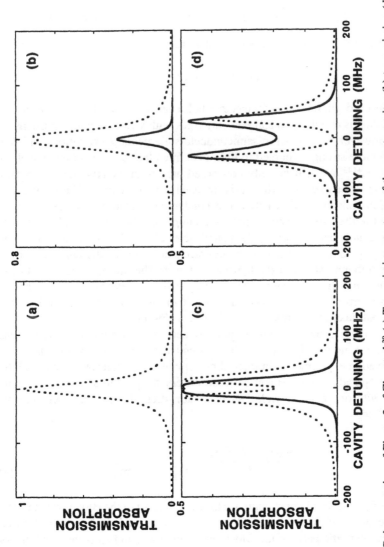

FIGURE 14 Reinterpretation of Figure 2 of Zhu et al.[39] (a) Transmission spectrum of the empty cavity; (b) transmission (dotted line) and absorption (solid line) of weak coupling regime; (c) weak coupling regime with a (nonrelevant) splitting in transmission but none in absorption; (d) strong coupling regime that exhibits a vacuum field Rabi splitting.

in the valence band is coupled to one momentum-matched state in the conduction band. Then, the Rabi frequency for such a transition is given by:

$$\hbar\Omega = e\vec{r} \cdot \vec{E}_{vac} = \langle 1|e \cdot \vec{r}|2\rangle \cdot \vec{E}_{vac} \qquad (34)$$

Given that $E_{vac} = \sqrt{\dfrac{\hbar\omega}{2\varepsilon_0 \varepsilon V}} \approx 6.10^4 \; V \cdot m^{-1}$ in a 1-μm λ–sized cavity, and that $\langle e\,\vec{r}\rangle$ is about an atomic dipole, of the order of $1.6 \; 10^{-19} \; Cb \times \mathring{A}$, one finds

$$\Omega \approx 10^{10} \, rad \cdot s^{-1} \qquad (35)$$

As the dephasing times for electron-hole pairs are in the 10^{-13} s range, the Rabi oscillation is overdamped and one only deals with weak-coupling situations.

However, there exist in semiconductors elementary excitations which are much more strongly coupled to light than free electron-hole pairs, namely the excitons.[44,45] We will be only concerned here with weakly bound, so-called Wannier-Mott excitons, which arise in usual semiconductors. They originate in the Coulomb interaction between a conduction electron and the hole it left behind in the valence band. That interaction leads both to bound states of the electron-hole pair, similar to the bound states of the electron and proton in the hydrogen atom, and to unbound but correlated states of the pair, also similar to the unbound states of the hydrogen atom.[46] Therefore, a first effect of the interaction is the appearance of an hydrogen-like series of peaks in the absorption spectrum, each corresponding to a hydrogenic level characterized by a principal quantum number n (n = 1, 2,. . .) (Figure 15).

Of major importance for our purpose here is the oscillator strength in each transition, and mainly in the exciton ground state n = 1, as it is the one mainly observed in experiments. In three-dimensional, bulk materials, the calculation of the optical matrix element between the crystal ground state and its excited state with one exciton yields an oscillator strength per unit volume, given by[44,45,47,48]

$$f = \frac{2 \, m_0 \omega \left| \langle u_c | \vec{\varepsilon}_0 \vec{r} | u_v \rangle \right|^2}{\hbar} \frac{1}{\pi a_B^3} = f_{at} \cdot \frac{1}{\pi a_B^3} \qquad (36)$$

where f_{at} is the oscillator strength for the periodic part of the Bloch wavefunction, of the order of an atomic oscillator strength, u_c and u_v are conduction and valence periodic parts of the Bloch wavefunctions, and a_B is the exciton Bohr radius.

Instead of detailing here the straightforward but lengthy calculation which leads to Equation (36), let us discuss its origin and significance: one has first to calculate the energy levels and wavefunctions of excitons, the elementary excitations of insulators which take into account the carrier-carrier interaction

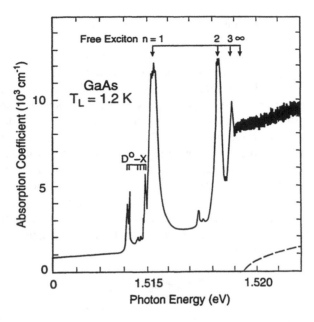

FIGURE 15 Low-temperature exciton absorption spectrum of hight-purity GaAs. The weaker sharp peaks correspond to bound exciton transitions. (R. G. Ulbrich and C. Weisbuch, unpublished).

(without that interaction the elementary excitations are those obtained in the single-electron approximation, i.e., the uncorrelated electron-hole pairs). One has then to calculate the interaction of the excitons with electromagnetic radiation. In such a sequential calculation, it is of course assumed that the electromagnetic field will only act in a perturbative manner on the excitons' energy levels and wavefunctions. Perturbation theory then gives Equation 36, which shows that excitons act "like" Lorentz oscillators in the description of the light-solid interaction. This comes from first-order perturbation theory, which always leads to an expression for the induced electric polarization of any medium, which in turn gives an equation like (36) for the dielectric function, i.e., which includes the momentum matrix element between the ground and excited states of the system. This "equivalent" description, appearing here in an equation form, has its origin in the constitutive equation, i.e., the relation between the electric polarization field and the applied electromagnetic field, which defines the dielectric coefficient in Maxwell's equation. The equivalence between such a macroscopic description and the effect of microcospic Lorentz oscillators is rarely carried out. We are not aware of a single report demonstrating the equivalence for excitons. Curious readers could use the following sequence of references to construct their own understanding of the relation between the microscopic and macroscopic descriptions. For the purely classical case, a good description can be found in the optics book by Rossi.[49] The equivalence of the classical Lorentz oscillator model and the quantum two-level atom model can be found in Section 6.4 of Eberly and Milonni.[42] The details of the description of excitons as Lorentz oscillators are given in various papers by Hopfield, reviewed in the book by Knox.[44]

We see from Equation 36 that the exciton oscillator strength per unit volume is given by an atomic oscillator strength times the number of excitons which can be closely packed in a unit volume. This can be justified by a close look at free electron-hole pair and exciton wavefunctions respectively. In the latter case, the electron-hole correlation in the hydrogenic state increases the overlap between the electron and hole, as they now have a relative extension of the order of a_B, instead of being delocalized in the whole crystal. By close inspection of the dipole moment in the two cases, one can convince oneself that the dipole moment is then increased by the factor $1/\pi\, a_B^3$.

In 2-D quantum wells, one has to resort to involved approximate methods to calculate exciton energy levels and oscillator strengths of these quasi-two-dimensional systems. In the limit of exact 2-D systems, the oscillator strength per unit surface would be:

$$f = \frac{2\, m_0\omega\left|\langle u_0|\bar{\varepsilon}\cdot\bar{r}|u_v\rangle\right|^2}{\hbar}\ \frac{8}{\pi a_B^2} \tag{37}$$

Where a_B has the usual 3-D value. Use has been made of the four times smaller value of the Bohr radius for exact 2-D excitons. Detailed calculations of exciton binding energies and oscillator strengths have been performed by Andreani and Pasquarello[50] and are shown in Figure 16. A very important feature of excitons in quantum wells is that the increase in binding energy from the 3-D case makes exciton features observable at room temperature. Although the dissociation time is quite short, as described by a linewidth of ~ 5 meV at 300 K, this leads to observable effects on those phenomena related to unrelaxed excitations such as absorption, electro-optical and nonlinear optical effects[3] (Figure 17). The importance of excitonic effects on long-lived phenomena such as luminescence is still a hotly debated issue, as it could have a very important impact on laser action.

To finish, let us briefly evaluate the mechanism and order of magnitude for the vacuum-field Rabi oscillations of excitons. Again, the Rabi frequency is given by:

$$\hbar\Omega = e\,\bar{r}\cdot\bar{E}_{vac} \tag{38}$$

Using a value of 5.10^{16} m^{-2} for the oscillator strength per unit surface (i.e., $\langle r\rangle \approx 3.10^{-2}$) (Figure 16) and $E_{vac} \approx 6.10^{-2}$ for a 1-μm-thick unit surface cavity, one finds $\hbar\Omega \approx 1,8\cdot10^{-3}$ eV ($\omega \approx 2.5\ 10^{12}$ rad·s$^{-1)}$ which can be larger than exciton linewidth at low temperatures, as excitons are less scattered than electrons or holes due to their neutral nature, and of comparable size at room temperature.

Experimental results to be described now will indeed show that the Rabi oscillation is observable up to room temperature.

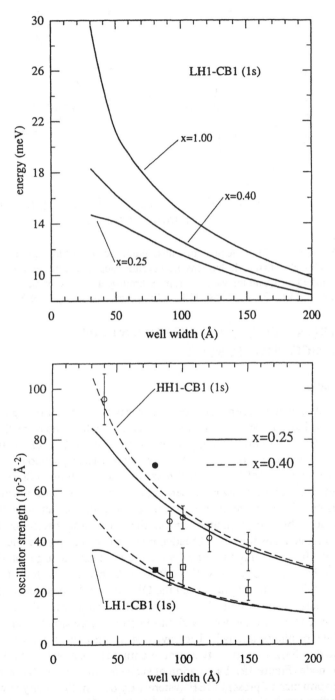

FIGURE 16 Binding energies (top) and oscillator strength per unit area (bottom) of excitons in GaAs-Ga$_x$Al$_{1-x}$As quantum wells, for various values of x, as a function of well thickness. (From Andreani, L. C. and Pasquarello, A., *Phys. Rev.*, B42, 8928, 1990. With permission ©American Physical Society.)

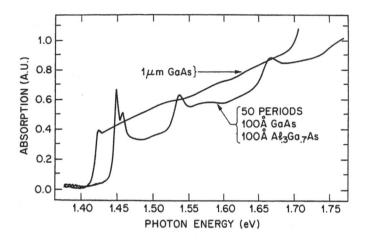

FIGURE 17   Room temperature absorption spectra of bulk and quantum well GaAs samples. Note that in the bulk material excitons appear as a bump whereas they are well-identified in quantum wells. (From Schmitt-Rink, S., Chemla, D. S., and Miller, D. A. B., *Adv. Phys.*, 38, 89, 1989. With permission ©Taylor & Francis.)

V.   STRONG COUPLING EXPERIMENTS IN SEMICONDUCTORS[51]

A.   SAMPLE

A sketch of a semiconductor microcavity is shown in Figure 18. The low (high) index material is usually AlAs (GaAs) where the refractive index is n = 2.96 (3.54). For optical pumping experiments it can be useful to use $Al_xGa_{1-x}As$ ($x \approx 10\%$) instead of GaAs. The alloy is grown as a pseudoalloy for molecular beam epitaxy (MBE) technical reasons. The quantum wells are $In_{0.13}Ga_{0.87}As$ and 75 Å thick, allowing experiments in reflection and transmission. In the design of the quantum well layer, a compromise between the limitations induced by the mismatch strain, the increasing oscillator strength, and inhomogeneous linewidth broadening when decreasing the thickness has to be made. In optimizing the optical structure to achieve a large splitting a compromise has also to be made with the fact that Ω increases with f but decreases with the cavity length. In practice all the QWs cannot be positioned where the field intensity is maximum. A structure factor arises from the phase difference between the different wells and thus Ω grows more slowly than the square root of the oscillator strength. The cavity of the structures that have been investigated is $3\lambda/2$ long (i.e., two usable antinode positions) with 3 QWs at each antinode (Figure 19). Fabrication of such structures is not far from the ultimate performance of present MBE systems. By design, the cavity is wedge-shaped, leading to a variation of the relative position of the QW exciton and the cavity mode across the sample. The splitting can then be studied as a function of the detuning between the FP mode and the exciton energy.

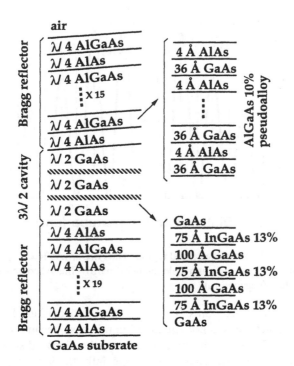

FIGURE 18   Sketch of a wedge-shaped semiconductor microcavity.

B.   REFLECTIVITY AND ABSORPTION

Figure 20 shows the reflectivity spectrum of the sample described above at 300K and 77K. The sample is designed to be resonant at low temperatures. At room temperature no resonance occurs and a single FP line is observed. Measurement at 77K on the same spot on the sample shows a split FP line. It is possible to demonstrate that this structure is not simply a dip formed by the absorption line of the excitonic line by looking at the position of both normal modes as a function of the detuning (in making use of the wedge shape of the cavity). As shown in Figure 21, taken from data in Reference 51, both lines exhibit an anticrossing behavior, as is expected by the theory.

Figure 22 shows the absorption spectrum at resonance at 80K of the structure described in Figure 18. The absorption (A) spectrum has been deduced from reflectivity (R) and transmitivity (T) measurements. Moreover it was checked that at resonance $T \ll R$ in this unbalanced Fabry-Pérot structure, then $A = 1-R-T \approx 1-R$. The splitting is 8.8 meV. The continuous line is a fit using a transfer matrix method including a linear dispersion model for the exciton, where the fitting parameters are the oscillator strength and the excitonic linewidth. The high energy line is much broader than the low energy one, probably because of the coupling to the continuum or excited states of the exciton. It should be noted that the observed Rabi splitting is of the order or even bigger than the exciton binding energy. In fact simulations including

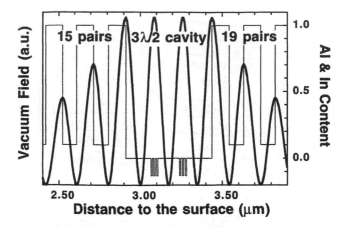

FIGURE 19   Sketch of the central region of an $Al_xGa_{1-x}As/In_yGa_{1-y}As$ semiconductor microcavity, thick line: vacuum-field intensity, narrow line Al content (positive values) and In content (negative values).

coupling to the 2s exciton state show a large broadening of the high energy line. A complete fit was not attempted in this case because of too many free parameters. Therefore the comparison to the linear dispersion model is satisfactory for the low energy line (especially the low energy tail) and the amplitude of the line. A linewidth that fits the high energy line would not have given the right amplitude and moreover would have given a low energy tail above the experimental one. The inhomogeneous linewidth (2.7 meV) and the oscillator strength ($4.8 \cdot 10^{12}$ cm^{-2} per QW) are in good agreement with the literature.[52] It can be noticed that this also gives one of the most accurate and reliable measurements of the oscillator strength of a QW exciton. Figure 23 shows the absorption spectrum of the same nominal structure tuned so that the cavity and the QWs are resonant at room temperature. The splitting is now 4.6 meV. An excellent fit is obtained with the same oscillator strength that was measured at 77K, by adding to the 77K inhomogeneous linewidth the homogeneous linewidth given by LO phonon scattering at 300K (≈ 6.5 meV). No broadening due to continuum effects is observed in this case because of the smaller Rabi splitting.

C.   EMISSION

While absorption can be understood in an dispersionless model, dispersion has to be taken into account in order to understand photoluminescence (PL). This is because an absorption process only involves states at a given value of momentum k while photoluminescence involves the creation of states at various nonresonant energies and momenta. Emission occurs at a given k and energy state after a relaxation process that can involve a large part of the dispersion curve. The situation here is quite different from atomic physics:[53] there, an atomic beam, having a monochromatic emission line, can only

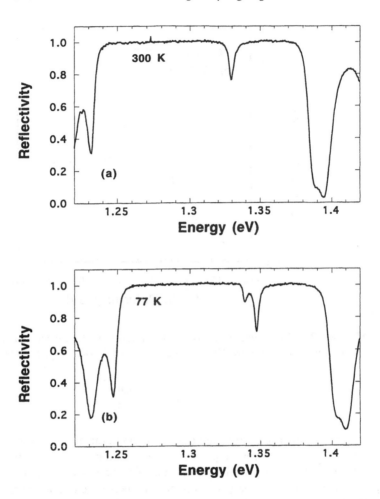

FIGURE 20 Reflectivity spectrum of a 3-quantum well, λ cavity sample (a) at 300K, where the exciton and the Fabry-Pérot are not resonant, (b) at 77K where both oscillators are resonant. The doublet structure due to the vacuum-field Rabi splitting is observed.

interact with a single mode of the cavity, while in a semiconductor microcavity photons and excitons have an in-plane dispersion. Each exciton state can couple with the photon state of the same in-plane wavevector $k_{//}$. During the photoluminescence process, a number of states with different $k_{//}$ are created, each interacting with the energy- and momentum-matched photon state. These considerations are the exact extension to 2-D systems of the well-known bulk exciton-photon polariton. In fact we believe that in the case of semiconductor microcavities, where in-plane dispersion is essential, the denomination cavity-polariton is much more suitable than vacuum-field Rabi splitting. The main differences, going from bulk polaritons to 2-D cavity polaritons, comes from the following considerations, starting from the bulk case: for 3-D polaritons the coupling strength is

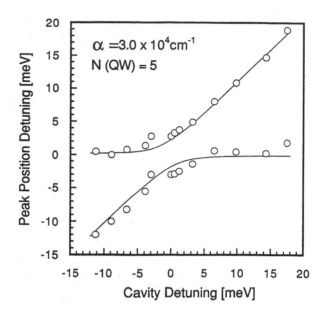

FIGURE 21 Reflectivity peak positions as a function of the cavity detuning for a 5-quantum well sample at 5K. Note the anticrossing behavior of the two normal modes.

$$2\hbar\omega_c \text{ where } \omega_c = \sqrt{\frac{\omega_0\omega_{LT}}{2}}, \tag{39}$$

ω_0 is the resonance energy, and ω_{LT} is the longitudinal-transverse splitting,[54,55] $2\hbar\omega_c$ is also the Rabi frequency of the exciton-photon system and is the frequency range over which eigenstates of wave function are strongly mixed. Because of the extremely steep photon dispersion curve as compared to the exciton one, the longitudinal-transverse splitting (ω_{LT}) which is not *stricto sensu* a polariton effect (i.e., it is an instantaneous effect) just remains accessible to experiment. For 2-D excitons it has been shown that $\omega_{LT} = 0$ and that the coupling energy is strongly reduced, mainly because the exciton states inside the light cone are radiative states.[56] Two-dimensional cavity-polaritons, which are not radiative states almost recover the 3-D coupling strength,[56] $\omega_c = \sqrt{\frac{\omega_0\Gamma_0}{2\pi}}$ (4 meV compared to 7.8 meV) but still with $\omega_{LT} = 0$. The main difference is that, due to the cavity, the photon dispersion curve has a vanishing slope close to the resonance, making the polariton effect the dominant one, even up to room temperature.[57]

In bulk material, it has been shown through a long sequence of important contributions[58-65] that excitonic polaritons have a luminescence process essentially different from the weak-coupling description where an exciton is transformed into a photon state through a mechanism that can be modeled with a perturbative approach such as Fermi's golden rule. In the exciton-polariton picture the emission process consists in the random-walk transfer of the polariton,

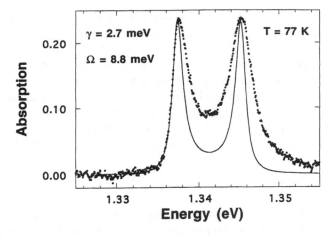

FIGURE 22   Absorption spectrum at 77K of a 6-quantum well, $3\lambda/2$ cavity sample. The solid line is a linear dispersion model fit. The oscillator strength per quantum well is $4.8 \cdot 10^{12}$ cm^{-2}.

regarded as a local property of the system, toward the crystal surface and its transmission as an outside photon at the surface. Also essential is the polariton thermalization process because both the propagation to the surface, which is characterized by the polariton group velocity and mean free path, and the transmission coefficient are very dependent on the exciton energy and wavevector. In addition, excitonic polaritons undergo a relaxation bottleneck at the resonance energy, where they can eventually connect to an outside photon state. The bottleneck leads to a peaked emission in this region.[58] It was evaluated theoretically[63] and demonstrated experimentally by resonant excitation

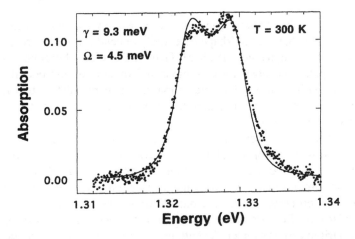

FIGURE 23   Absorption spectrum at 300K of a 6-quantum well, $3\lambda/2$ cavity sample. The solid line is a linear dispersion model fit. The oscillator strength per quantum well is $4.8 \cdot 10^{12}$ cm^{-2}.

experiments[64,65] that the strong coupling regime of the exciton polariton leads to quite inefficient luminescence, even in the purest semiconductors, which can be seen as a paradox: this is a system where the transformation of a crystal excitation into a photon is as fast as it can be but the price to pay is that this transformation is not an irreversible process anymore. Moreover, the coherent eigenstate is badly coupled to the outside photon states, because the random walk to the surface is very long compared to exciton destructive events and the transmission coefficient is most often very weak.

The bulk polariton luminescence line shape is given by:[62]

$$I\left(E,\ \theta_{out}\right)d\Omega_{out} = \sum_{i=up,lp} f_i\left(E,\ \theta_{in},\ z=0\right)\rho_i(E)v_{g\perp,i}(E)T_i\left(E,\ \theta_{in}\right)d\Omega_{in} \qquad (40)$$

where the sum is over the upper and lower branch of the polariton, f is the polariton energy distribution function at the surface, r is the polariton density of states, $v_{g\perp}$ is the component of the polariton group velocity perpendicular to the surface, T the transmission coefficient of a polariton incident at the surface into an outside photon, and the propagation directions θ_{in} and θ_{out} and solid angles $d\Omega$ are related by Snell-Descartes law. Although the energy and $k_{//}$ selection rule are already included in θ_{in}, θ_{out}, and $T(E, \theta)$ it is instructive to include them in the following form:

$$I\left(E,\ \theta_{out}\right)d\Omega_{out} = \sum_{i=up,lp} f_i\left(E,\ \theta_{in},\ z=0\right)\rho_i(E)v_{g,i}(E)T_i\left(E,\ \theta_{in}\right)$$

$$\delta\left[E - E\left(\vec{k}\right)\right]\delta\left[\vec{k}_{//} - \left(\vec{q}\cdot\vec{u}_{//}\right)\vec{u}_{//}\right]d\Omega_{in}$$

$$= \sum_{i-lp,up} I_i\left[E\left(k_\perp + q\ \sin\left[\theta_{out}\right]\right)\right]d\Omega_{in} \qquad (41)$$

where θ is the photon wavevector and $u_{//}$ a unitary vector in the direction of **k**. It follows that the line shape essentially depends on geometrical, static (r, v_g, T, etc.), and dynamic (f [E, θ, z]) factors. Therefore photoluminescence of the bulk polaritons does *not* give direct information on the polariton dispersion curve without an explicit calculation of the energy distribution function. It is known that this point makes the analysis of bulk polariton luminescence somewhat difficult.

The cavity-polariton situation is quite different:[66] as cavity polaritons are a macroscopic property of the microcavity, the luminescence process does not involve any transport of excitation. Therefore, we can describe the luminescence process just by knowing the distribution of the cavity polariton along its dispersion curve and by the outside transmission coefficient of such cavity polaritons. Because of the absence of a perpendicular wavevector the cavity-polariton case is simpler. As both the exciton and the cavity mode energy only depend on $k_{//}$, the in-plane **k** selection rule reduces the photoluminescence spectrum to a sum of two delta functions:

$$I(E, \theta_{out})d\Omega_{out} = \sum_{i=lp,up} f_i(E, \theta_{in})\rho_i(E)T_i(E, \theta_{in})$$

$$\delta\left[E - E(\vec{k}_{/\!/})\right]\delta\left[\vec{k}_{/\!/} - (\vec{q}\cdot\vec{u}_{/\!/})\vec{u}_{/\!/}\right]d\Omega_{in}$$

$$= \sum_{i-lp,up} I_i\left[(E, \theta_{in})\delta(E - E(q\,\sin[\theta_{out}]))\right]d\Omega_{in} \qquad (42)$$

and therefore the emission spectrum for a fixed incidence angle should exhibit two lines whose positions are *directly* related to the cavity-polariton dispersion curve and the relative intensities of which are function of static and dynamical factors. In the presence of homogeneous or inhomogeneous broadening the delta functions are replaced by Lorentzian- or Gaussian-type lines, but their energy position is not changed as long as the broadening is smaller than the energy separation of both lines. This is in agreement with PL spectrum shown in Figure 24. Moreover, it follows from the previous consideration that measurement of the PL at different emission angles allows us to measure dispersion curves of the cavity polariton.[66]

Cavity-polariton dispersion curves can be calculated in different ways: (i) A classical local Lorentz oscillator model[39] can be included inside the FP cavity as a dispersive medium. Although applicable to atomic physics, this model is not suitable for a semiconductor microcavity: it is independent of the Lorentz oscillator position inside the cavity, it neglects the reflections from the Lorentz oscillator layer, and it underestimates the splitting by a factor[56] of 2. (ii) The classical Lorentz oscillator of model (i) that takes into account the reflected waves at the quantum well interfaces[57] is sufficient to solve the weaknesses

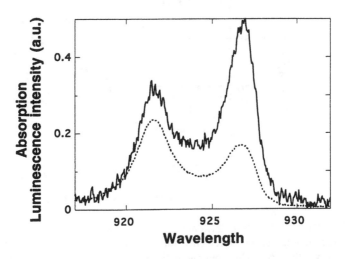

FIGURE 24 Absorption at 77K (dashed line) and photoluminescence (continuous line) spectrum of a 6-quantum well, $3\lambda/2$ cavity sample.

mentioned for (i). (iii) Model (ii) can be extended to use the transfer matrix of the quantum well layer in the nonlocal susceptibility framework.[56] (iv) Full quantum mechanical treatment,[67,68] which is usually done for a perfect lossless (i.e. closed) metallic cavity, can be given. The full quantum mechanical treatment directly gives the cavity-polariton dispersion curve vs. the in-plane wavevector $\mathbf{k}_{//}$. Models (ii) and (iii) give the dispersion curves of reflectivity, transmission, or absorption resonances in an open cavity as a function of the angle of incidence θ, with $\mathbf{k}_{//} = \theta \sin(\theta)$. Models (ii), (iii), and (iv) are in very good agreement.

The continuous lines in Figure 25 are theoretical calculations of the dispersion curve using model (ii). The fitting parameters are the Rabi splitting energy Ω (which is directly related to the exciton oscillator strength f) and the resonance energy between the exciton and cavity mode. The exact resonance condition is determined when the energy separation of both lines is minimum. The dashed lines are for the uncoupled dispersion curves. In this wavevector range the exciton can be regarded as dispersionless and the cavity mode has the usual dependency:

$$E\left(k_{//}\right) = \sqrt{E_0^2 + \frac{c^2 \hbar^2}{n_{eff}^2} k_{//}^2} \tag{43}$$

As can be observed, an excellent fit is obtained for $\Omega = 7.3 \pm 0.3$ meV (i.e., $f = 4.6 \cdot 10^{12}$ cm^{-2}).

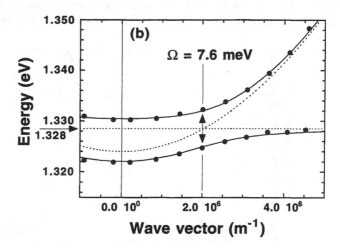

FIGURE 25 Cavity-polariton dispersion curves, deduced from angle resolved photoluminescence measurements, for a resonance occurring at an oblique angle of incidence ($\theta = 29°$). The continuous lines are theoretical calculations and the dashed lines are the uncoupled exciton and cavity dispersion curves.

VI. CONCLUSION

As can be seen, the strong coupling regime is well achieved in semiconductor microcavities and observable at room temperature. The future directions in this remarkable area are set with several challenges if one wants to take full advantage of this unique optical material/system. These are:

• Increase of the coupling, to be well beyond any thermal broadening effect while operating at or above room temperature.
• Master the growth homogeneity so that over a whole wafer, and wafer after wafer, the resonance condition is fulfilled. This implies a layer thickness control better than ω_0/Ω, i.e., better than 0.3%.
• Define new device operating modes. It is not clear that the strong-coupling regime is the best for devices. It might well be that the critical coupling regime, where Ω is equal to the damping rates, is the optimal one. Of course, when exciting resonantly under strong-coupling conditions, one only excites one mode of the system, which therefore has a spontaneous emission factor[6-8] of $\beta = 1$. However, "real world" excitation such as electrical injection populate many strongly coupled states, which means that the overall system has $\beta \leq 1$.

These opportunities and challenges make that the field of semiconductor strongly coupled microcavities will remain a very open area of fundamental and applied studies.

REFERENCES AND NOTES

1. Weisbuch, C. and Vinter, B., *Quantum Semiconductor Devices*, Academic Press, Boston, 1991.
2. Burstein, E. and Weisbuch C., Eds., Confined electrons and photons: *New Physics and Applications*, Plenum, New York, 1995.
3. Chemla, D. S. and Miller, D. A. B., Room-temperature excitonic nonlinear optical effects in semiconductor quantum wells, *J. Opt. Soc. Am.*, B2, 1155, 1985.
4. Benisty, H., Sotomayor-Torres, C. M., and Weisbuch, C., Intrinsic mechanism for the poor luminescence properties of quantum-box systems, *Phys. Rev.*, B44, 10945, 1991.
5. See, e.g., the contributions by M. Bawendi, Y. Arakawa, J.-M. Gérard, and Böckelmann in Ref. 2.
6. Yokoyama, H., Physics and device applications of optical microcavities, *Science*, 256, 66, 1992.
7. Bjork, G., Machida, So, Yamamoto, Y., and Igeta, K., Modification of spontaneous emission rate in planar dielectric microcavity structures, *Phys. Rev. A*, 44, 669, 1991; Bjork, G., Heitmann, H., and Yamamoto, Y., Spontaneous emission coupling factor and mode characteristics of planar dielectric microcavity lasers, *Phys. Rev. A*, 47, 4451, 1993.

8. Baba, T., Hamano, T., Koyama, F., and Iga, K., Spontaneous emission factor of a microcavity DBR surface-emitting laser, *IEEE J. Quantum Electronics*, QE-27, 1347, 1991.

9. See, e.g., the contributions by H. Yokoyama, G. Björk, and S. John in Ref. 2.

10. See, e.g., Yablonovitch, E., Photonic bandgap structures, *J. Opt. Soc. Am. B*, 10, 283, 1993 and the other papers in the same special issue of *J. Am. Soc. Am.*, on Photonic Bandgap Materials.

11. See, e.g, the special issue on Photonic Bandgap Materials, *J. Mod. Opt.*, 41, (2), 1994.

12. Everitt H. O., Applications of photonic bandgap structures, *Opt. Photon. News*, Nov. 1992, p. 18.

13. See also the contributions by E. Yablonovitch, S. John, and P. Russel in Ref. 2.

14. Hunt, N. E. J., Schubert, E. F., Kopf, R. F., Sivco, D. L., Cho, A.Y., and Zydzik, G. J., *Appl. Phys. Lett.*, 19, 2600, 1993.

15. Born, M. and Wolf, E., *Principles of Optics*, Pergamon Press, Oxford, 1970.

16. Yariv, A. and Yeh, P., *Optical Waves in Crystals*, John Wiley & Sons, New York, 1984.

17. Stanley, R. P., Houdré, R., Oesterle, U., and Ilegems, M., Impurity modes in one-dimensional periodic-systems: the transition from photonic band-gaps to microcavities, *Phys. Rev. A*, 48, 2246, 1993.

18. Yablonovitch, E., Inhibited spontaneous emission in solid state physics and electronics, *Phys. Rev. Lett.*, 58, 2059, 1987.

19. Jewell, J. L., Lee, Y. H., and McCall, S. L., High-finesse (Al,Ga)As interference filters grown by molecular beam epitaxy, *Appl. Phys. Lett.*, 53, 640, 1988.

20. Oudar, J. L., Kuszelewicz, R., Sfez, B., Pellat, D., and Azoulay, R., Quantum well nonlinear microcavities, *Superlatt. Microstruct.*, 12, 89, 1992.

21. Stanley, R. P., Houdré, R., Oesterle, U., Gailhanou, M., and Ilegems, M., Ultra-high finesse microcavity with distributed Bragg reflectors, *Appl. Phys. Lett.*, 65, 1883, 1994.

22. Cohen-Tannoudji, C., Diu, B., and Laloë, F., *Mécanique Quantique*, (Hermann, Paris, 1973); *Quantum Mechanics*, John Wiley & Sons, New York, 1976.

23. Purcell, E. M., Spontaneous emission probabilities at radio frequencies, *Phys. Rev.*, 69, 681, 1946.

24. Kleppner, D., Inhibited spontaneous emission, *Phys. Rev. Lett.*, 47, 233, 1981.

25. Haroche, S., Rydberg atoms and radiation in a resonant cavity, in *New Trends in Atomic Physics*, 193, Grynberg, G. and Stora, R., Eds., North-Holland, Amsterdam, 1983.

26. Kaluzny, Y., Goy, P., Gross M., Raimond J. M., and Haroche S., Observation of self-induced Rabi oscillations in two-level atoms excited inside a resonant cavity: the ringing regime of superradiance, *Phys. Rev. Lett.*, 51, 1175, 1983.

27. Carmichael, H. J., Brecha, R. J., Raizen, M. G., Kimble, H. J., and Rice, P. R., Subnatural linewidth averaging for coupled atomic and cavity-mode oscillators, *Phys. Rev. A*, 40, 5516, 1989.

28. Rempe, G., Schmidt-Kaler, F., and Walther, H., Observation of sub-Poissonian statistics in a micromaser, *Phys. Rev. Lett.*, 64, 2783, 1990.

29. Rempe, G. , Thompson, R. J., Brecha, R. J., Lee, W. D., and. Kimble, H. J., Optical bistability and photon statistics in cavity quantum electrodynamics, *Phys. Rev. Lett.*, 67, 1727, 1991.

30. Haroche, S., Cavity quantum electrodynamics, in *Fundamental Systems in Quantum Optics,* 769, 1. Dalibard, J., Raimond, J. M., and Zinn-Justin, J., Eds., Elsevier, Amsterdam, 1992.
31. Berman, P. R., Ed., *Cavity Quantum Electrodynamics,* Academic Press, Boston, 1994.
32. Haroche, S. and Raimond, J. M., Manipulation of nonclassical field states in a cavity by atom interferometry, in *Cavity Quantum Electrodynamics,* p. 123, Berman, P. R., Ed., Academic Press, Boston, 1994.
33. Rabi, I. I., Space quantization in a gyrating field, *Phys. Rev.,* 51, 652, 1937.
34. Mollow, B. R., Power spectrum of light scattered by two-level systems, *Phys. Rev.,* 188, 1969, 1969.
35. Sanchez-Mondragon, J. J., Narozhny, N. B., and Eberly, J. H., Theory of spontaneous emission line shape in an ideal cavity, *Phys. Rev. Lett.,* 51, 550, 1983.
36. Agarwal, G. S., Vacuum-Field Rabi splittings in microwave absorption by Rydberg atoms in a cavity, *Phys. Rev. Lett.,* 53, 1732, 1984.
37. Thompson, R. J., Rempe, G., and Kimble, H. J., Observation of normal mode splitting for an atom in an optical cavity, *Phys. Rev. Lett.,* 68, 1132, 1992.
38. Rempe, G., Walther, H., and Klein, N., Observation of quantum collapse and revival in a one atom maser, *Phys. Rev. Lett.,* 58, 353, 1987.
39. Zhu, Y., Gauthier, D. J., Morin, S. E., Wu, Q., Carmichael H. J., and Mossberg, T. W., Vacuum Rabi splitting as a feature of linear-dispersion theory: analysis and experimental observations, *Phys. Rev. Lett.,* 64, 2499, 1990.
40. Stern, F., Elementary optical properties of solids, *Solid State Physics,* vol. 15, Seitz, F. and Turnbull, D., Eds., Academic Press, New York, 1963, p. 300.
41. Ziman, J., *Principles of the Theory of Solids,* 2nd ed. Cambridge University Press, Cambridge, UK, 1970.
42. Eberly, J. H. and Miloni, P. W., *Lasers,* John Wiley & Sons, New York, 1988.
43. Yariv, A., *Quantum Electronics,* 3rd ed., John Wiley & Sons, New York, 1989.
44. Knox, R., Theory of excitons, *Solid State Physics,* suppl. 5, Seitz, F. and Turnbull, D., Eds., Academic Press, New York, 1963.
45. See the very complete review by Andreani in Ref. 2.
46. See, e.g., Bethe, H. and Sommerfeld, A., *The Theory of One and Two Electron Atoms,* Plenum Press, New York, 1970.
47. Elliott, R. J., Intensity of optical absorption by excitons, *Phys. Rev.,* 108, 1384, 1957.
48. Dimmock, J. O., Theory of exciton states, in *Semiconductors and Semimetals,* Willardson, A. K. and Beer, A. C., Eds., Academic Press, New York, 1967, p. 259.
49. Rossi, B., *Optics,* Addison-Wesley, Reading, MA, 1957; see also Ref. 15.
50. Andreani, L. C. and Pasquarello, A., Accurate theory of excitons in GaAs-GaAlAs quantum wells, *Phys. Rev.,* B42, 8928, 1990.
51. Weisbuch, C., Nishioka, M., Ishikawa, A., and Arakawa, Y., Observation of the coupled exciton-photon mode splitting in a semiconductor quantum microcavity, *Phys. Rev. Lett.,* 69, 3314, 1992.
52. Andreani, L. C. and Pasquarello, A., Accurate theory of excitons in GaAs-Ga(1-x)AlxAs quantum wells, *Phys. Rev.,* B 42, 8928, 1990.
53. Childs, J. J., An, K., Dasari, R. R., and Feld, M. S., Single atom emission in an optical resonator, in *Cavity Quantum Electrodynamics,* Berman, P. R., Ed., Academic Press, Boston, 1994, 123.

54. Hopfield, J. J., Resonant scattering of polaritons as composite particles, *Phys. Rev.*, 182, 945, 1969.

55. Andreani, L. C., Optical transitions, excitons and polaritons in bulk and low-dimensional semiconductor structures, in *Confined Electrons and Photons,* Weisbuch, C. and Burstein, E., Eds., Plenum, Boston, 1995.

56. Andreani, L. C. , unpublished.

57. Houdré, R., Stanley, R. P., Oesterle, U., Ilegems, M., and Weisbuch, C., Room-temperature exciton-photon Rabi splitting in a semiconductor microcavity, *J. Phys. IV (Paris)* 3, 51, 1993.

58. Toyozawa, Y., On the dynamical behavior of an exciton, *Prog. Theor. Phys. Suppl.*, 12, 111, 1959.

59. Tait, W. C. and Weiher, R. L., Contributions of scattering of polaritons by phonons to emission of radiation by solids, *Phys. Rev.*, 178, 1404, 1969.

60. Benoit à la Guillaume, C., Bonnot, A., and Debever, J. M., Luminescence from polaritons, *Phys. Rev. Lett.*, 24, 1235, 1970.

61. Sell, D. D., Stokowski, S. E., Dingle, R., and DiLorenzo, J. V., Polariton reflectance and photoluminescence in high-purity GaAs, *Phys. Rev.*, B 10, 4568, 1973.

62. Bonnot, A. and Benoit à la Guillaume, C., Polariton luminescence line shape, in *Polaritons,* Burstein, E. and Martini, F. D., Eds., Pergamon, New York, 1974.

63. Sumi, H., On the exciton luminescence at low temperatures: importance of the polariton viewpoint, *J. Phys. Soc. Jpn.*, 41, 526, 1976.

64. Weisbuch, C. and Ulbrich, R. G., Resonant polariton fluorescence in gallium arsenide, *Phys. Rev. Lett.*, 39, 654, 1977.

65. Weisbuch, C. and Ulbrich, R. G., Spatial and spectral features of polariton fluorescence, *J. Luminesc.*, 18/19, 27, 1979.

66. Houdré, R., Weisbuch, C., Stanley, R. P., Oesterle, U., Pellandini, P., and Ilegems, M., Measurement of cavity-polaritons dispersion curve from angle resolved photoluminescence experiments, *Phys. Rev. Lett.*, 73, 2043, 1994.

67. Citrin, D. S., Effect of an optical cavity on spontaneous emission, Rabi oscillations and radiative shifts of excitons in quantum wells, in *Confined Electrons and Photons,* Weisbuch, C. and Burstein, E., Eds., Plenum, Boston, 1995.

68. Savona, V., Hadril, Z., Quattropani, A., and Schwendimann, P., Quantum theory of quantum well polaritons in semiconductor microcavities, *Phys. Rev.*, B49, 8774, 1994.

5 Electromagnetic Field Mode Density Calculated via Mode Counting

Stuart D. Brorson

TABLE OF CONTENTS

I. INTRODUCTION

The spontaneous emission lifetime of an excited atom may be changed by placing it into a cavity having dimensions close to the atom's transition wavelength.[1] The presence of the cavity alters the mode density of the electromagnetic field coupled to the atom, thereby changing the atomic transition rate as calculated via Fermi's golden rule. This phenomenon is best understood using a simple example: a closed, lossless rectangular cavity, as is shown in Figure 1. This structure has a series of discrete resonant frequencies; its frequency-dependent mode density is a series of Lorentzians centered at the cavity resonances, such as is depicted schematically in Figure 2. We consider placing an atom having a very narrow transition frequency into this cavity. When the atomic transition frequency does not lie near a cavity resonance

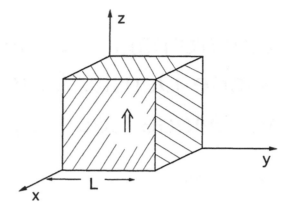

FIGURE 1   The simplest microcavity: a closed box having perfectly reflecting walls. The wide double arrow symbolizes the presence of an atomic dipole in the center of the cavity.

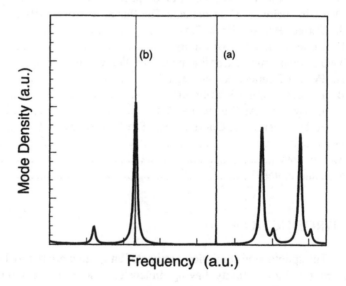

FIGURE 2   The resonant modes of a box cavity form a series of Lorentzians in frequency as indicated by the darker curve. Depicted is the case when the atom (thin line) is off resonance (a), and on resonance (b).

(Figure 2(a)), the atom cannot radiate and decay, since it cannot give off a photon of the appropriate energy. In this case, the atom remains in its excited state forever; the radiative lifetime of the atom is infinite. Conversely, if the atomic line overlaps a cavity resonance (Figure 2(b)), then the spontaneous emission rate is resonantly enhanced by the cavity. The spontaneous emission lifetime can be quite short — shorter indeed than the corresponding free space lifetime. The basic effect is **the presence of a cavity changes the mode density of the electromagnetic field,** and thereby, the lifetime of the excited atom. The occurrence of this phenomenon has been amply investigated over the past several years.[2]

The purpose of this chapter is to present calculations of the frequency-dependent mode density in several different model microcavity structures. The calculations are performed using the "mode counting" method,[3] which offers a simple and intuitive prescription for calculating the mode density in cavities of many different shapes and dimensionalities. Mode counting assumes that the cavity's interior is free space and is bounded by conductive (metallic) walls. By concentrating on such model cavities we can develop intuition about the general importance e.g., of dimensionality and loss on the mode density, without having to worry about the complexities inherent to real microcavity structures. Calculations pertaining to realistic dielectric planar microcavities are presented in Chapter 6 of this volume, while Chapter 7 is concerned with the technologically important "dielectric post" microcavity.

The mode counting calculations yield the mode density seen by a hypothetical two-level atom, which we model as a Hertzian dipole of infinitesimal size. We start in Section II by deriving the mode density seen by an atom sitting in free space, where there is **no confinement** of the electromagnetic field. In Section III we consider the mode density inside a cavity formed by coplanar metal mirrors. This is the case of **one-dimensional confinement.** Both the case of perfectly reflecting mirrors, and a lossy cavity are discussed. In Section IV we consider cavities giving **two dimensions of confinement:** a rectangular waveguide having perfectly reflecting walls, as well as a cylindrical waveguide cavity having both perfect and lossy walls. Finally, in Section V we briefly discuss the case of **three-dimensional confining cavities,** i.e., resonant boxes. The most important result of this work is that the critical factor in determining the amount of increase or decrease of the mode density at one particular frequency is the dimensionality of the cavity. **Cavities confining in more dimensions** (i.e., waveguide cavities or resonant boxes) **give larger effects.** In the discussion in Section VI, we point out how the mode counting method might be extended to experimentally relevant structures such as dielectric slab cavities, as well as cavities incorporating multilayer dielectric mirrors.

II. NO CONFINEMENT: A DIPOLE IN FREE SPACE

To start, we imagine an excited two-level atom coupled to the quantized radiation field. The coupling can be described by a perturbation term in the atom-field Hamiltonian,

$$V = e\mathbf{d} \cdot \mathbf{E},$$

where $-e\mathbf{d}$ is the (vector) dipole moment, and $\mathbf{E}$ is the electric field seen by the atom. The atom starts in its excited state and drops into the ground state upon spontaneously emitting a photon of energy $\hbar\omega = \hbar c k$ and momentum $\hbar\mathbf{k}$. The spontaneous emission rate is given by the transition rate $W_{e\rightarrow g}$ between the excited ($|e\rangle$) and the ground ($|g\rangle$) state of the atom-photon system. We have, from Fermi's golden rule,

$$W_{e\rightarrow g} = \frac{2\pi}{\hbar^2 c} \sum_{\mathbf{k}} \left| \langle g | e\mathbf{d} \cdot \mathbf{E} | e \rangle \right|^2 \rho(\mathbf{k}).$$

(1)

The sum is taken over all available wave vectors $\mathbf{k}$ satisfying $|\mathbf{k}| = k = \omega/c$. $\rho(\mathbf{k})$ is the mode density of the field in the $\mathbf{k}$ direction at frequency $\omega = c|\mathbf{k}|$. For the rest of this chapter we shall always write k to denote the frequency variable in order to eliminate the spurious factor of c which would otherwise appear in the derived expressions for the mode density.

In mode counting, we re-express Equation 1 in the form

$$W_{e\rightarrow g} = \frac{2\pi}{\hbar^2 c} |M|^2 g(k),$$

(2)

where $g(k)$ plays the role of an effective mode density of the optical field. M is the dipole-field matrix element. This expression separates Equation 1 into two parts — a quantum mechanical part: the matrix element M, and a "classical" part: the effective mode density $g(k)$.

Our interest is in calculating the classical part, $g(k)$. It can be factored as

$$g(k) = \frac{1}{V} \sum_{|\mathbf{k}|=k} P(\mathbf{k})\rho(\mathbf{k}),$$

(3)

where $P(\mathbf{k})$ describes the angular dependence of the dipole-field matrix element $|\mathbf{d} \cdot \mathbf{E}|^2$, and $\rho(\mathbf{k})$ is determined by the availability of states in k-space. In a cavity, $\rho(\mathbf{k})$ will in general depend on the direction of the $\mathbf{k}$ vector. The sum is performed over all allowed $\mathbf{k}$ and is normalized by the volume of the box in which the field is presumed to exist (the quantization volume V). In the case of free space, the quantization volume is traditionally taken to be a rectangular box of infinite size.

In this context, mode counting means that we interpret Equation 3 as the sum over the surface of a sphere in k-space having radius $k = \omega/c$, weighted by the polarization factor $P(\mathbf{k})$ and the — cavity-dependent — density of states $\rho(\mathbf{k})$. The important feature of mode counting is that the structure of the allowed modes in k-space is very simple and easy to visualize for many important cavities.

Since all the direction-dependent information has been subsumed into $g(k)$, the matrix element M, which holds all the quantum mechanical information,

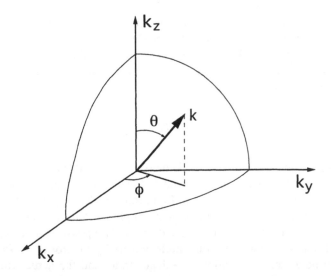

FIGURE 3   Mode counting of a free space dipole. In k-space, the allowable modes lie on the surface of a sphere of radius $k = \omega/c$. The mode density is given by integrating over this spherical surface. Shown here is only the first octant.

can be factored into two *scalar* parts: the atomic dipole matrix element $\langle g|ed|e \rangle$: and the electric field matrix element corresponding to the creation of a photon in a previously empty cavity, $\langle 1|\sqrt{\hbar\omega/2\varepsilon_0}\ \alpha^\dagger|0\rangle$. For our purposes, we regard this factor as a constant proportional to the dipole moment, d, and henceforth ignore it in mode counting. In the present form, the quantization volume V is *not* included in the matrix element M because it has been moved inside $g(k)$ as is shown explicitly in Equation 3.

Now we perform mode counting for a dipole in free space. Imagine that the dipole is oriented vertically; choose a spherical coordinate system for simplicity. We assume the dipole sits in the center of a very large box having sides of length L, giving a quantization volume $V = L^3$. (We can later take the limit $L \to \infty$.) In k-space, the accessible states into which the decaying atom can radiate form a thin spherical shell of radius k and thickness dk centered on the origin. This is schematically illustrated in Figure 3. Following the prescription outlined above, the effective mode density $g(k)$ is found by summing the "background" mode density $\rho(\mathbf{k}) = (L/2\pi)^3$ over this spherical shell, multiplied by the polarization factor $P(\mathbf{k})$. For a vertically oriented dipole, this dipole-field coupling factor is $P(\mathbf{k}) = \sin^2\theta$, where θ is the angle $\mathbf{k}$ makes with respect to the vertical axis. The volume element of k-space is $d^3k = \sin\theta\ k^2 d\phi d\theta dk$. Incorporating these factors into the mode counting sum gives

$$g_{\text{free}}(k)dk = \frac{1}{V}\int_0^{2\pi} d\phi \int_0^{\pi} d\theta\ \sin\theta\ k^2 dk \left(\frac{L}{2\pi}\right)^3 \sin^2\theta. \qquad (4)$$

In accord with the discussion above, the integral in Equation 4 can be regarded as counting up all the available modes in k-space into which the atom can radiate, multiplied by some weight factor depending on the polarization of the dipole and the outgoing photon. Hence we use the name "mode counting". Upon evaluation we get

$$g_{free}(k)dk = \frac{1}{3\pi^2} k^2 dk.$$

(5)

As expected, the effective mode density is proportional to k^2 since the number of available states is proportional to the surface of the sphere in k-space.[4]

A subtle feature hidden in Equation 5 is that $g_{free}(k)$ is the mode density seen by an *oriented* dipole, which couples to only one polarization state of the field for each k vector. Since there are two possible polarizations for each k, (5) undercounts the actual mode density by a factor of $1/2$. In free space the two polarization directions are degenerate, and the actual free space mode density can be gotten by multiplying $g_{free}(k)$ by 2 to compensate for the undercounting. In a cavity, however, the mode density depends on the polarization state of the field and the orientation of the dipole. In the following sections we wish to compare $g_{free}(k)$ with mode densities calculated for oriented dipoles in cavities. Therefore, we will keep (5) in its present form, and henceforth refer to the mode density seen by an *oriented* dipole as the "effective mode density".[5]

III. ONE DIMENSION OF CONFINEMENT: THE PLANAR MIRROR CAVITY

A. PERFECT PLANAR MIRROR CAVITY

We now consider the problem of a dipole in a cavity formed by two perfectly reflecting planar mirrors made of metal. We call this the "perfect planar mirror microcavity". It is analogous to the cavity used in many of the microcavity experiments performed to date.[2] The physical system is depicted in Figure 4. The dipole is situated at the origin of a cylindrical coordinate system, as shown. The mirrors are defined by the planes $z = \pm L_z/2$. We must consider two possible orientations for the dipole: parallel to the $\hat{z}$ axis ($d \parallel \hat{z}$) (Figure 4(a)), and perpendicular ($d \perp \hat{z}$) (Figure 4(b)). All other dipole orientations can be represented as a superposition of these two cases.

Consider the structure of k-space seen by the dipole between two planar mirrors. In the $\hat{r}$ direction, the cavity is unbounded. Under Fourier transform, k-space in the $\hat{k}_r$ direction is also unbounded and continuous. In the $\hat{z}$ direction, however, the mirrors at $z = \pm L_z/2$ impose periodicity in the z direction on the real-space fields. Thus, in k-space, k_z exists only at a series of points centered at $k_z = 0$ and separated by a distance of π/L_z. The allowed k values form a set of planes in the $\hat{k}_r$ direction intersecting the k_z axis at $k_z = n\pi/L_z$,

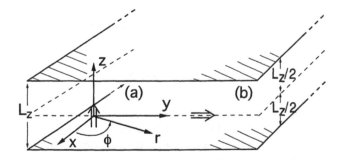

FIGURE 4 Atomic dipole in a planar mirror microcavity. The dipole is indicated by the doubled arrow, and sits exactly halfway between the mirrors. The parallel case ($d \parallel \hat{z}$) is shown in (a), and the perpendicular case ($d \perp \hat{z}$) is shown in (b). (From Brorson et al., *IEEE J. Quantum Electron.*, 26, 1492, 1990. With permission © 1990 IEEE.)

where $n = \dots, -2, -1, 0, 1, 2, \dots$. The k-space volume element is $k_r d\phi dk_r$, **on the plane**. Moreover, since the planes are continuous and unbounded in two dimensions, we must include a factor $(L/2\pi)^2$ accounting for the "background" mode density on the plane.

The effective mode density for a given k is found by again summing the allowed **k** vectors over a spherical shell of radius k and width dk, as in the free space case. This time, however, the allowed modes occur where the spherical shell intercepts the planes, as shown in Figure 5. There, the allowed modes form rings (annuli) on the planes having radii

$$k_r^{(n)} = k \sin \theta_n,$$

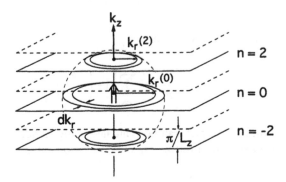

FIGURE 5 The k-space of the planar mirror microcavity. The boundary conditions imposed by the mirrors constrain the possible values of k to lie on a series of planes perpendicular to the $\hat{k}_z$ axis, separated by $n\pi/L_z$. For a given frequency $k = \omega/c$, the allowed modes form rings on the k-space planes. The mode density is accordingly found via a weighted *sum* over the rings — hence the name "mode counting". (From Brorson et al., *IEEE J. Quantum Electron.*, 26, 1492, 1990. With permission © 1990 IEEE.)

where n designates the plane. The width of each ring is given by

$$dk_r^{(n)} = \frac{dk}{\sin \theta_n}.$$

The number of allowed modes is then obtained by integrating over each ring and summing over all planes for which $|n\pi/L_z| < k$.

We must include the dipole weighting factors in the mode sum. For the parallel dipole, we have

$$P_{\parallel}(\mathbf{k}) = \sin^2 \theta \tag{6}$$

for the dipole-field coupling factor. In addition, a parallel dipole positioned halfway between the mirrors couples only to **even** order (TM) modes in a perfectly reflecting cavity. This occurs because the $\hat{z}$ component of the electric field varies as $\cos[n\pi(z/L_z - 1/2)]$ in order to match boundary conditions. Thus, only even n appears in the mode sum. The k-space mode counting geometry is depicted in Figure 5. Incorporating all these factors into the sum, we get

$$g_{\parallel}(k)dk = \frac{1}{V} \sum_{n \text{ even}} 2\pi k_r^{(n)} dk_r^{(n)} \sin^2 \theta_n \left(\frac{L}{2\pi} \right)^2,\tag{7}$$

where the sum is taken over all integer $|n| < kL_z/\pi$. In the present case, the quantization volume is $V = L^2 L_z$. Defining the "cut-off" wavevector $k_c = \pi/L_z$, and using

$$k_r^{(n)} = \sqrt{k^2 - (nk_c)^2}$$

we can re-express Equation 7 as

$$g_{\parallel}(k) = \frac{1}{2\pi^2} k_c k \sum_{n \text{ even}} \left(1 - \left(n\frac{k_c}{k} \right)^2 \right).$$

This is a finite sum which can be evaluated using standard identities giving a closed-form expression for the effective mode density:

$$g_{\parallel}(k) = \frac{1}{2\pi^2} k_c k \left[1 + 2\left[p - \frac{1}{6}\left(\frac{2k_c}{k} \right)^2 p(p+1)(2p+1) \right] \right],\tag{8}$$

where $p = \lfloor k/2k_c \rfloor$ denotes the largest integer less than or equal to $k/2k_c$. This function is plotted in Figure 6. The effective mode density normalized to its free space value, $g_{\parallel}(k)/g_{\text{free}}(k)$ is shown in Figure 7. Since the spontaneous

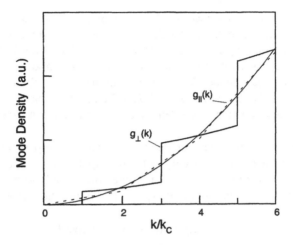

FIGURE 6   Planar mirror microcavity mode density $g(k)$ vs. frequency k. The dashed line is the mode density seen by the parallel dipole ($g_{\parallel}$); the full "staircase" is the perpendicular dipole ($g_{\perp}$). The parabolic curve is the free space mode density (g_{free}).

emission rate is determined from the mode density via (2), the curve shown in Figure 7 is directly proportional to the **change** in spontaneous emission rate caused by the cavity confinement.

Examine the limiting behavior of Equation 8. The first case to consider is the high-frequency limit $k/k_c \to \infty$. There, we may approximate $\lfloor k/2k_c \rfloor$ by $k/2k_c$, yielding

$$g_{\parallel}(k \to \infty) = \frac{k^2}{3\pi^2},$$

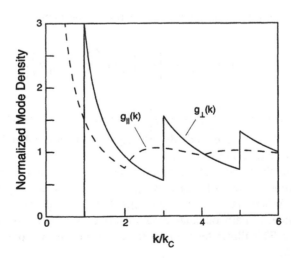

FIGURE 7   Normalized mode density in a planar mirror microcavity. Shown here is $g_{\parallel}/g_{free}$ (dashed line) and $g_{\perp}/g_{free}$ (full line).

equivalent to the free space mode density in (5). This makes sense, since in this limit we have $L_z \gg \lambda$, and the dipole does not "feel" the presence of the cavity. The other limiting case occurs when $k \to 0$. This is the low frequency limit; cavity confinement effects are expected to be important. In this case, the allowed k-space modes form the circumference of a circle in the $n = 0$ plane. Because $g_\parallel(k)$ varies linearly with k, although it varies as k^2 in free space, the normalized effective mode density diverges in the low frequency limit. (See Figure 7.)

Next we consider the perpendicular dipole. The perpendicular dipole differs from the parallel case only in that the dipole-field coupling factor is rotated by 90° and that the perpendicular dipole couples only to the odd-order modes. We get $P_\perp(\mathbf{k})$ upon rotating $P_\parallel(\mathbf{k})$ by 90° around the x axis giving

$$P_\perp(\mathbf{k}) = \sin^2 \theta \, \cos^2 \phi + \cos^2 \theta. \tag{9}$$

The dipole couples only to modes that have a maximum in either H_ϕ or E_r at $z = 0$; since both H_ϕ and $E_r \propto \sin[n\pi(z/L_z - 1/2)]$, only odd-order (TE) modes need be included in the sum.

Now we count modes as before. This time, we must integrate explicitly over ϕ. We have

$$g_\perp(k)dk = \frac{1}{V} \sum_{n \text{ odd}} \int_0^{2\pi} d\phi \, k_r^{(n)} dk_r^{(n)} \left(\sin^2 \theta_n \cos^2 \phi + \cos^2 \theta_n \right) \left(\frac{L}{2\pi} \right)^2$$

$$= \frac{1}{4\pi^2} k_c k \sum_{n \text{ odd}} \left(1 + \left(n \frac{k_c}{k} \right)^2 \right). \tag{10}$$

Summing for $|n| < k/k_c$ gives the effective mode density

$$g_\perp(k) = \frac{1}{2\pi^2} k_c k \left[q + \left(\frac{k_c}{k} \right)^2 \left(\frac{4}{3} q^3 - \frac{1}{3} q \right) \right], \tag{11}$$

where this time $q = \left\lfloor \frac{1}{2}(k/k_c + 1) \right\rfloor$. This function is also plotted in Figure 6. In Figure 7 the effective mode density Equation 11 normalized to the free space value (5) is also shown.

Using the same kinds of arguments as in the parallel case, we find that the high frequency limit of (11) converges to the free space expression in (5) as it should. When $k/k_c < 1$, we have $g_\perp(k) = 0$ — all modes are cut off the waveguide. A perpendicular dipole cannot spontaneously emit at these frequencies. This illustrates the importance of the cavity's presence for λ longer than $\approx L_z$.

B.   Lossy Planar Mirror Cavity

Thus far, we have always taken the mirror reflectivity to be perfect, i.e., $|r| = 1$, where r is the *field* reflection coefficient. As a consequence, the locus of allowed **k** vectors in the cavity is a series of planes in k-space, and the mode density $g(k)$ is found by *summing* over the intersection of these planes with a sphere having radius $k = \omega/c$.

In reality, all microcavities are lossy, so it is inappropriate to sum the allowed modes over sharply defined surfaces in k-space. The question arises, "How do we calculate the effective mode density of a lossy cavity?" For the purposes of modeling, we can distinguish two types of loss: dissipation in the cavity walls and radiation away from the cavity through partially transmitting mirrors. From the dipole's point of view, there is no difference between these two processes. Both dissipation and radiation represent sinks for the dipole's energy. Therefore, it is immaterial which process is taking place to the calculation of the emission rate, and hence the mode density. Since most cavities are fabricated from dielectric elements, and we are furthermore only interested in a model calculation, we can ignore the possibility of dissipative loss and concentrate on calculating the effective mode density of a cavity having walls which are partially transmissive. Furthermore, we shall assume that the reflectivity and transmissivity of the walls is constant for all angles of incidence, i.e., the cavity walls are "partially silvered mirrors". With these assumptions, we can easily derive expressions for the electric fields inside and outside the cavity. Upon inserting these expressions into (3) we find that in the case of a lossy planar mirror microcavity, the effective mode density may be expressed as a mode counting integral over the surface of a k-space sphere of radius $|\mathbf{k}| = \omega/c$.

The strategy for computing $g(k)$ in a lossy cavity will be as follows: We will first compute the electromagnetic field modes in a symmetric **dielectric** slab cavity using a method similar to DeMartini et al.[6] A similar calculation is presented in Chapter 6 in the present volume. The propagation directions of the fields are shown schematically in Figure 8. The cavity walls separate space into three regions, as is shown in the figure. The microcavity (region 1) is assumed to have an index n_1, whereas the regions outside the cavity (regions 0 and 2) are free space. Plane-wave solutions for the wave equation are assumed, with the field incident from $-\infty$ as shown, impinging upon the cavity at an angle θ_0. Inside the cavity the solutions are also plane waves traveling at an angle θ_1. The transmitted wave exits to the right at an angle θ_2 ($= \theta_0$ since both regions are free space). Upon matching the boundary conditions at the cavity walls, we will be able to express the field inside the cavity (region 1) in terms of the Fresnel reflection and transmission coefficients r and t at a dielectric interface.[7] Now, we are ultimately interested in modeling a free space cavity with metallic walls (half-silvered mirrors). Therefore, upon obtaining expressions for the electric field in terms of r and t, we will simply ignore their angular dependences and

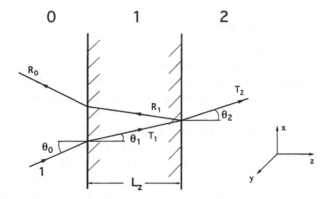

FIGURE 8 Fields in a lossy planar mirror cavity. The incoming wave has unit amplitude and enters from the left (region 0) at angle θ_0. A reflected wave (amplitude R_0) leaves to the left. Inside the cavity (region 1) are two counterpropagating waves having amplitudes T_1 and R_1. The transmitted wave (T_2) is shown exiting at θ_2 from region 2 on the right.

take them to be constants fixed by the properties of the mirror. Furthermore, since region 1 is free space, we may set $n_1 = 1$. Then, the mode density may be found by integrating the dipole-field coupling factor $\mathbf{d} \cdot \mathbf{E}$ over the surface of the sphere in k-space.

As in the case of perfectly reflecting walls, there are two possible orientations of the dipole: parallel ($\parallel$) and perpendicular ($\perp$) to the $\hat{\mathbf{z}}$ axis. We start with the parallel dipole. Coupling can occur only with the TM modes of the cavity, which are characterized by $\mathbf{H} \parallel \hat{\mathbf{y}}$. The $\mathbf{H}$ fields in each region can be expressed as:

$$\mathbf{H}_0^{\mathrm{TM}} = \hat{\mathbf{y}}\left\{1^{\mathrm{TM}} \exp\left[ik\left(x \sin \theta_0 + z \cos \theta_0\right)\right]\right.$$

$$\left. -R_0^{\mathrm{TM}} \exp\left[ik\left(x \sin \theta_0 - z \cos \theta_0\right)\right]\right\},$$

$$\mathbf{H}_1^{\mathrm{TM}} = \hat{\mathbf{y}}\left\{T_1^{\mathrm{TM}} \exp\left[in_1 k\left(x \sin \theta_1 + z \cos \theta_1\right)\right]\right.$$

$$\left. -R_1^{\mathrm{TM}} \exp\left[in_1 k\left(x \sin \theta_1 - z \cos \theta_1\right)\right]\right\},$$

$$\mathbf{H}_2^{\mathrm{TM}} = \hat{\mathbf{y}}\left\{T_2^{\mathrm{TM}} \exp\left[ik\left(x \sin \theta_2 + z \cos \theta_2\right)\right]\right\}. \tag{12}$$

(The $-$ sign in front of R^{TM} is chosen for consistency with the electric fields defined below in Equation 16.) The $\mathbf{E}$ fields in each region can be determined via the relation $\mathbf{E} = -(\eta_0/n)\,\hat{\mathbf{n}} \times \mathbf{H}$, where $\hat{\mathbf{n}}$ is the unit normal pointing in the direction of $\mathbf{k}$, η_0 is the impedance of free space, and n is the index in the cavity region under consideration. The coefficients giving the fields in the microcavity region are T_1^{TM} and R_1^{TM}. They are determined by the boundary conditions at the cavity walls. We have

$$T_1^{TM} = \frac{t}{1 - r^2 e^{i2w}}$$

and

$$R_1^{TM} = \frac{rt}{1 - r^2 e^{2iw}} e^{2iw},$$

where $w = n_1 k L_z \cos\theta_1$, and r and t are the field reflection and transmission coefficients at each boundary given by the Fresnel relations:[7]

$$r = \frac{n_1 \cos\theta_0 - \cos\theta_1}{n_1 \cos\theta_0 + \cos\theta_1}$$

and

$$t = \frac{2n_1 \cos\theta_0}{n_1 \cos\theta_0 + \cos\theta_1}$$

Following our prescription above, we shall henceforth assume that r and t are constants determined by the properties of the mirror, and assume free space in the interior of the cavity ($n_1 = 1$).

Having written down the fields in each region, it is necessary to get the dipole-field coupling $\mathbf{d} \cdot \mathbf{E}$ for use in (3). To do so, we again consider the case where the dipole sits exactly halfway between the mirrors at $z = L_z/2$. As mentioned above, the parallel dipole at $z = L_z/2$ couples only to the TM polarized mode, giving

$$|\mathbf{d} \cdot \mathbf{E}|^2 \propto \sin^2\theta_1 \left| \frac{t}{1 + re^{iw}} \right|^2, \tag{13}$$

where the $\sin^2\theta_1$ dependence arises from the dot product. This is the dipole radiation pattern $P_\parallel(\theta,\phi)$ defined in (6) of Section III.A multiplied by a "resonance factor"

$$A_z^{TM}(w) = \left| \frac{t}{1 + re^{iw}} \right|^2, \tag{14}$$

which depends on the reflectivity of the cavity walls. For $|r|, < 1$, $A^{TM}(w)$ is a Lorentzian, peaking periodically with varying w.

Mode counting now involves inserting the dipole-field coupling (13) into (3) and integrating over the surface of a sphere in k-space. We get

$$g_{\parallel}(k) = \left(\frac{1}{2\pi}\right)^3 \int_0^\pi d\theta \int_0^{2\pi} d\phi \, \sin \theta \, k^2 \left\{\sin^2 \theta \, A_z^{TM}\left(kL_z \cos \theta\right)\right\}. \qquad (15)$$

(The prefactor $(2\pi)^{-3}$ is found by comparing Equation 15 to the free space form (4) in the $r \to 0$ limit.) No closed form expression exists for $g_{\parallel}(k)$. However, (15) may be evaluated numerically. Shown in Figure 9 is the normalized effective mode density $g_{\parallel}(k)/g_{\text{free}}(k)$ for a metal mirror with $|r| = 1.0, 0.8, 0.6, 0.4, 0.2,$ and 0. As can be seen, as the mirror reflectivity approaches 0, the mode density inside the microcavity tends to its free space value.

The expression (15) is reminiscent of the free space mode density integral (4), except that the points in k-space are weighted by the resonance factor $A_z^{TM}(k_z)$. This observation provides an appealing mental picture of the integral (15). When $|r| = 1$, $A_z^{TM}(w)$ is a "picket fence" of delta functions at frequencies given by $kL_z \cos \theta = \pm n\pi$, where n is an even integer. (Remember that $r = -1$ for a metal.) Physically, for $|r| = 1$ the available states in k-space are constrained to lie on the planes defined by $k_z = kL_z \cos \theta = \pm n\pi$. In this limit, (15) reduces to the mode counting sum (7). Upon the introduction of loss, $|r|$ moves away from 1, and the planes smear out to become Lorentzians. The mode counting sum becomes a **mode** counting integral over a spherical surface intersecting a series of Lorentzians corresponding to the cavity resonances. This is shown schematically in Figure 10.

Now we turn to the case of the perpendicularly oriented ($\perp$) dipole. We employ the same calculational strategy as for the parallel dipole above. In the

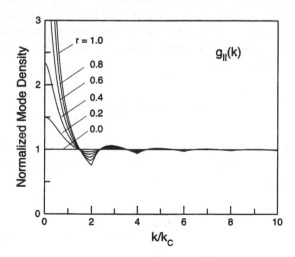

FIGURE 9 Normalized mode density $g_{\parallel}(k)$ computed in Equation 15 for a parallel dipole in a lossy planar mirror microcavity. Shown are curves for different values of the field reflection coefficient r.

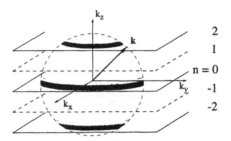

FIGURE 10   Upon introducing radiative loss to a planar mirror cavity, the k-space planes (as shown in Figure 5) smear out. The mode density is then given by *integration* over the surface of a sphere having radius $k = \omega/c$. Compare this picture to that of the perfect cavity case, Figure 5, where the mode density is found via a mode counting *sum*.

presence of partially transmitting mirrors the perpendicular dipole can couple to *both* TE and TM modes. This is in contrast to the case of the perfect planar mirror cavity (Section III.A) where coupling is only possible to the TM modes. For TE we have $\mathbf{E} \parallel \hat{\mathbf{y}}$. The electric fields are

$$\mathbf{E}_0^{TE} = \hat{\mathbf{y}}\left\{1^{TE} \exp\left[ik\left(x \sin \theta_0 + z \cos \theta_0\right)\right]\right.$$

$$\left. + R_0^{TE} \exp\left[ik\left(x \sin \theta_0 - z \cos \theta_0\right)\right]\right\},$$

$$\mathbf{E}_1^{TE} = \hat{\mathbf{y}}\left\{T_1^{TE} \exp\left[in_1 k\left(x \sin \theta_1 + z \cos \theta_1\right)\right]\right.$$

$$\left. + R_1^{TE} \exp\left[in_1 k\left(x \sin \theta_1 - z \cos \theta_1\right)\right]\right\},$$

$$\mathbf{E}_2^{TE} = \hat{\mathbf{y}}\left\{T_2^{TE} \exp\left[ik\left(x \sin \theta_2 + z \cos \theta_2\right)\right]\right\}. \tag{16}$$

As in the TM case, the $\mathbf{H}$ fields in each region are given by Maxwell's equation $\mathbf{H} = (n/\eta_0)\,\hat{\mathbf{n}} \times \mathbf{E}$. The coefficients in the microcavity region (region 1) are found to be

$$T_1^{TE} = \frac{t}{1 - r^2 e^{i2w}}$$

and

$$R_1^{TE} = \frac{rt}{1 - r^2 e^{2iw}}\, e^{2iw},$$

where the Fresnel relations governing r and t for TE incidence are[7]

$$r = \frac{\cos \theta_0 - n_1 \cos \theta_1}{\cos \theta_0 + n_1 \cos \theta_1}$$

and

$$t = \frac{2n_1 \cos \theta_0}{\cos \theta_0 + n_1 \cos \theta_1}.$$

(Note the sign difference between R_1^{TM} and R_1^{TE}.) As before, we henceforth take r and t to be constants, and assume that the interior (region 1) is free space (i.e., $n_1 = 1$).

When calculating the dipole-field coupling $\mathbf{d} \cdot \mathbf{E}$ for the perpendicular case we recall that the general dipole orientation is given by $\mathbf{d} = \hat{\mathbf{x}} \cos \phi + \hat{\mathbf{y}} \sin \phi$. That is, we must consider all orientations for the dipole since the incoming field may impinge at all angles ϕ. We again take the dipole position as $z = L_z/2$. The dipole-field interaction term is

$$|\mathbf{d} \cdot \mathbf{E}|^2 \propto \left(\cos^2 \theta \sin^2 \phi \, A_x^{\text{TM}}(kL_z \cos \theta) + \cos^2 \phi \, A^{\text{TE}}(kL_z \cos \theta)\right), \tag{17}$$

with resonance factors

$$A_x^{\text{TM}}(w) = A^{\text{TE}}(w) = \left|\frac{t}{1 - re^{iw}}\right|^2. \tag{18}$$

We have two resonance factors, since the perpendicular dipole may couple to both TE and TM waves. The trigonometric functions of the angles in Equation 17 correspond to the angle-dependent dipole radiation patterns $P(\theta, \phi)$. Integrating these expressions over the spherical shell in k-space gives

$$g_\perp(k) = \left(\frac{1}{2\pi}\right)^3 \int_0^\pi d\theta \int_0^{2\pi} d\phi \, \sin \theta \, k^2$$

$$\left(\cos^2 \theta \sin^2 \phi \, A_x^{\text{TM}}(kL_z \cos \theta) + \cos^2 \phi \, A^{\text{TE}}(kL_z \cos \theta)\right), \tag{19}$$

where the radiation factor $P_\perp(\theta, \phi)$ again appears because of the dot product. The normalized function $g_\perp(k)/g_{\text{free}}(k)$ is plotted in Figure 11 for $|r| = 1.0, 0.8, 0.6, 0.4, 0.2$, and 0. Again, as can be seen, for $|r| \to 0$, $g_\perp(k)$ reduces to its free space value.

IV. TWO DIMENSIONS OF CONFINEMENT: THE WAVEGUIDE CAVITY

A. PERFECT RECTANGULAR WAVEGUIDE

Here we apply the mode counting method to calculate the mode density of structures providing confinement of the optical field in *two* dimensions. To

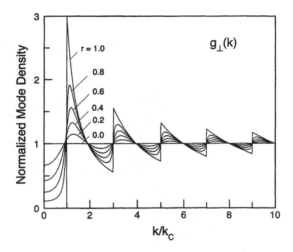

FIGURE 11 Normalized mode density given by (19) for a perpendicular dipole in a lossy planar mirror microcavity. Shown are curves for different values of the field reflection coefficient, r.

begin, we take the case of a rectangular waveguide cavity. The dipole sits exactly in the center of a long metallic waveguide having a rectangular cross section. The physical system is shown in Figure 12. The metal forming the waveguide is assumed lossless. In k-space, this forces the x and y components of $\mathbf{k}$ to be discrete. If we take the length of the cavity L to be much larger than a wavelength, then the z component of $\mathbf{k}$ can be regarded as a continuous quantity. Accordingly, the allowed k-space modes form a series of parallel lines in the k_z direction separated by π/L_x in the k_x direction and by π/L_y in the k_y direction, as is depicted in Figure 13. The lines are indexed by m in the k_x direction and by n in the k_y direction. The "background" mode density on each line is $(L/2\pi)$.

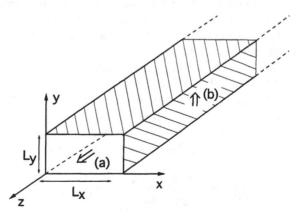

FIGURE 12 Dipole in a rectangular waveguide cavity. Two polarization states are possible, one parallel to the cavity axis ($\mathbf{d} \parallel \hat{\mathbf{z}}$) (a), and the other perpendicular ($\mathbf{d} \perp \hat{\mathbf{z}}$) (b).

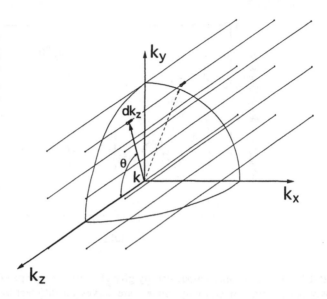

FIGURE 13 k-space of a rectangular waveguide cavity. The possible **k** vectors form a series of lines running in the $\hat{\mathbf{k}}_z$ direction. The lines are separated by π/L_x in the $\hat{\mathbf{k}}_x$ direction and by π/L_y in the $\hat{\mathbf{k}}_y$ direction. Mode counting takes place by summing the contribution of each line segment formed by the intersection of a line with the surface of a sphere of radius $k = \omega/c$ (shown highlighted here).

The effective mode density is again found by counting the number of allowed modes contained within a spherical shell of width dk and radius k. Each allowed line is intersected by the sphere's surface twice, giving a factor of two. This situation is schematically shown in Figure 13. The number of modes included in each intersection is determined by the length of the line $dk_z^{(m,n)}$ lying inside the shell of width dk. A dipole-field coupling factor $P(\mathbf{k})$ must again be included. For a waveguide cavity, there are three orientations for the dipoles: two perpendicular ($\perp$) and one parallel ($\parallel$) to the long ($\hat{\mathbf{z}}$) axis. Depending on the dipole's orientation, only certain of the intersected modes need be included in the sum.

We first treat the parallel dipole (Figure 12(a)). In this case, there can only be coupling to the TM modes of the waveguide. Accordingly, both m and n take on only *odd* values in the mode counting sum to satisfy the boundary conditions. For convenience, we define the angle which the **k** vector makes with respect to the $\hat{\mathbf{z}}$ axis to be $\theta_{m,n}$, as indicated in the figure. The coupling factor is then $P_\parallel(\theta,\phi) = \sin^2\theta_{m,n}$. The mode counting sum is

$$g_\parallel(k)dk = \frac{k_{cx}k_{cy}}{2\pi^3}\sum_{m\,\text{odd}}\sum_{n\,\text{odd}} 2\cdot P_\parallel\left(\theta_{m,n}\right)\cdot dk_z^{(m,n)}.$$

As before, the sum proceeds over all integer m and n satisfying $k^2 \geq (mk_{cx})^2 +$ $(nk_{cy})^2$. The quantization volume is $V = LL_xL_y$. We have defined cut-off wavevectors $k_{cx} = \pi/L_x$ and $k_{cy} = \pi/L_y$.

By inspecting the figure, it is easy to see that

$$\cos \theta_{m,n} = \frac{\sqrt{k^2 - (mk_{cx})^2 - (nk_{cy})^2}}{k} \tag{20}$$

and

$$\sin \theta_{m,n} = \frac{\sqrt{(mk_{cx})^2 + (nk_{cy})^2}}{k}. \tag{21}$$

Furthermore, we have $dk_z^{(m,n)} = dk/\cos\theta_{m,n}$. Inserting all of these factors into the mode counting sum yields

$$g_\parallel(k)dk = \frac{k_{cx}k_{cy}}{2\pi^3} \sum_{m \text{ odd}} \sum_{n \text{ odd}} \frac{2dk\, k}{\sqrt{k^2 - (mk_{cx})^2 - (nk_{cy})^2}} \left\{ \frac{(mk_{cx})^2 + (nk_{cy})^2}{k^2} \right\}. \tag{22}$$

In contrast to the formula for the planar mirror cavity, no closed form expression exists for $g_\parallel(k)$. However, Equation 22 may be readily evaluated by computer, given appropriate values for L_x and L_y. We have evaluated (22) for the case $L_x = L_y$. The result is plotted in Figure 14. In Figure 15 is plotted the effective mode density normalized by the free space density $g_\parallel(k)/g_{\text{free}}(k)$.

Both Figure 14 and Figure 15 show that $g_\parallel(k)$ is singular at many points. Indeed, the effective mode density has a singularity of form $(\omega^2 - \omega_c^2)^{-1/2}$ at a countably infinite number of points, whenever the frequency passes through the cut-off of a mode, and that mode turns on. This behavior is different to that of the planar mirror microcavity, which is **never** singular (except at $k = 0$ for the parallel dipole). This exemplifies the importance of dimensionality on the effective mode density: the more confinement in a microcavity, the larger the change of the mode density. Importantly for applications, this implies that an atom having a suitably narrow spectral width could have its spontaneous emission rate *dramatically* enhanced if it were resonant with the cavity. In this case, the ultimate limit to the enhancement is determined by the linewidth of the atom itself. This is discussed more fully in Chapter 8 by Yokoyama. Such behavior is not possible using the smaller mode density enhancement afforded by a planar mirror cavity.

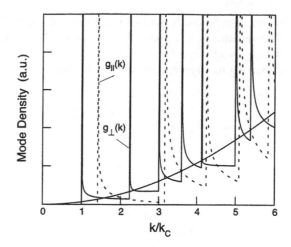

FIGURE 14 Mode density in a rectangular waveguide microcavity. The dashed line is the mode density seen by the parallel dipole ($\mathbf{d} \parallel \hat{\mathbf{z}}$), the solid line is the case of the perpendicular dipole ($\mathbf{d} \perp \hat{\mathbf{z}}$). Note that the mode density diverges at the cut-off frequencies of the modes. Also shown is the free space mode density.

The other dipole orientation is the perpendicular ($\hat{\mathbf{y}}$ oriented) dipole (Figure 12(b)). This time, the allowed modes are odd m and even n because of the boundary conditions imposed by the perfect conductor. The dipole-field coupling factor is $P_\perp(\theta,\phi) = \cos^2\theta_{m,n} + \sin^2\theta_{m,n}\cos^2\phi_{m,n}$, where $\phi_{m,n}$ is the angle in the $k_x - k_y$ plane. We have

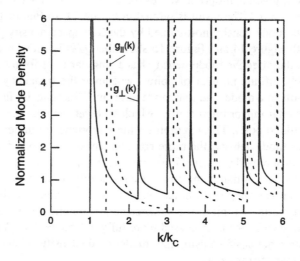

FIGURE 15 Normalized mode density in a rectangular waveguide microcavity. The dashed line corresponds to the parallel dipole ($\mathbf{d} \parallel \hat{\mathbf{z}}$), the solid line to the perpendicular dipole ($\mathbf{d} \perp \hat{\mathbf{z}}$).

$$\cos \phi_{m,n} = \frac{mk_{cx}}{\sqrt{\left(mk_{cx}\right)^2 + \left(nk_{cy}\right)^2}}.$$

With this information, the mode counting sum becomes

$$g_\perp(k)dk = \frac{k_{cx}k_{cy}}{2\pi^3} \sum_{m \text{ odd}} \sum_{n \text{ even}} \frac{2dk\,k}{\sqrt{k^2 - \left(mk_{cx}\right)^2 - \left(nk_{cy}\right)^2}}\left\{1 - \left(nk_{cy}/k\right)^2\right\}. \tag{23}$$

As before, the sum proceeds over all integer m and n satisfying $k^2 \geq (mk_{cx})^2 + (nk_{cy})^2$. This function is also plotted in Figure 14 for the special case $L_x = L_y$. Finally, in Figure 15 we plot the effective mode density normalized by the free space density $g_\perp(k)/g_{\text{free}}(k)$. Again, dramatic enhancements of the spontaneous emission rate from atoms having suitably narrow linewidths are feasible in this cavity geometry.

B.   PERFECT CIRCULAR WAVEGUIDE

The mode counting method can also be applied to the case of a cavity having a circular cross section. Again we assume perfectly reflecting cavity walls and take the dipole to lie exactly on the cavity axis. The cavity geometry is shown schematically in Figure 16. The cavity radius is a, and the cavity length is assumed infinite. Two types of modes are supported by a circular waveguide: TM and TE. The radial dependence of the field is described by Bessel functions $J_n(k_r r)$. In this case, the k-space seen by the dipole is a series

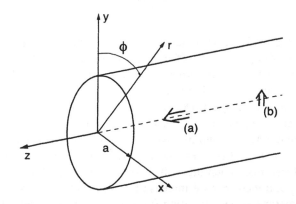

FIGURE 16 Schematic of circular waveguide cavity. The radius of the cavity is a. The two dipole orientations are shown: parallel ($\mathbf{d} \parallel \hat{\mathbf{z}}$) (a) and perpendicular ($\mathbf{d} \perp \hat{\mathbf{z}}$) (b).

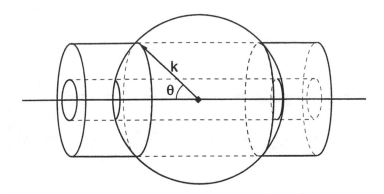

FIGURE 17   The circular waveguide cavity in k-space. The locus of allowed **k** vectors form a series of concentric cylinders. The mode density is found by a weighted sum over the intersection of the cylinders with a sphere of radius $k = \omega/c$.

of concentric cylinders whose radii are determined by the cut-off frequencies of the corresponding modes. This is shown in Figure 17.

We examine first the case of a parallel ($\hat{z}$ oriented) dipole (Figure 16(a)). Because of the dipole's orientation, coupling can only take place to the TM modes of the waveguide. Furthermore, inspection of the form of the electric and magnetic fields (given in Equations 28 to 38 below) shows that only the zeroth-order mode (i.e., $J_0(k_r r)$) couples to a $\hat{z}$ dipole, since this mode is rotationally symmetric about the $\hat{z}$ axis. Higher-order modes have angular dependencies which prohibit coupling to the rotationally symmetric parallel dipole.

The radii in k-space of the concentric cylinders, $k_r^{(n)}$ (see Figure 17), are determined by the condition

$$J_0\left(k_r^{(n)}a\right) = 0.$$

This relation is satisfied at an infinite number of points, $n = 1, 2, 3,...,$ corresponding to the cut-off frequencies of the J_0 modes.

Note that the higher-order modes (i.e., those given by $J_1, J_2,...$) are absent, since they cannot couple to the dipole by reason of symmetry. Since the only modes included in the mode counting sum are the J_0 modes, the mode counting sum picks up only a small subset of all the possible modes present in a circular waveguide! Mode counting gives the wrong mode density since it ignores the $J_1, J_2,...$ modes. However, these modes first turn on at higher frequencies. Therefore, mode counting will give correct results for the first few allowed modes, and then become increasingly unreliable as $k \to \infty$. This is a simple mathematical artifact resulting from our choice of placing the dipole **exactly** on the $\hat{z}$ axis, and has no physical content. Indeed, if we assume that the dipole has finite size or displace it off the cavity axis, then the higher-order modes will also couple to the dipole.[8] This caveat must be borne in mind when considering the $k \rightarrow \infty$ limit of the mode counting calculation.

The polarization factor is $P_{\parallel}(\theta) = \sin^2 \theta$. Reference to Figure 17 shows that the effective mode density is found by summing over all the rings in k-space representing the intersection of the concentric cylinders with a sphere of radius k. We have

$$g_{\parallel}(k)dk \propto \sum_{k_r^{(n)} < k} 2 \int_0^{2\pi} d\phi \; k_r^{(n)} dk_z P_{\parallel}(\theta_n).$$

Define

$$\cos \theta_n = \frac{k_z}{k} = \frac{\sqrt{k^2 - \left(k_r^{(n)}\right)^2}}{k} \tag{24}$$

and

$$\sin \theta_n = \frac{k_r^{(n)}}{k}. \tag{25}$$

We have $dk_z = dk/\cos \theta_n$ and $k_r^{(n)} = k \sin \theta_n$, giving the mode counting sum

$$g_{\parallel}(k)dk \propto \sum_{k_r^{(n)} < k} 4\pi k \left(\frac{\sin^3 \theta_n}{\cos \theta_n} \right) dk = \sum_{k_r^{(n)} < k} \frac{4\pi}{k} \frac{\left(k_r^{(n)}\right)^3}{\sqrt{k^2 - \left(k_r^{(n)}\right)^2}} dk. \tag{26}$$

The effective mode density $g_{\parallel}(k)$ is plotted in Figure 18. The resemblance of $g_{\parallel}(k)$ in Figure 18 to that of the rectangular waveguide shown in Figure 14 is apparent. In both cases, the mode density "blows up" as $(\omega^2 - \omega_c^2)^{-1/2}$ at the cut-off frequencies of the cavity. As in the rectangular waveguide case, the divergent mode density will significantly enhance the spontaneous emission rate of an atom placed in such a cavity. The only difference between the two cases occurs in the position of the cut-off frequencies and their relative sizes. This point is illustrated in Figure 19, which shows the effective mode density normalized to its free space value. This mode density increase in this figure should be contrasted with that in Figure 7, which is manifestly smaller.

The case of a perpendicular dipole is similar (Figure 16(b)). The dipole couples only to the first-order TE modes for reasons of symmetry. The cut-off frequencies, $k_r^{(n)}$, are determined by

$$J_1'\left(k_r^{(n)}a\right) = 0,$$

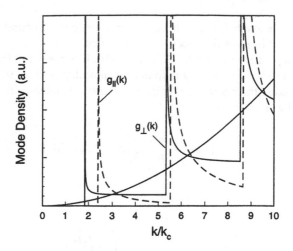

FIGURE 18 Mode density in a circular waveguide cavity having perfectly reflecting walls. The dashed line is the mode density seen by a parallel dipole, $g_{\parallel}(k)$, and the solid line is the mode density seen by a perpendicular dipole, $g_{\perp}(k)$. As in the case of the rectangular waveguide cavity, the mode density diverges at the cut-off frequencies of the cavity. The quadratically rising free space mode density is also shown.

where prime denotes differentiation with respect to the *entire* argument. The polarization factor is $P_{\perp}(\theta, \phi) = \sin^2\theta_n \cos^2\phi + \cos^2\theta_n$. For the effective mode density we get

$$g_{\perp}(k)dk \propto \sum_{k_r^{(n)}<k} 2\int_0^{2\pi} d\phi \; k_r^{(n)}dk_z \left\{\sin^2\theta_n \cos^2\phi + \cos^2\theta_n\right\}$$

$$= \sum_{k_r^{(n)}<k} 2\pi k \frac{k_r^{(n)}dk}{\sqrt{k^2 - \left(k_r^{(n)}\right)^2}} \left\{2 - \left(k_r^{(n)}/k\right)^2\right\}.$$

$$(27)$$

This function is also shown in Figure 18.

C.   LOSSY CIRCULAR WAVEGUIDE

It is of interest to see what effect the introduction of lossy mirrors has on the alteration of the effective mode density in a waveguide cavity. Unfortunately, for the **rectangular** waveguide cavity, there is no simple closed form expression for the modes in the event that the cavity walls are partially transmissive. This difficulty is well known from the theory of integrated optical waveguides.[9] Accordingly, the lossy rectangular waveguide cavity cannot be

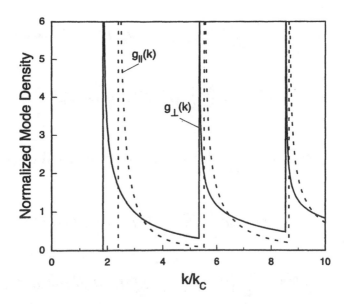

FIGURE 19 Normalized mode density in a circular waveguide microcavity. The dashed line corresponds to the parallel dipole, and the solid line to the perpendicular dipole.

treated. However, in the case of a **circular** waveguide cavity, closed form expressions do exist both inside and outside the cavity, regardless of the reflectivity of the cavity walls. Inside the cavity the modes are Bessel functions, $J_n(k_r r)$ corresponding to standing waves. Outside the cavity the modes are the Hankel functions $H_n^{(1)}(k_r r)$ and $H_n^{(2)}(k_r r)$ corresponding to outward and inward propagating waves, respectively.

We will treat the lossy cylindrical waveguide cavity differently from the lossy planar mirror cavity. The cavity geometry is shown in Figure 16. In contrast to the strategy for dealing with the planar mirror cavity, we assume from the beginning and there is free space inside the cavity. We let the cavity have metallic walls of vanishing thickness δ and finite conductivity σ. The regions inside and outside the cavity are free space. We imagine that a **cylindrically symmetric** wave of amplitude 1 propagates inward from infinity towards the cavity. Part of the wave is reflected back outward at the cylinder wall with reflection coefficient R. Inside the cylinder, the field is taken to have amplitude T. The goal is to calculate the dipole-field coupling, $|\mathbf{d} \cdot \mathbf{E}|$, and integrate it over a sphere in k-space to get the effective mode density via Equation 3.

We start by examining the parallel ($\parallel$) dipole case. As usual, we consider a dipole situated exactly on the cavity axis. The boundary conditions imposed by the cavity walls yield two possible polarizations for the field, TM and TE. The parallel dipole can couple only to the TM modes. In complete generality, the TM fields inside and outside the cavity can be expressed by the following modal sums:[10]

$$E_r^{\mathrm{TM,o}} = iC\frac{k_z}{k_r}\sum_n\left\{1_n^{\mathrm{TM}}\,H_n^{(2)'}\!\left(k_r\mathrm{r}\right)+R_n^{\mathrm{TM}}H_n^{(1)'}\!\left(k_r r\right)\right\}\cos(n\phi)e^{ik_z z},$$
$$\tag{28}$$

$$E_r^{\mathrm{TM,i}} = iC\frac{k_z}{k_r}\sum_n\left\{T_n^{\mathrm{TM}}\,J_n'\!\left(k_r r\right)\right\}\cos(n\phi)e^{ik_z z},$$
$$\tag{29}$$

$$E_\phi^{\mathrm{TM,o}} = -iC\frac{k_z}{k_r^2 r}\sum_n n\left\{1_n^{\mathrm{TM}}\,H_n^{(2)}\!\left(k_r r\right)+R_n^{\mathrm{TM}}H_n^{(1)}\!\left(k_r r\right)\right\}\sin(n\phi)e^{ik_z z},$$
$$\tag{30}$$

$$E_\phi^{\mathrm{TM,i}} = -iC\frac{k_z}{k_r^2 r}\sum_n n\left\{T_n^{\mathrm{TM}}\,J_n\!\left(k_r r\right)\right\}\sin(n\phi)e^{ik_z z},$$
$$\tag{31}$$

$$E_z^{\mathrm{TM,o}} = C\sum_n\left\{1_n^{\mathrm{TM}}\,H_n^{(2)}\!\left(k_r\mathrm{r}\right)+R_n^{\mathrm{TM}}H_n^{(1)}\!\left(k_r r\right)\right\}\cos(n\phi)e^{ik_z z},$$
$$\tag{32}$$

$$E_z^{\mathrm{TM,i}} = C\sum_n\left\{T_n^{\mathrm{TM}}\,J_n\!\left(k_r r\right)\right\}\cos(n\phi)e^{ik_z z},$$
$$\tag{33}$$

$$H_r^{\mathrm{TM,o}} = iC\frac{k}{\eta_0 k_r^2 r}\sum_n n\left\{1_n^{\mathrm{TM}}\,H_n^{(2)}\!\left(k_r r\right)+R_n^{\mathrm{TM}}H_n^{(1)}\!\left(k_r r\right)\right\}\sin(n\phi)e^{ik_z z},$$
$$\tag{34}$$

$$H_r^{\mathrm{TM,i}} = iC\frac{k}{\eta_0 k_r^2 r}\sum_n n\left\{T_n^{\mathrm{TM}}\,J_n\!\left(k_r r\right)\right\}\sin(n\phi)e^{ik_z z},$$
$$\tag{35}$$

$$H_\phi^{\mathrm{TM,o}} = iC\frac{k}{\eta_0 k_r}\sum_n\left\{1_n^{\mathrm{TM}}\,H_n^{(2)'}\!\left(k_r r\right)+R_n^{\mathrm{TM}}H_n^{(1)'}\!\left(k_r r\right)\right\}\cos(n\phi)e^{ik_z z},$$
$$\tag{36}$$

$$H_\phi^{\mathrm{TM,i}} = iC\frac{k}{\eta_0 k_r}\sum_n\left\{T_n^{\mathrm{TM}}\,J_n'\!\left(k_r r\right)\right\}\cos(n\phi)e^{ik_z z},$$
$$\tag{37}$$

$$H_z^{\mathrm{TM,o}} = H_z^{\mathrm{TM,i}} = 0.$$
$$\tag{38}$$

The superscripts "o" and "i" indicate the fields outside and inside the cavity, respectively. C is a constant which will be determined by normalizing the field. As usual, k denotes the frequency of the wave, while k_r is the wavevector in the radial direction. The field propagates in the z direction with wavevector k_z.

We have written down the general expressions for a TM wave inside and outside the cavity. The parallel dipole sitting at the exact center of the cavity will couple only to those modes which are nonzero at the center of the cavity, and have the proper symmetry. Inspection of Equations 28 to 38 shows that the only nonzero term in the modal sum occurs for $n = 0$. All higher-order terms have oscillatory ϕ dependencies, and hence do not couple. This has been discussed previously in Section IV.B. Therefore, we are left with only three nonzero field components, E_r, E_z, and H_ϕ.

To find the amplitude of the electric field inside the cavity, we need boundary conditions at the cylinder wall (i.e., at $r = a$) to account for the nonzero transmission of the field. For the **E** field, we use Faraday's law,

$$\nabla \times \mathbf{E} = - \partial \mathbf{B}/\partial t \tag{39}$$

which implies that the tangential component of **E** remains constant across the boundary,

$$E_z^o(r = a) = E_z^i(r = a). \tag{40}$$

The boundary condition for the **H** field is found from Ampere's law,

$$\nabla \times \mathbf{H} = \mathbf{j} + \partial \mathbf{D}/\partial t \tag{41}$$

and Ohm's law $\mathbf{j} = \sigma \mathbf{E}$. The boundary condition becomes

$$H_\phi^o(r = a) - H_\phi^i(r = a) = \delta \sigma E_z(r = a). \tag{42}$$

Because we are interested in a **dissipationless** wall, we take the conductivity to be pure imaginary, i.e., $\sigma = i|\sigma|$, so that no power loss occurs. Inserting the expressions (32), (33), (36), and (37) into (40) and (42), and solving for the amplitude of the internal field T_0^{TM}, we get

$$T_0^{\mathrm{TM}} = \frac{H_o^{(1)'}(k_r a)H_o^{(2)}(k_r a) - H_o^{(1)}(k_r a)H_o^{(2)'}(k_r a)}{H_o^{(1)'}(k_r a)J_o(k_r a) - H_o^{(1)}(k_r a)J_o'(k_r a) + i\delta|\sigma|\eta_o(k_r/k)J_o(k_r a)H_o^{(1)}(k_r a)}. \tag{43}$$

Next, we recall that, for $\delta \to 0$, the *field* transmission coefficient of an infinitely thin conducting layer is[11]

$$t = \frac{2}{2 + \delta \sigma \eta_o}. \tag{44}$$

Then, using standard identities for the Wronskians of the Cylinder functions,[12] can write T_0^{TM} as

$$T_0^{TM} = \frac{2t}{t + i\pi a(1-t)(k_r^2/k)J_o(k_r a)H_o^{(1)}(k_r a)}. \tag{45}$$

This is the desired expression giving the magnitude of the electric field inside the cavity.

We have determined the magnitude of the field up to a constant normalization C. We determine the normalization by demanding that the total energy of the field in the cavity,

$$\int dV \frac{1}{2}\varepsilon_0 \left(|E_r|^2 + |E_z|^2 \right) + \int dV \frac{1}{2}\mu_0 |H_\phi|^2 = \text{constant},$$

independent of k_r or k_z. Physically, we are demanding that the energy contained in a mode should not depend on its "direction of propagation", but only on its absolute frequency. Imposing this condition yields $C = k_r/k$.

Since k_r and k_z refer to the wavevectors in the $\hat{r}$ and $\hat{z}$ directions respectively, and we have $k^2 = k_z^2 + k_r^2$, we can define, as usual, an angle θ such that

$$k_z = k \cos \theta, \tag{46}$$

$$k_r = k \sin \theta, \tag{47}$$

which measures the direction of propagation of the field with respect to the z axis. With this convention, and assuming a z oriented dipole at the origin, the dipole-field coupling factor $P = |\mathbf{d} \cdot \mathbf{E}|^2$ entering into (3) becomes

$$|\mathbf{d} \cdot \mathbf{E}|^2 = |dE_z|^2 \propto \sin^2 \theta |T_0^{TM}(k, \theta)J_0(0)|^2,$$

where the $\sin^2 \theta$ comes from the normalization condition. Inserting this expression into (3), we find the expression for the effective mode density,

$$g_\parallel(k) \propto \int_0^{\pi/2} d\theta \, k^2 \sin \theta \cdot \sin^2 \theta \cdot |T_0^{TM}(\theta)|^2, \tag{48}$$

where $T_0^{TM}(\theta)$ is given by

$$T_0^{TM}(\theta) = \frac{2t}{t + i\pi ka(1-t)\sin^2 \theta \, J_0(ka \sin \theta)H_0^{(1)}(ka \sin \theta)}. \tag{49}$$

The computed mode density is shown in Figure 20 for several different values of the reflectivity, $|r|$. We have taken $a = 1$ as the cavity radius. As can be seen, in the $|r| \to 1$ limit, the effective mode density converges to that already deduced in Section IV.B via the mode counting method. In the opposite limit, the effective mode density goes to its free space value, as it should.

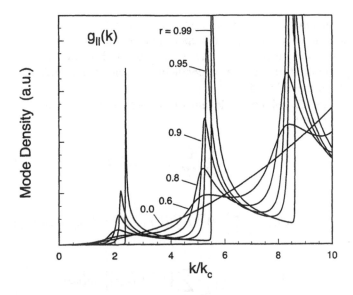

FIGURE 20   Mode density seen by a parallel dipole in a lossy circular waveguide cavity. The field reflection coefficient of the walls r is indicated for each curve. In the $r \to 1$ limit, the results of the perfect cavity case (Figure 18) are recovered. For $r \to 0$ the mode density tends to the k^2 dependence of free space as given by Equation 5.

The interpretation of (48) is very similar to that of the mode counting integral in the planar mirror case. In the $r \to 1$ ($t \to 0$) limit, the function $|T_0^{TM}(k,\theta)|^2$ converges to a sequence of delta functions at points corresponding to the zeros of $J_0(k_r)$. Accordingly, the integration (48) reduces to the mode counting sum over concentric cylinders in k-space, as depicted in Figure 17. As r moves away from 1, the sharply defined k-space cylinders "smear out", causing the peaks in the mode density to broaden. In the opposite limit ($r \to 0$, $t \to 1$), $T_0^{TM}(\theta) \to 2$ and Equation 48 reduces to the free space integral (4).

An important point to notice in Figure 20 is that even a moderately lossy waveguide cavity (e.g., $r = 0.9$) still offers roughly the same mode density enhancement (factor of 3) as a perfect planar mirror cavity, in spite of the loss. (Compare the lossy cylinder in Figure 20 to the perfect planar mirror mode density in Figure 6.) The conclusion is that confinement of the field in two dimensions is a preferred method to achieve an altered mode density.

We now turn to the perpendicular dipole ($\perp$). In principle, it can be handled in exactly the same way at the TM case. The dipole is located exactly on the cavity axis and points in the $\hat{y}$ direction, as depicted in Figure 16. However, in contrast to the parallel dipole, the perpendicular dipole couples to *both* TM and TE modes of the field when the cavity walls are not perfectly reflecting. In this case, analytical solution of the boundary value problem for the electric fields becomes difficult; the resulting expressions are extremely complicated, and it is impossible to develop intuition about the nature of the solutions. Since mode counting is first and foremost meant to be a simple and intuitive computational prescription, unwanted analytical complexity is to be avoided, particularly if it hinders a deeper understanding of the physics of the mode density in one particular cavity or another. For this reason, we adopt the **approximation** that the perpendicular dipole in a cylindrical cavity couples only to the TE modes.[13]

The TE solutions relevant to the propagation of waves in a cylindrical cavity are:[10]

$$E_r^{\text{TE,i}} = iC\eta_0 \frac{k}{k_r^2 r} \sum_n n\{T_n^{\text{TE}} J_n(k_r r)\} \cos(n\phi) e^{ik_z z},$$ (50)

$$E_r^{\text{TE,o}} = iC\eta_0 \frac{k}{k_r^2 r} \sum_n n\{1_n^{\text{TE}} H_n^{(2)}(k_r r) + R_n^{\text{TE}} H_n^{(1)}(k_r r)\} \cos(n\phi) e^{ik_z z},$$ (51)

$$E_\phi^{\text{TE,i}} = -iC\eta_0 \frac{k}{k_r} \sum_n \{T_n^{\text{TE}} J_n'(k_r r)\} \sin(n\phi) e^{ik_z z},$$ (52)

$$E_\phi^{\text{TE,o}} = -iC\eta_0 \frac{k}{k_r} \sum_n \{1_n^{\text{TE}} H_n^{(2)'}(k_r r) + R_n^{\text{TE}} H_n^{(1)'}(k_r r)\} \sin(n\phi) e^{ik_z z},$$ (53)

$$E_z^{\text{TE,o}} = E_z^{\text{TE,i}} = 0$$ (54)

$$H_r^{\text{TE,i}} = iC \frac{k_z}{k_r} \sum_n \{T_n^{\text{TE}} J_n'(k_r r)\} \sin(n\phi) e^{ik_z z},$$ (55)

$$H_r^{\text{TE,o}} = iC \frac{k_z}{k_r} \sum_n \{1_n^{\text{TE}} H_n^{(2)'}(k_r r) + R_n^{\text{TE}} H_n^{(1)'}(k_r r)\} \sin(n\phi) e^{ik_z z},$$ (56)

$$H_\phi^{\text{TE,i}} = iC \frac{k_z}{k_r^2 r} \sum_n n\{T_n^{\text{TE}} J_n(k_r r)\} \cos(n\phi) e^{ik_z z},$$ (57)

$$H_\phi^{\text{TE,o}} = iC \frac{k_z}{k_r^2 r} \sum_n n\{1_n^{\text{TE}} H_n^{(2)}(k_r r) + R_n^{\text{TE}} H_n^{(1)}(k_r r)\} \cos(n\phi) e^{ik_z z},$$ (58)

$$H_z^{\text{TE,i}} = C \sum_n \{T_n^{\text{TE}} J_n(k_r r)\} \sin(n\phi) e^{ik_z z},$$ (59)

$$H_z^{\text{TE,o}} = C \sum_n \{1_n^{\text{TE}} H_n^{(2)}(k_r r) + R_n^{\text{TE}} H_n^{(1)}(k_r r)\} \sin(n\phi) e^{ik_z z}.$$ (60)

The subscripts and superscripts have the same meanings as in the TM case in Equations 28 to 38 above.

The modal sums (50) to (60) can be used to express any arbitrary TE field pattern present in the cavity. A perpendicular dipole, however, can couple only to the first-order modes ($n = 1$) for symmetry reasons. This point has been discussed previously. We therefore discard all solutions for which $n \neq 1$.

The fields are found by matching the boundary conditions at the cavity walls. Since the tangential component of **E** remains constant across the walls from Faraday's law (39), we have

$$E_\phi^i(r = a) = E_\phi^o(r = a). \tag{61}$$

We again model the cavity walls as lossless metals of infinitesimal thickness δ and conductivity σ. Applying Ampere's law (41), the boundary condition for the **H** field is

$$H_z^o(r = a) - H_z^i(r = a) = \delta\sigma E_\phi(r = a). \tag{62}$$

Solving the two resulting equations for the internal field amplitude coefficient T_1^{TE}, and assuming a pure imaginary conductivity as before, we find

$$T_1^{TE} = \frac{2t}{t + i\pi ka(1 - t)H_1^{(1)'}(k_r a)J_1'(k_r a)}. \tag{63}$$

Here, again, t is the field transmission coefficient of the cavity walls, as given by Equation 44.

The amplitude of the dipole-field coupling is found by computing the magnitude of the $\hat{y}$ directed electric field in the $r \to 0$ limit. The normalization coefficient $C = k_r/k$ as before. Using the angle θ defined in Equations 46 and 47 above, the dipole-field coupling factor $|\mathbf{d} \cdot \mathbf{E}|^2$ becomes

$$|\mathbf{d} \cdot \mathbf{E}|^2 = |dE_r(r \to 0)|^2 \propto |T_1^{TE}(\theta) \cos \phi|^2.$$

Inserting this expression into the mode counting integral (3), we get the expression for the mode density seen by a perpendicular dipole in a cylindrical cavity,

$$g_\perp(k) \propto \int_0^\pi d\theta \int_0^{2\pi} d\phi k^2 \sin \theta \cdot \cos^2 \phi \cdot |T_1^{TE}(\theta)|^2, \tag{64}$$

where $T_1^{TE}(\theta)$ is given by

$$T_1^{TE} = \frac{2t}{t + i\pi ka(1 - t)H_1^{(1)'}(ka \sin \theta)J_1'(ka \sin \theta)}. \tag{65}$$

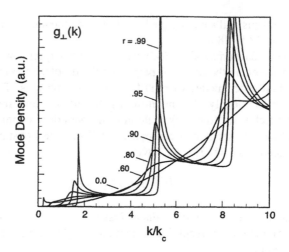

FIGURE 21   Mode density seen by a perpendicular dipole in a lossy circular waveguide cavity. We approximate that coupling occurs only to the TE modes of the cavity, as discussed in the text. As in the parallel dipole case, the results of the perfect cavity case (Figure 18) are recovered in the $r \to 1$ limit. For $r \to 0$ the mode density tends to the k^2 dependence of free space (5).

This expression may be evaluated numerically, and is shown in Figure 21 for several different values of the wall reflectivity $|r| = \sqrt{1-|t|^2}$. The cavity radius $a = 1$. As can be seen, the results of the perfect cavity case (Figure 18) are recovered in the $r \to 1$ limit. In the opposite limit, $r \to 0$, the mode density tends to the k^2 dependence of free space as given by Equation 5, as it should.

V.   THREE DIMENSIONS OF CONFINEMENT: THE BOX MICROCAVITY

The ultimate limit of confinement is provided by a box microcavity, where the field is bounded by reflecting walls in all three dimensions. The effective mode density in this case is a series of delta functions at frequencies corresponding to the resonances of the box. It should be noted here that any completely closed structure, such as a sphere or a closed cylinder, as well as a rectangular box, will serve as a resonant cavity. Indeed, an active area of experimentation has been in microdroplets, which are spherical dielectric microcavities created by ejecting liquids through small vibrating orifices.[14] However, for simplicity, we will here consider only the mode density of a cubic box having sides of length L.

The structure has already been depicted in Figure 1. The resonant frequencies of the structure are given by the relation

$$k^2 = \left(mk_{cx}\right)^2 + \left(nk_{cy}\right)^2 + \left(pk_{cz}\right)^2,$$

(66)

where $k_{cx} = k_{cy} = k_{cy} = \pi/L$ and m, n, and p are integers. In k-space the cavity resonances form a lattice of points determined by $\mathbf{k}_{mnp} = (\pi/L)(m,n,p)$.

Since the box is closed on all three sides, there is no distinction between a parallel and a perpendicularly oriented dipole. As usual, we take the dipole to sit *exactly* in the center of the cavity. Then, for a vertically ($\mathbf{d} \parallel \hat{\mathbf{z}}$) oriented dipole, coupling can occur only to modes for which p is an **even** integer and m and n are **odd** integers. The polarization factor for a vertical dipole is

$$P(\theta) = \sin^2 \theta = \frac{p^2}{m^2 + n^2 + p^2}. \tag{67}$$

Combining all these factors together gives the expression for the effective mode density,

$$g(k)dk \propto \sum_{m \text{ odd}} \sum_{n \text{ odd}} \sum_{p \text{ even}} \frac{p^2}{m^2 + n^2 + p^2} \delta\left(k - \left|\mathbf{k}_{mnp}\right|\right), \tag{68}$$

where $\delta(k - |\mathbf{k}_{mnp}|)$ signifies the presence of the cavity resonances at frequencies given by (66).

Since depicting the mode density of a *lossless* box is very difficult pictorially (how should one draw a true delta function?), it is helpful to introduce a small amount of cavity loss. Incorporating the effects of arbitrary amounts of loss into (68) is not possible in a rigorous way. Similarly to the case of the rectangular waveguide cavity, no proper eigenfunctions exist for the box microcavity if the cavity walls are partially transmissive. Hence, no mode counting integral can be derived for the box cavity. (This situation is different for a spherical microcavity, where the modes inside and outside the cavity are spherical Bessel functions, and an approach similar to that in Sections III.B and IV.C can be used to get the effective mode density for arbitrary wall reflectivity.) Nonetheless, *small* amounts of loss can be easily introduced into (68) in an *ad hoc* way by replacing the delta functions in (68) with Lorentzians,

$$\delta(k) \rightarrow \frac{1}{2\pi} \frac{1/\tau}{(1/\tau)^2 + k^2},$$

where τ is the cavity lifetime. Then, in the limit of perfectly reflecting walls ($r = 1$), we have $\tau \rightarrow \infty$ and the delta functions are recovered. As the reflectivity of the walls moves away from $r = 1$, the cavity lifetime decreases, and the delta functions broaden out. A plot of the effective mode density for a vertical dipole with $\tau = 10^4$ (normalized units) is shown in Figure 22. As can be seen, with increasing k the number of modes per unit frequency increases quadratically, as in the free space case.

As is also evident, the heights of the Lorentzians are different at different frequencies. This is a real effect and has nothing to do with the presence of loss

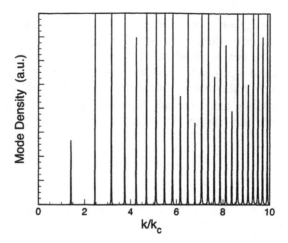

FIGURE 22   Mode density seen by a dipole in a box microcavity as given by (68). Here we have assumed sides of equal length L for simplicity. The number of modes per unit frequency interval manifestly increases with increasing frequency.

in the cavity or finite resolution in the numerical results. Indeed, even if the cavity was lossless and the peaks were true delta functions, their "areas" would be different. Mode counting provides an easy way to understand this phenomenon and clarifies the two contributions to the effect: First, the mode counting sum picks up different numbers of modes in k-space depending on the frequency k. Physically, many of the modes are frequency degenerate, since we have chosen to calculate the case of a cubic cavity. An easy way to see this is to imagine the sphere of possible $\mathbf{k}$ to sit at the center of a cube, which represents a possible cell defined by the lattice in k-space. When the sphere is tangent to the faces of the cube, there are six points of intersection. If the modes lie on the faces of the cube, then six modes are included in the sum. When the sphere intersects the corner vertices of the cube, there are eight points of intersection. If these points correspond to modes, then eight modes must be included in the sum. The second contributing factor is the $\sin^2\theta$ polarization factor given in Equation 67. This gives a weight to each of the k-space modes which depends on its angle with respect to the $\hat{\mathbf{k}}_z$ axis. Both factors cause different Lorentzians to have different heights.

VI.   DISCUSSION

We have seen that the effective mode density seen by an atomic dipole in a microcavity may be calculated using a simple, intuitive prescription called "mode counting". For cavities having rectangular geometries and perfectly reflecting mirrors, this prescription may be summarized as follows:

1. Find the available modes in k-space imposed by boundary conditions at the cavity walls. For example, the planar mirror structure gives a series of equally spaced planes, the rectangular waveguide gives a series of lines, and so on.
2. Find the (k-space) volume of intersection between the allowed modes and a spherical surface of radius $k = \omega/c$ and width dk.
3. Express the effective mode density $g(k)$ as a sum over all the available k-space modes. In the sum, include the following factors:
   - The radiation pattern of the dipole $P(\theta,\phi)$.
   - For rectangular cavities, include a "background" mode density factor $(L\ 2\pi)^D$, where D is the dimension of the surface of available modes, e.g., $D = 2$ for the planar mirror cavity, $D = 1$ for the rectangular waveguide. (Note that this simple prescription does not apply for cavities having shapes other than rectangular, such as the circular waveguide cavity.)
   - Normalize by the cavity volume V.
4. In certain cases, the resulting mode counting sum may be evaluated analytically. Normally, however, one must resort to numerical calculation to evaluate the effective mode density.

In the case of lossy cavity walls, we have seen that the mode counting sum can be generalized to a mode counting integral, at least for certain cavity geometries. Deriving the appropriate integral is usually rather involved. Interpreting the resulting expression, however, reveals the underlying simplicity of mode counting in k-space. The mode counting integral is weighted by a polarization-dependent "resonance factor" which is essentially a series of Lorentzians whose width depends on the reflectivity of the cavity walls. In the limit of perfectly reflecting walls, the Lorentzians become to delta functions, and the integral reduces to the mode counting sum.

A very important extension of the mode counting method would be to cavities having complex reflection and transmission coefficients dependent on the angle of incidence of the light ray, as well as on its polarization state. This generalization would enable one to deal with calculating the effective mode density in, for example, a dielectric microcavity, which would be important for real-world applications.

In dielectrics, the reflection coefficient depends upon both the angle of incidence and the polarization state of the light ray. The reflection coefficient for rays impinging at angles larger than the critical angle has unit magnitude and an angle-dependent phase, i.e., $r = e^{i\psi(\theta)}$. The mode counting prescription suggests that we simply replace the constants r and t with angle- and polarization-dependent quantities in the mode counting integrals (15) and (19). Although the resulting expressions are fairly complicated, they are susceptible to numerical evaluation. Dielectric microcavities are discussed further in Chapter 6.

It is important to note that this procedure would also be valuable in calculating the mode density of a cavity having dielectric stack mirrors. In this case, one must use other methods to compute the angle and polarization dependent reflection coefficient r of the mirror. Using this as the input, the effective mode density $g(k)$ may be again found via Equations 15 and 19. Such a procedure is similar to that described in Reference 15 to treat the dielectric post microcavity, which is discussed further in Chapter 7.

In conclusion, an important lesson to be learned from the mode counting calculations is simply that as the amount of confinement increases, so does the mode density alteration. Starting with free space (no singularities), the frequency-dependent mode density progresses to the planar mirror case (one singularity at $k \to 0$), to the waveguide cavity (infinitely many $(\omega^2 - \omega_c^2)^{-1/2}$ singularities), and finally to the box cavity (infinitely many $\delta(\omega - \omega_c)$ singularities). **For device applications where the magnitude of the mode density alteration is important, it is important to confine the field in as many dimensions as possible.** Calculations in lossy structures also support this observation since even a moderately lossy cavity of low dimension offers a substantial mode density alteration. Realizing this concept in real devices, however, remains a challenging task since it requires the development of microstructures having high quality mirrors on several walls at once.

ACKNOWLEDGMENTS

I acknowledge with pleasure continuing fruitful correspondence with H. Yokoyama and G. Björk. I am also grateful to C. Svendsen for carefully proofreading this manuscript as well as to T. Nyborg for valuable assistance with the artwork.

REFERENCES AND NOTES

1. E. M. Purcell, "Spontaneous emission probabilities at radio frequencies," *Phys. Rev.,* vol. 69, p. 681, 1946.
2. H. Yokoyama, "Physics and device applications of optical microcavities," *Science,* vol. 256, p. 66, 1992, and references therein.
3. S. D. Brorson, H. Yokoyama, and E. P. Ippen, "Spontaneous emission rate alteration in optical waveguiding structures," *IEEE J. Quantum Electron.,* vol. 26, p. 1492, 1990.
4. R. Loudon, *The Quantum Theory of Light.* Oxford: Clarendon Press, 1972.
5. A different — but complementary — perspective on this issue is offered by G. Björk, "On the spontaneous lifetime change in an ideal planar microcavity — transition from a mode continuum to quantized modes," *IEEE J. Quantum Electron.,* vol. 30, p. 2314, 1994.
6. F. DeMartini, F. Cairo, P. Mataloni, and F. Verzegnassi, "Thresholdless microlaser," *Phys. Rev. A,* vol. 46, p. 4220, 1992.
7. M. Born and E. Wolf, *Principles of Optics,* ch. 1.5.2. Oxford: Pergamon, 6th ed., 1980.
8. D. Kleppner, "Inhibited spontaneous emission," *Phys. Rev. Lett.,* vol. 47, p. 233, 1981.
9. D. Marcuse, *Theory of Dielectric Optical Waveguides.* New York: Academic Press, 1974.

10. J. A. Stratton, *Electromagnetic Theory*. New York: McGraw-Hill, 1941.
11. M. Born and E. Wolf, *Principles of Optics,* ch. 1.6.4. Oxford: Pergamon, 6th ed., 1980. In particular, it can be shown that Born and Wolf's Equation 58 reduces to our form in the $\delta \to 0$, $\sigma \to \infty$ limit.
12. M. Abramowitz and I. A. Stegun, *Handbook of Mathematical Functions,* ch. 9. Washington, DC: National Bureau of Standards, 1964.
13. It should be noted that calculations including coupling to both TE and TM modes show that the effect of the TM modes is not large. In particular, no divergences in the mode density are caused by the TM modes.
14. A. J. Campillo, J. D. Eversole, and H. B. Lin, "Cavity quantum electrodynamic enhancement of stimulated emission in microdroplets," *Phys. Rev. Lett.,* vol. 67, p. 437, 1991.
15. T. Baba, T. Hamano, F. Koyama, and K. Iga, "Spontaneous emission factor of a microcavity DBR surface-emitting laser," *IEEE J. Quantum. Electron.,* vol. 27, p. 1347, 1991.

6 Spontaneous Emission in Dielectric Planar Microcavities

Gunnar Björk and Yoshihisa Yamamoto

TABLE OF CONTENTS

I. INTRODUCTION

It is now more than thirty years since the first demonstration of a semiconductor laser.[1-4] During these years there have been steady improvements in semiconductor laser performance regarding output power, quantum efficiency, dynamic single mode output, tunability, etc. During this development, new laser designs of increasing complexity have created a never-ending demand on improved fabrication technology. We now find ourselves in the position of being able to deposit semiconductor films that are smooth down to the atomic scale, and it is possible to etch features as small as tens of nanometers. Hence the era of quantum confined optical devices has begun. In these devices the photon is confined to a volume of about a wavelength cubed, and hence the energy difference between the adjacent quantized modes is comparable with quantization energy itself. This should be compared to the photoemission linewidths of semiconductor electron-holes or excitons which is on the order of $\frac{1}{1000}$ to $\frac{1}{50}$ of the transition energy. Hence, it would seem that in semiconductor microcavity devices only one, or at most a few, optical modes should be able to interact with the transition line. This is in stark contrast with the conventional macroscopic semiconductor laser with a cavity volume of perhaps 10^4 cubic wavelengths. In these lasers the energy difference between adjacent modes is only about 10^{-4} times the photon energy so that ten or more cavity modes can interact with the gain material. In macroscopic lasers multimode behavior is always prevalent and sometimes, especially under direct modulation, dominant.

A major reason it is desirable to make a truly single mode optical device is that, more often than not, spontaneous emission into spurious modes is simply a loss process. One exception is multimode light emitting diodes for flat panel displays, where the multimode characteristic is actually beneficial, since it increases the possible viewing angle of the panel. In most applications, however, a single mode, narrow emission lobe device is clearly preferred. Above threshold, a semiconductor diode laser approximately meets these requirements. Most of the emission goes into the lasing mode, and in a well-designed laser the laser oscillates in a stable single mode. Nonetheless, some power is still radiated as spontaneous emission into unwanted modes. Even if nonradiative recombination is completely suppressed, the power dissipated by unwanted spontaneous emission is a fundamental loss process, and it sets a lower limit to the threshold pump power. A deeper discussion of the threshold power and external quantum efficiency of microcavity lasers can be found in Chapter 8 by Yokoyama. Why even such relatively small dissipated power may have significant impact in densely packed optical interconnects is discussed in Chapter 10 by Hayashi.

In reality, today's microlasers are not quite as good as we would want them to be. Specifically, it is very difficult to design a cavity which excludes all but one cavity mode. Since metals have limited reflectivity (Au or Al films have maximum reflectivities of about 0.95 in the visible to near infrared range) and are lossy, small cavities with reasonably long photon lifetimes cannot be fabricated using metal films. Dielectric cavities, on the other hand, almost invariably support leaky modes and hence couple to the free space mode continuum. One research direction in the microcavity field today is the design and evaluation of various dielectric cavity geometries. Ideally one would like to find a cavity which couples weakly to the free space continuum, which has a high quality factor (Q-factor), which supports mainly one single mode, and which is relatively easy to fabricate. So far no clear winner has emerged, although the dielectric post cavity, originally devised for optical bistability devices, is gaining popularity. References 5 to 10 review and summarize some of the recent developments in the microcavity laser field.

In this chapter we will focus on the planar dielectric cavity, Figure 1, one of the least complex cavity geometries. In addition to simplicity of fabrication, the planar cavity also has the advantage that one does not have to employ any etching. This is advantageous in that etching the active layer in these small structures usually introduces surface defects that increase the nonradiative recombination rate quite significantly.[5]

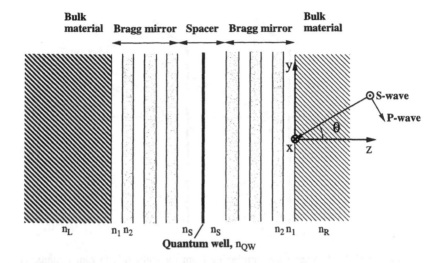

FIGURE 1   Schematic picture of a planar dielectric microcavity. (From Weisbuch, C. and Burstein, E., *Confined Electronics and Photons: New Physics and Devices*, Plenum Press, London, 1995. With permission.)

II.   THE IDEAL PLANAR CAVITY

We will start the discussion about spontaneous emission in planar dielectric microcavities by quickly deriving some results for an atom suspended in free space, and for the same atom placed in an ideal planar cavity. This is an old theoretical problem addressed by a number of authors over the years.[11-23] Our justification for repeating some of the calculations here is that the results will serve as a platform to better understand the physical limitations of the dielectric planar cavity. The discussion also serves as an introduction to the calculation method we will use for the more involved computation of the dielectric planar cavity. Throughout the chapter we will calculate the modification of the spontaneous emission pattern and the corresponding rate change using classical physics. In the weak coupling regime, defined as the regime where both the atomic dipole dephasing time and the cavity photon lifetime are much smaller than the Rabi-flopping cycle time, the modification of spontaneous emission rates and patterns can be seen as the result of the modified zero-field mode patterns and/or mode density. In the following we will take the view that spontaneous emission is "stimulated" by zero-point fluctuations. This is one but not the only interpretation of the cause of spontaneous emission. In the weak coupling regime, with proper precautions, this view leads to the correct results. (For a complete discussion on the various views of the cause of spontaneous emission, see Reference 24.) In the strong coupling regime, which is treated in Chapter 1 by Goldstein and Meystre, additional phenomena such as reversible spontaneous decay and energy-level splitting will also occur. (For a few recent reviews of the field of cavity quantum electrodynamics, please consult References 25 to 28). However, to date, most of the interest in microcavity

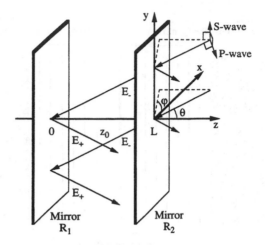

FIGURE 2 A plane wave impinging from the right side toward a planar cavity of length L. The wave is propagating in the φ, θ direction. The electric field at any position z_0 can be computed as the superposition of the E_+ and E_- waves.

physics from the device point of view has been focused on the ability to effectively channel the spontaneous emission into one or a few selected modes. For this end the classical analysis is sufficient.

The ideal cavity we use as our "benchmark" is an idealized planar cavity with lossless mirrors of fixed reflectivity R. The reflectivity is assumed constant not only as a function of wavelength, but also as a function of incidence angle. The mirrors are assumed to be spaced a distance L apart, and they are assumed to extend infinitely far in the mirror plane. Our coordinate system is oriented so that the x- and y-axes lie in the mirror plane, and the z-axis is normal to the mirrors, Figure 2.

We will also assume that the gain material can be modeled as a two-level atomic dipole transition. In contrast to the usual bulk gain material case we will often not assume that the dipole moment has a random spatial orientation, for in practice the gain material in the dielectric microcavities are often thin quantum wells in which the transition dipole moment is not isotropic. In general, the dipole moment of excitonic transitions in quantum wells is substantially stronger in the quantum-well plane, i.e., parallel to the mirror plane. Hence we could use a vectorial representation of the dipole moment $\mathbf{d}(\omega)$, e.g., $\mathbf{d}(\omega) = d(\omega)(1,1,0)/\sqrt{2}$ for a quantum well with the dipole moment isotropically oriented in the plane. However, in the remainder of the text we will use the terminology y-dipole (z-dipole) for any emitter, be it an exciton, atom, or electron-hole pair, with its transition dipole moment oriented in the y-direction (z-direction). Likewise, an emitter with $\mathbf{d}(\omega) = d(\omega)(1,1,0)/\sqrt{2}$ will be denoted an xy-dipole, and an emitter with $\mathbf{d}(\omega) = d(\omega)(1,1,1)/\sqrt{3}$ will be called an xyz-dipole.

The rate of spontaneous emission Γ_{sp} (the inverse of the spontaneous emission lifetime τ_{sp0}) can be calculated from Fermi's golden rule[29]

$$\Gamma_{sp} = \frac{1}{\tau_{sp}} = \frac{2\pi}{\hbar^2} |\langle|\hat{\mathbf{d}} \cdot \hat{\mathbf{E}}_{\mathbf{k},\sigma}^{\dagger}|\rangle|^2 \rho(\omega), \tag{1}$$

where $\hat{\mathbf{d}}$ is the dipole moment operator of the emitter, $\hat{\mathbf{E}}$ is the electric field operator of the mode $\mathbf{k},\sigma$, where $\mathbf{k}$ is the plane-wave k-vector, $\sigma = 1,2$ is the polarization index of the mode, and $\rho(\omega) = \omega^2 V / c_0^3 \pi^2$ is the density of states of the electromagnetic field per unit angular frequency in a quantization volume V.

The quantities in (1) we are going to focus our attention on are the electric field operator and the mode density. All other parameters are cavity independent. The electric field operator and the mode density always appear as a product. In a real cavity (e.g., with dissipation), it is impossible, by any measurement, to separate the influence of the electric field from that of the mode density in (1). The reason is that the quasi-modes in a dissipative cavity are not power orthogonal. Therefore, it is not possible to decompose the electric field inside the cavity in the components of the quasi-modes in a unique way. For this reason it is customary to attribute the changes in spontaneous emission characteristics to *either* the electric field *or* the mode density. In this chapter we will take the former view; in Chapter 5 by Brorson the opposite viewpoint is taken, and the changes in emission characteristics are attributed to a modified density of states. Which viewpoint one wishes to take is irrelevant and largely a matter of convenience, since, treated correctly, both lead to the same result.

To compute the rate of spontaneous emission of an atom suspended in free space we use (1), but we need to average the bracket, which depends on $\mathbf{k},\sigma$, and, in general, ω over all modes, which we will take as plane wave modes in the volume V, which is very large compared to our cavity. The smallness of the planar cavity compared to V ensures that the modes in V are perturbed by a negligible amount by the cavity, although the local electric field inside the cavity may significantly differ from that outside of the cavity as the mirror reflectivity R approaches unity. We will therefore use a first-order perturbation and use the unperturbed external modes and the unperturbed density of states to calculate the field inside the planar cavity.

To facilitate the averaging over $\mathbf{k}$ and σ, we will use a spherical coordinate system in which we can replace the k-vector index with the plane-wave angular frequency ω and the angular coordinates θ and φ. In addition, we will replace the polarization index σ by the notation S and P for the S- and P-polarized waves respectively. The P-polarized wave always has its electric field in the plane defined by $\mathbf{k}$ and the z-axis, whereas the S-polarized mode has its electric field vector normal to that plane, Figure 2. (When there is no cavity present the choice of the S and P directions has no physical significance.)

In making the transition from a Cartesian to a spherical coordinate system it will also prove useful to introduce the concept of density of states per unit frequency and unit solid angle $\zeta(\omega, \theta, \varphi)$. In a sufficiently large cavity it is reasonable to assume that the number of modes with k-vectors falling within some fixed solid angle is independent of the direction. Therefore we will assume that ζ is isotropic so that

$$\zeta(\omega,\ \theta,\ \varphi) = \zeta(\omega) = \rho(\omega)/4\pi = \omega^3\, V/4c_0^3\pi^3. \tag{2}$$

(The assumption preceding (2) is, in fact, not true. The mode density is only isotropic in a roughly spherical cavity irrespective of volume. In an oblong cavity the mode density per unit solid angle along the "long" direction would be higher than the mode density in the "short" directions. However, this higher mode density is exactly compensated for by a smaller electric field for a fixed mode energy (the zero-point energy). We are interested in the product of the two and therefore it is permissible to assume that both are isotropic.)

We write the electric field operator $\hat{E}$ in the volume V as $(\hbar\omega/2\varepsilon_0 V)^{1/2}(a+a^\dagger)$ where ε_0 is the permittivity of free space. For a vacuum mode the electric field expectation value $\langle 0|\,\hat{E}\,|0\rangle$ vanishes, while the electric field square has the expectation value $\langle 0|(\,\hat{E}\,)^2|0\rangle = \hbar\omega/2\varepsilon_0 V$. In the following we will loosely use the term the "electric field intensity" for the mean square of the zero-point fluctuation electric field. In the weak coupling regime we are interested in, if the electric field and dipole moment vectors are parallel, the calculation of the bracket in (1) yields

$$\left|\langle|\hat{\mathbf{d}}\cdot\hat{\mathbf{E}}_{\mathbf{k},\sigma}^\dagger|\rangle\right|^2 = \frac{d(\omega)^2\,\hbar\omega}{2\varepsilon_0 V}, \tag{3}$$

where d represents the matrix element between the emitter and the electric field. We see that the factor $\hbar\omega/2\varepsilon_0 V$, which may be attributed to the mode function, drops out of Equation 1. This is the factor we eventually want to modify by enclosing the atom by a cavity.

With these preparations, we can compute the rate of spontaneous emission per unit angle, $\gamma_{sp}(\omega,\theta)$ to within some constant factor. For a y-dipole, the result is

$$\gamma_{sp}(\omega,\ \theta) = \frac{2\pi}{\hbar^2}\,\frac{d(\omega)^2\,\hbar\omega}{2\varepsilon_0 V}\,\frac{\omega^2 V}{4c_0^3\pi^3}\cos^2(\varphi) = \frac{3\Gamma_0\cos^2(\varphi)}{8\pi} \tag{4}$$

for the S-polarized radiation, and

$$\gamma_{sp}(\omega,\ \theta) = \frac{3\Gamma_0\sin^2(\varphi)\cos^2(\theta)}{8\pi} \tag{5}$$

for the P-polarized radiation, where the constant $\Gamma_0 = 2\omega^3 d^2(\omega)/3\pi\hbar\varepsilon_0 c_0^3$ has been chosen for later convenience. The trigonometrical factors in (4) and (5) are due to the dot product in (1). The spontaneous emission radiation pattern for an x-dipole is given by interchanging $\sin^2(\varphi)$ and $\cos^2(\varphi)$ in (4) and (5) above. The radiation from an z-dipole is zero for the S-polarization, and

$$\gamma_{sp}(\omega,\ \theta,\ \varphi) = \frac{3\Gamma_0 \sin^2(\theta)}{8\pi},\tag{6}$$

for the *P*-polarization. The total emission rate of a *y*-dipole thus becomes

$$\Gamma_{sp0} = \frac{3\Gamma_0}{8\pi} \int_0^{2\pi} d\varphi \int_0^{\pi} d\theta\ \sin\ (\theta)\big(\cos^2(\varphi) + \sin^2(\varphi)\cos^2(\theta)\big)$$

$$= \frac{3\Gamma_0}{4}\left(1 + \frac{1}{3}\right) = \Gamma_0.\tag{7}$$

In (7) the left term in the bracket on the second line corresponds to the spontaneous emission carried away by the *S*-polarized waves, while the right term is that carried away by the *P*-waves. Since we have assumed that the space surrounding our atom is isotropic, the decay rate for an *x*- or *z*-dipole must also be the same. For the *x*-dipole 3/4 of the radiation is also carried away by the *S*-polarized modes, while for the *z*-dipole, all the radiation is carried away by the *P*-waves as suggested by Equation 6.

In the derivation of (4) to (7) it was assumed that neither the mode density nor the zero-point fluctuations of the electric field varied appreciably with frequency within the atomic transition linewidth. As mentioned above this is usually a good approximation in free space since a typical transition line width for the semiconductor systems we consider here are at most a few percent of the transition center energy. When the atom is placed in a cavity this assumption is usually not true anymore, since the cavity electric field resonances may have any width depending on the specific cavity. Therefore we will assume that the dipole moment has some distribution given by the distribution function $f_n(\omega)$. The function is normalized so that $\int_0^\infty f_n(\omega)d\omega = 1$. To derive the spontaneous emission patterns or rates we must average the emission rate at each frequency over this distribution function. It is important to understand that this assumption is only valid if the emission line is homogeneously broadened (albeit not necessarily at its natural linewidth limit due to the radiative decay). In most semiconductor cavities this is a good assumption since the emission lifetime from electron-hole recombination is of the order of 0.1–1 ns while the intraband phonon scattering time is of the order of a few ps. The phonon scattering time is roughly the same for excitons, but the recombination is faster. At temperatures of 4 K, the recombination time can be as fast as 25–100 ps.

Now we will move on and see how the inclusion of a planar cavity will modify the spontaneous emission characteristics. In order to figure out the modification of the spontaneous emission rate into any of our defined modes it suffices to compute the modification of the mode function at the location of the atom(s). Putting the atom between two mirrors with reflectivities R_1 and R_2, Figure 2, the vacuum-field mode inside the cavity will be modified by subsequent reflections in the mirrors. The two fields $E_+(z_0)$ and $E_-(z_0)$ in the cavity, originating from the incident wave E_0 impinging from the right, can be computed by summing the wave transmitted through the right mirror and all subsequent internal reflections:

$$
\begin{aligned}
E_- &= E_0\sqrt{T_2}\,\exp\!\big(ik_z(L-z_0)\big)\big[1+\exp(i2\pi)\sqrt{R_1R_2}\,\exp(i2kzL)+\cdots\big] \\
&= \frac{\sqrt{1-R_2}\,\exp\!\big(ik(L-z_0)\cos(\theta)\big)}{1-\sqrt{R_1R_2}\,\exp(i2kL\,\cos(\theta))}E_0,
\end{aligned}
\tag{8}
$$

where $T_2 = 1 - R_2$ is the transmittivity of mirror 2 and we have used the fact that $k_z = k\cos(\theta) = 2\pi\cos(\theta)/\lambda$. It is worth noticing that the resonance condition depends only on the z-component of the k-vector. In a similar fashion we get

$$
\begin{aligned}
E_+ &= E_0\,\exp(i\pi)\sqrt{T_2R_1}\,\exp\!\big(ik_z(L+z_0)\big)\big[1+\exp(i2\pi)\sqrt{R_1R_2}\,\exp(i2kzL)+\cdots\big] \\
&= \frac{\sqrt{R_1(1-R_2)}\,\exp\!\big(i\big[k(L+z_0)\cos(\theta)+\pi\big]\big)}{1-\sqrt{R_1R_2}\,\exp(i2kL\,\cos(\theta))}E_0.
\end{aligned}
\tag{9}
$$

The in-parallel $E_{//}$ and cavity normal E_z electric field mean intensities can be computed as follows from (8) and (9). For the S-wave we have

$$
E_{//}^2 = |E_+ + E_-|^2 = \frac{(1-R_2)\big[1+R_1-2\sqrt{R_1}\,\cos(2kz_0\cos(\theta))\big]}{\big(1-\sqrt{R_1R_2}\big)^2+4\sqrt{R_1R_2}\,\sin^2(kL\,\cos(\theta))}|E_0|^2
\tag{10}
$$

and

$$
E_z^2 = 0.
\tag{11}
$$

For the P-wave we get

$$
\begin{aligned}
E_{//}^2 &= |E_+ + E_-|^2\cos^2(\theta) \\
&= \frac{(1-R_2)\big[1+R_1-2\sqrt{R_1}\,\cos(2kz_0\cos(\theta))\big]\cos^2(\theta)}{\big(1-\sqrt{R_1R_2}\big)^2+4\sqrt{R_1R_2}\,\sin^2(kL\,\cos(\theta))}|E_0|^2
\end{aligned}
\tag{12}
$$

and

$$
\begin{aligned}
E_z^2 &= |E_+ - E_-|^2\sin^2(\theta) \\
&= \frac{(1-R_2)\big[1+R_1+2\sqrt{R_1}\,\cos(2kz_0\cos(\theta))\big]\sin^2(\theta)}{\big(1-\sqrt{R_1R_2}\big)^2+4\sqrt{R_1R_2}\,\sin^2(kL\,\cos(\theta))}|E_0|^2.
\end{aligned}
\tag{13}
$$

Equations 10 to 13 will serve as the basis for all our calculations for the modification of spontaneous decay in the ideal planar cavity. Instead of using the no-mirror value $E_0^2 \equiv \hbar\omega/2\varepsilon_0 V$ for the electric field square we will use the equations above to compute the electric field "intensity". Using time reversibility arguments, one can easily convince oneself that the emission due to the interaction with the zero-point field incident from the right will leave the cavity through the right mirror. The zero-point fluctuations impinging from the left will likewise be modified by the presence of the mirrors. The expressions will be identical to Equations 10 to 13 above, with $L - z_0$ substituted for z_0 and R_1 for R_2.

The first thing to notice about the equations is that the denominators are resonant when

$$k_z = k \, \cos(\theta) = m\pi/L, \tag{14}$$

where $m = 0, 1, 2, 3, \ldots$. An important conclusion from Equation 14 that we will come back to, is that the resonance condition can be fulfilled for any wave with a $k > \pi/L$, but at angles

$$\theta_m = \arccos(m\pi/L), \tag{15}$$

where m is a positive integer $\leq kL/\pi$. In the following we will usually fix the cavity length so that kL/π is an integer. The wavelength corresponding to this resonance in the $\theta = 0$ (cavity normal) direction will be denoted λ_r.

An important observation from (10) to (13) is that if $R_1, R_2 \approx 1$, the FWHM (full width at half maximum) of the resonance is determined by the equation

$$kL \, \cos(\theta) \approx \pm \frac{1 - \sqrt{R_1 R_2}}{2(R_1 R_2)^{1/4}}. \tag{16}$$

From this equation the following three important relations follow:

$$\Delta k_{FWHM} = \frac{\left(1 - \sqrt{R_1 R_2}\right)}{L(R_1 R_2)^{1/4}}, \tag{17}$$

$$\Delta\lambda_{FWHM} = \frac{\lambda_r^2 \left(1 - \sqrt{R_1 R_2}\right)}{2\pi L(R_1 R_2)^{1/4}}, \tag{18}$$

and

$$\Delta\theta_{FWHM} = \sqrt{\frac{2\lambda_r \left(1 - (R_1 R_2)^{1/2}\right)}{\pi L(R_1 R_2)^{1/4}}}. \tag{19}$$

Equations 17 to 19 express the mode spread in k-vector, wavelength, and angle, respectively. The resonant cavity quasi-mode is not quite a plane wave, but rather a TEM_{00}-like mode with near Lorentzian far-field mode profile. As the dissipation decreases (R_1 and R_2 increases), the spread in all three quantities become smaller and the mode becomes more and more plane-wave-like.

We will derive one more important feature from Equation 17 before calculating any spontaneous emission patterns or rates. Assume we have placed an atom with a free space transition linewidth substantially smaller than the cold cavity linewidth given by Equation 18. Also assume that placing the atom in the cavity does not introduce significant linewidth broadening. (As will be shown shortly, this assumption is justified in the planar cavity case.) Having such a narrow transition line means that we can assume that the k-vector of the emitted line has a constant length k_0. However, both the cavity mode k-vector z-component, and the radial component k_r can still have some uncertainty as manifested by Equations 17 and 19. The relation between the uncertainty of the two components is

$$\Delta k_r = \sqrt{2k_0 \Delta k_z},\qquad(20)$$

where Δk_z is to be taken as the mean square uncertainty of k_z. Invoking Heisenberg's uncertainty principle $\Delta p_r \Delta r \geq \hbar/2$ where r is the radial coordinate (perpendicular to z), we arrive at the equation

$$\Delta r \geq \frac{\hbar}{2\hbar\Delta k_r} = \sqrt{\frac{\lambda_{r0}L(R_1 R_2)^{1/4}}{8\pi n_{cav}(1-\sqrt{R_1 R_2})}},\qquad(21)$$

where λ_{r0} is the resonant mode vacuum wavelength. This result tells us that although the cavity has infinite lateral extent, the cold-cavity modes of the planar cavity are finite in extent. As the mirror reflectivity increases the mode(s) spread out more and more. This is an unwanted peculiarity of the planar cavity that is absent in cavities with transverse confinement. The result (21) agrees to within a factor of order of unity to results obtained in different fashions.[30-32] The implication of Equation 21 is that as the reflectivity of the mirrors become higher, the mode volume of the cavity becomes larger. As long as the cold-cavity linewidth is wider than the atomic linewidth an increased mirror reflectivity in general increases the atom-zero-point fluctuation coupling, thereby increasing the rate of decay. In this particular cavity geometry this effect is effectively canceled by the increase of the mode volume, which decreases the zero-point fluctuation "electric field intensity". Therefore, the planar cavity will not enable one to increase the spontaneous decay rate by any substantial amount.

To calculate the spontaneous emission pattern (i.e., the emission rate per unit solid angle) of a dipole emitter in a planar cavity, we essentially use (4), (5), and (6) but insert the appropriate electric field modification factors given

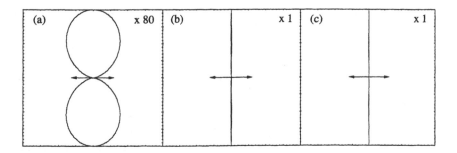

FIGURE 3 The radiation patterns in the xz-plane from a z-dipole emitter. In (a) the emission pattern is drawn for an emitter sitting in free space. In (b) and (c) the emitter is sitting at the center of an $R = 0.95$ cavity. In (b), the normalized cavity length is $L/\lambda_{em} = 0.5$, while in (c), it is $L/\lambda_{em} = 1.0$. In all three plots the radiation patterns have rotational symmetry about the z-axis. Plot (a) has been magnified 80 times with respect to (b) and (c).

by (10), (12), and (13). In Figure 3 the radiation patterns for a dipole emitter in free space and the emission patterns for a z-dipole in the center of a half-wavelength-long, and a one-wavelength-long cavity are shown. It is seen that the cavity selectively enhances the radiation rate per solid angle in some of the cavity resonant modes. For the z-dipole, only the P-polarized even modes couple, with the exception of the mode with $k_z = k$, e.g., the resonant mode propagating parallel to the z-axis. For the half lambda cavity, the $m = 1$ and $m = 0$ modes are resonant, but only the $m = 0$ mode couples to the dipole emitter. This mode radiates parallel to the cavity mirror plane. For the one-wavelength-long cavity the $m = 0$, 1, and 2 modes are resonant, but again only the $m = 0$ mode couples to the dipole emitter. The $m = 1$ mode does not couple because the emitter is at the node position of the mode function, and the $m = 0$ mode does not couple because this mode has its electric field perpendicular to the dipole moment.

In Figure 4 the radiation patterns for a dipole emitter oriented in the y-direction placed in the center of an $R = 0.95$ cavity are shown. The cavity length in (a) is slightly shorter than half a wavelength, in (b) the length is exactly half

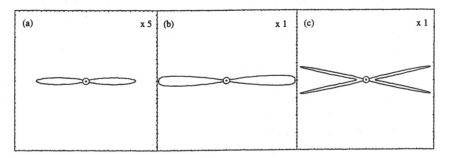

FIGURE 4 The radiation patterns in the xz-plane from a y-dipole emitter sitting in the center of an $R = 0.95$ cavity. The normalized cavity length has been assumed to be $L/\lambda_{em} = 0.49, 0.50$, and 0.51 in (a), (b), and (c) respectively. The patterns are rotationally symmetric about the z-axis to within the resolution of the plots.

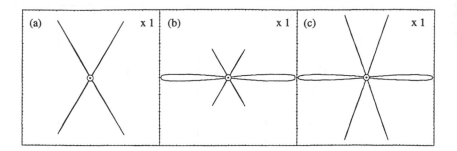

FIGURE 5   The radiation patterns in the xz-plane from a y-dipole emitter sitting in an $R = 0.95$ cavity. The normalized cavity length has been assumed to be $L/\lambda_{em} = 1$ in (a) and (b), and 1.5 in. (c). In (a) and (c), $z_0 = L/2$, while in (b), $z_0 = L/4$. The emission into the mode at $\theta = 60$ (70) degrees does not have rotational symmetry about the z-axis. Its intensity in the yz-plane is only 1/4 (1/9) of the intensity in the xz-plane.

a wavelength, and in (c) the cavity is slightly longer than the (vacuum) emission wavelength $\lambda_{em}/2$. For this cavity only the $m = 1$ mode is resonant, and as the cavity length approaches $\lambda_{em}/2$, the radiation intensity in the normal direction increases quite rapidly. When the cavity is slightly longer than $\lambda_{em}/2$ the mode splits, and the spontaneous emission far-field pattern is ring-shaped. Such patterns have been experimentally observed.[33,34] In Figure 5, finally, the corresponding patterns for a one-wavelength-long cavity are plotted in (a) and (b), and the pattern for a $3\lambda_{em}/2$-long cavity is plotted in (c). In (a), the dipole emitter is sitting in the center of the cavity, so it does not couple to the even modes. The only significant emission is emitted in the $m = 1$ lobe at 60 degrees. In (b), the dipole is sitting at $z_0 = L/4$. At this position the emitter couples both with the $m = 1$ and $m = 2$ modes. The mode function of the $m = 1$ mode is only a factor $1/\sqrt{2}$ as large at this position as at the cavity center. Therefore the emission intensity at 60 degrees is only half as great in (b) as in (a). The $3\lambda_{em}/2$-long cavity supports the $m = 1, 2$, and 3 modes. Since the emitter is assumed to be placed in the cavity center, virtually no emission goes into the $m = 2$ mode. Since the cavity is longer than in (a) and (b), the $m = 1$ mode now emits at an angle $\arccos(1/3) \approx 70.5$ degrees. It should also be noted that the emission lobe width of the mode propagating in the forward direction is smaller than the corresponding width for the $\lambda_{em}/2$ long cavity in Figure 4(b).

We are now ready to calculate the decay rate for a y-dipole using Equations 10 and 12. We will still assume that the line shape of the atomic transition is much narrower than the cold-cavity linewidth. We can therefore write the emission lineshape function as a delta function centered at the emission wavelength λ_{em} so that $f_n(\omega) = \delta(\omega - 2\pi c/\lambda_{em})$. We will also assume that the cavity is symmetric, i.e., $R_1 = R_2 = R$ and $z_0 = L/2$. In this case the decay rate becomes

$$\Gamma_{sp} = \frac{3\Gamma_0}{8\pi} \int_0^\infty d\omega \int_0^{2\pi} d\varphi \int_0^\pi d\theta \ \sin(\theta)$$

$$\frac{(1-R)\left[1+R-2\sqrt{R}\ \cos\left(2\omega z_0 \cos(\theta)/c\right)\right]\left[\cos^2(\varphi)+\sin^2(\varphi)\cos^2(\theta)\right]\delta\left(\omega - 2\pi c/\lambda_{em}\right)}{(1-R)^2+4R\ \sin^2\left(\omega L\ \cos(\theta)/c\right)}$$

$$\approx \frac{3\Gamma_0}{4} \int_0^{\pi/2} d\theta \ \sin(\theta) \frac{(1-R)\left[1+R-2\sqrt{R}\ \cos\left(k_{em}L\ \cos(\theta)\right)\right]\left[1+\cos^2(\theta)\right]}{(1-R)^2+4R\ \sin^2\left(k_{em}L\ \cos(\theta)\right)}, \tag{22}$$

where $k_{em} = 2\pi/\lambda_{em}$. We can see that quite expectedly, when $R = 0$, the integral above identically reduces to (7). If, on the other hand, the reflectivity is sufficiently high, say $R > 0.99$, the integral can be solved by making the substitution $k_{em}L \cos(\theta) = t$, and (22) can be rewritten

$$\Gamma_{sp} = \frac{3\Gamma_0}{4} \int_0^{k_{em}L} \frac{dt}{k_{em}L} \frac{(1-R)\left[1+R-2\sqrt{R}\ \cos(t)\right]}{(1-R)^2+4R\ \sin^2(t)} \left[1+\left(\frac{t}{k_{em}L}\right)^2\right]. \tag{23}$$

The integrand in Equation 23 is sharply peaked at $t = \pi, 3\pi, 5\pi,\ldots$. These resonances correspond to the cos-type resonant modes (or in the nomenclature of Brorson et al.,[20] the even modes), see Figure 6. The denominator is also resonant at $t = 0, 2\pi, 4\pi,\ldots$ (sin-type, or odd modes), but at these t values the integrand vanishes as $R \rightarrow 1$. The reason is that with our choice of emitter position, the emitter is always located at the node position of the odd modes (see Figure 6) and therefore, although the cavity is resonant, the emission rate into these modes is quenched.

For the moment neglecting the factor $1 + (t/k_{em}L)^2$ in Equation 23 it can be verified that for any positive integer m

$$\int_{(m-1/2)\pi}^{(m+1/2)\pi} dt \frac{(1-R)\left[1+R-2\sqrt{R}\ \cos(t)\right]}{(1-R)^2+4R\ \sin^2(t)} \approx \begin{cases} \pi & \text{when } R = 0 \\ 0 & \text{when } m \text{ is even and } R \approx 1 \\ 2\pi & \text{when } m \text{ is odd and } R \approx 1. \end{cases} \tag{24}$$

If R is sufficiently close to unity, that is, the cavity has very little dissipation, the integrand in (24) approaches $2\pi\sum_{m=1}^\infty \delta(t - [2m-1]\pi)$ where $\delta(t)$ is the delta function, taking out 2π times the value of any slowly varying function at $t = \pi, 3\pi, 5\pi,\ldots$ Hence, the integral (23) approaches the sum

$$\Gamma_{sp} = \frac{3\Gamma_0}{4} \frac{2\pi}{k_{em}L} \sum_{m=1}^{\text{Int}\left(k_{em}L/\pi\right)} \frac{1-(-1)^m}{2} \left(1+\left(\frac{m\pi}{k_{em}L}\right)^2\right), \tag{25}$$

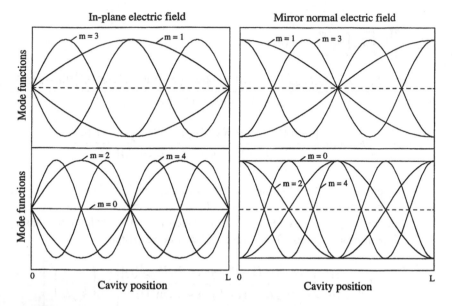

FIGURE 6   Schematic drawing of the electric field modes of the planar cavity. The in-plane electric field (left side) has cos-type modes for odd m:s and sin-type modes for even m:s. The $m = 0$ mode vanishes for the in-plane field. For the mirror normal electric field (right side) the cos-type modes correspond to even m:s, and sin-type modes correspond to odd m:s.

where Int(x) is the largest integer smaller or equal to x. This is effectively the same expression as that for the effective mode density for a perpendicular dipole derived in Chapter 5 by Brorson. We can immediately conclude that in the limit of a large cavity length ($L \rightarrow \infty$), the sum due to the left term in the bracket of the right-hand side of (25) approaches $(3\Gamma_0\pi/2k_{em}L) \cdot (k_{em}L/2\pi) = 3\Gamma_0/4$, and, using the formula $\sum_{m=1}^{N}(2m-1)^2 = N(4N^2-1)/3$, the right-hand term approaches $\Gamma_0/4$. Hence, (25) approaches the free space (no-mirror) value Γ_0 in the limit $L \rightarrow \infty$ as it should. In Figure 7 the total emission rate has been plotted as a function of cavity length and reference plane position, using the generalized form of (25), which reads

$$\Gamma_{sp} = \frac{3\Gamma_0}{4}\frac{\pi}{k_{em}L}\sum_{m=1}^{\text{Int}(k_{em}L/\pi)}\left(1-\cos\left(2\pi m z_0/L\right)\right)\left(1+\left(\frac{m\pi}{k_{em}L}\right)^2\right),$$

(26)

for a xy-dipole emitter and

$$\Gamma_{sp} = \frac{3\Gamma_0}{2}\frac{\pi}{k_{em}L}\left[1+\sum_{m=1}^{\text{Int}(k_{em}L/\pi)}\left(1+\cos\left(2\pi m z_0/L\right)\right)\left(1-\left(\frac{m\pi}{k_{em}L}\right)^2\right)\right],$$

(27)

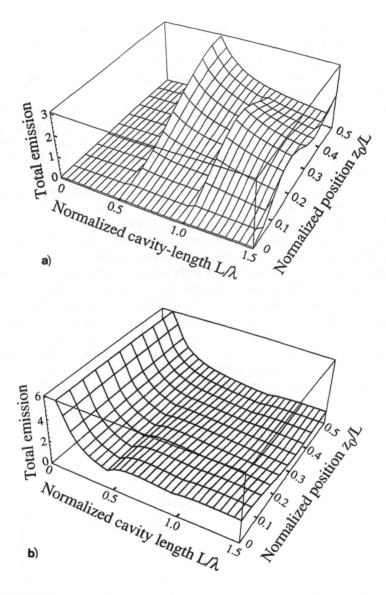

FIGURE 7   The total integrated spontaneous emission rate as a function of normalized cavity length and normalized emitter position. In (a) the dipole moment lies in the xy-plane (mirror plane) while in (b), the dipole is oriented in the z-direction (normal to the mirror plane). (Figure 7a from Weisbuch, C. and Burstein, E., *Confined Electronics and Photons: New Physics and Devices*, Plenum Press, London, 1995. With permission.)

for a z-dipole. It is clear that these expressions could just as easily have been derived by mode counting, rather than by looking at the modification of the mode functions. From Equation 26 it is seen that the maximum enhancement of the total spontaneous emission rate is a factor three as compared to the no cavity case for an xy-dipole emitter. For a z-dipole emitter, the total emission rate is also enhanced by an factor of three for the half-wavelength-long cavity, but in this case the total emission rate diverges and goes to $2\Gamma_0(1 +3/\pi^2)/(1 - R)$ as L goes to zero. In the opposite limit, when $L \to \infty$, the emission rate will approach Γ_0 if $z_0 \neq 0, L$ and will approach $2\Gamma_0$ when $z_0 \ll \lambda_r$ or when $L - z_0 \ll \lambda_r$, i.e., when the emitter is sitting very close to one of the mirrors. The last result is due to the presence of surface modes in the ideal planar cavity. The result is intimately connected with the underlying assumptions about the mirrors, for which we have assumed that all obliquely reflected P-waves interfere positively with the incident P-waves. At grazing mirror incidence this will set up surface plasmon modes. However, due to loss in real metal mirrors the emission rate for emitters close to the mirror surface will probably be closer to Γ_0 rather than to $2\Gamma_0$ in spite of our theoretical result.

Going back to the xy-dipole, the emission rate vanishes if the emitter is placed much closer than a wavelength from the mirrors, since the electric field parallel to the mirrors must vanish at the mirror surface. Again, this result will only be approximately true in practice due to the finite conductivity of metal mirrors. The maximum emission rate enhancement takes place for an approximately half-wavelength-long cavity with the emitter placed in the center of the cavity. However, any dissipative cavity actually has to be slightly longer than half a wavelength to achieve this maximum emission rate, and the mode pattern will have a conical shape, not the often desirable pencil-like beam. In Figure 8 we have solved the integral (23) numerically for a symmetric cavity for lengths around half a wavelength. Three different reflectivities, $R = 0.9, 0.95$, and 0.99 have been used. The trivial result for $R = 0$ is also included, and finally, the limiting result (25) is plotted for comparison. The plot and Equation

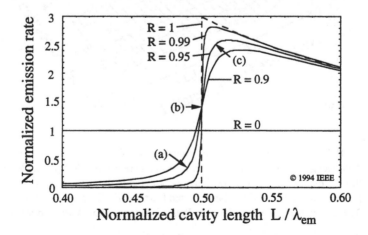

FIGURE 8 The total integrated spontaneous emission rate as a function of normalized cavity length L/λ_{em} for different mirror reflectivities. The dipole moment is distributed isotropically in the xy-plane. (From Björk, G., On the spontaneous lifetime change in an ideal planar microcavity — transition from a mode continuum to quantized modes, *IEEE J. Quantum Electron.*, 30, 10, 1994. With permission; © 1994 IEEE.)

24 very clearly show the transition from the quasi-mode regime to the quantized regime where $k_z \equiv \pi m/L$ (no uncertainty in k_z) as $R \to 1$. The formalism used here allows us to get arbitrarily close to the true mode regime and reach the regime as a limiting value. From the plot it is seen that for a dissipative cavity ($R < 1$), the spontaneous emission rate is about three halves the no-mirror rate for $L/\lambda_{em} = 1/2$. This is regardless of the mirror reflectivity (as long as the reflectivity is larger than, say, 0.9). The reason is clear from Equation 23. The integrand is sharply and symmetrically peaked at $t = \pi$, and for a cavity exactly half a wavelength long the integral is only taken over half of that peak. To illustrate this point more clearly, we have indicated by arrows labeled (a), (b), and (c) the points corresponding to the three radiation patterns (a), (b), and (c) shown in Figure 4. Remember that the integral in Equation 22 only runs over positive angles, so in (a) and (b) only two half lobes (corresponding to cavity resonances) are integrated, whereas in (c), the integral runs over two complete lobes. Hence, to get the maximum emission rate, the emission must be split up into a cone-shaped lobe. This lobe shape corresponds to one of the higher-order transverse modes in a three-dimensional cavity. In many applications this lobe shape is undesired, and therefore one may not wish to operate the device at this otherwise favorable point.

An important figure of merit for any microcavity light emitting device is its spontaneous emission coupling ratio β. It is defined as the ratio between the spontaneous emission rate into a single cavity mode and the total emission rate. In principle this means that every mode has its own β, but in practice only one mode (e.g., the one that eventually will lase) is interesting, so by convention, the β of this mode is referred to as the β of the cavity. Some authors use the letter C to denote the same ratio.[35] The reason β is important, especially for a laser, is that the threshold pump rate is proportional to[36,37] $1/\beta\tau_p$, where τ_p is the cavity photon lifetime, if all nonradiative recombination mechanisms are negligible. Since β in a macroscopic semiconductor laser is a small number, typically of the order of 10^{-4}, a drastic reduction in threshold pump rate (and hence in threshold pump power) is possible for microcavity lasers, which in principle can have a β of order unity. This possibility is one of the driving forces in the microcavity field. In Chapters 9 and 10, by Yokoyama and Hayashi, this topic will be probed further.

In dissipative cavities β cannot be precisely defined since, as mentioned above, the modes are not power orthogonal. Therefore it is impossible to tell where one mode "ends" and another one "begins" in k-space (which has a one-to-one correspondence to wavelength and far-field solid angle). For the specific cavity geometry discussed in this chapter, it is clear that the radiation pattern undergoes a smooth transition from a nearly isotropic pattern when $R = 0$ to a narrow beam pattern when $R \to 1$. To define a mode, we will conservatively define that all radiation emitted within the forward lobe FWHM angle $\Delta\theta_{FWHM}$ (given by Equation 19) and within the wavelength band $\Delta\lambda_{FWHM}$ (given by Equation 18), centered around λ_r, belongs to the cavity mode. For a high reflectivity cavity this is a conservative estimate since virtually all radiation is emitted in a very narrow lobe and could be used in, e.g., an interferometry

experiment or could be launched into a single mode fiber. The very definition
of the mode, given above, limits β for any planar cavity to about 0.5, even if
the cavity mode polarization degeneracy is removed.

To calculate β for a symmetric, half-wavelength-long cavity in which a y-
dipole emitter resides, one needs to integrate the emission within the emission
lobe which defines the mode. This power should subsequently be divided by
the total emission given by Equation 22. The emission into the resonant mode
will be polarized in the y-direction, so the emitter itself breaks the rotational
symmetry of the device around the z-axis. Since all constants in Equation 22
cancel in the division, this amounts to calculating

$$\beta = \frac{1}{2} \int_0^{\Delta\theta_{FWHM}} d\theta \frac{(1-R)\left[1+R-2\sqrt{R}\cos(\pi\cos(\theta))\sin(\theta)\right]}{(1-R)^2 + 4R\sin^2(\pi\cos(\theta))}$$

$$\approx \frac{1}{2} \int_0^{\sqrt{2\lambda_r(1-R)\cdot(\pi L)}} d\theta \frac{(1-R)\left[1+\sqrt{R}\right]^2 \theta}{(1-R)^2 + \pi^2 R\theta^4}$$

$$= \frac{2\arctan\left(\pi^{-1/2}\right)}{\sqrt{\pi}} \approx 0.58.$$

$\qquad\qquad\qquad\qquad\qquad\qquad\qquad\qquad\qquad\qquad\qquad\qquad\qquad\qquad$ (28)

As can be guessed from Figure 4 and Figure 8, β is slightly reflectivity
dependent. A numerical solution of the integral for $R = 0.95$ yields the more
accurate value β ≈ 0.50. If we instead place an xy-dipole emitter in the same
cavity, β drops by a factor of two due to polarization degeneracy. For an xyz-
dipole emitter the β drops by an additional factor of two. As mentioned above,
the upper limit (28) to β is somewhat artificial, and is a result of our definition
of the cavity mode. One could argue that in the case considered, a y-dipole with
an intrinsic emission linewidth much narrower than the cavity resonance, in a
high finesse cavity, β could be made arbitrarily close to unity. However, as
soon as the emission linewidth becomes wider than the cavity linewidth, the
spontaneous emission coupling ratio will start to drop,[32] because, as seen from
Figure 7, even in the $\lambda/2$-cavity, the total emission rate is not strongly dependent on the emission wavelength for wavelengths shorter than λ_r. Hence, on the
short wavelength side of the spectrum, spontaneous emission is going to be
emitted at angles greater than $\Delta\theta_{FWHM}/2$. This radiation corresponds to higher-
order transverse modes and is undesired. In Figure 9 we have plotted the
spontaneous emission rate per unit solid angle as a function of both angle and
emission wavelength. The emission line has been assumed to have a Gaussian
shape with an FWHM of 0.1 λ_{em} centered at λ_r. As predicted by Equation 15,
the cavity is resonant for all emission wavelengths shorter than $2L$ (for all
waves with $k > \pi/L$), albeit at an angle given by (15) from the cavity normal.
For emission wavelengths longer than $2L$ the cavity is off resonance. If an

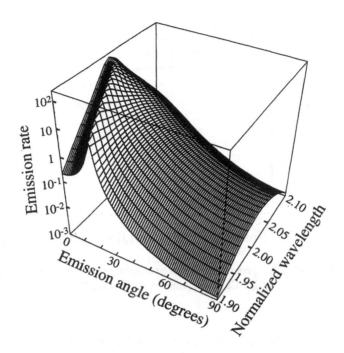

FIGURE 9 The spontaneous emission rate per unit solid angle as a function of normalized emission wavelength λ/L and emission angle for an $R = 0.95$ symmetric cavity. The emitter is located at $z_0 = L/2$.

emitter with a broad emission linewidth is placed in a planar cavity, the emission lobe will be broad, with "red" emission in the cavity normal direction and "blue" emission at larger angles. This "blue" emission can be seen as a sloping "ridge" in Figure 9. In the figure, only the emission between $2 - 0.016 < \lambda_r/L < 2 + 0.016$ and angles up to 7.2 degrees belongs to the resonant mode. It appears that this is actually a substantial part of all the radiation, but this is an illusion. The solid angle available as a function of θ is proportional to $\sin(\theta)$, so actually most of the radiation, about 87%, is radiated into unwanted modes. In any careful measurement or computation of the spontaneous emission rate into the cavity resonant mode this undesired emission should be rejected by simultaneous angular and wavelength filtering. The expression for β of the cavity mode can be written analytically as

$$\beta = \frac{\displaystyle\int_{\lambda_r - \Delta\lambda_{FWHM}/2}^{\lambda_r + \Delta\lambda_{FWHM}/2} d\lambda \, d(\lambda)^2 \int_0^{\Delta\theta_{FWHM}/2} d\theta \, \sin(\theta) \frac{\left[1 + R - 2\sqrt{R}\cos\left(4\pi z_0 \cos(\theta)/\lambda\right)\right]}{(1-R)^2 + 4R\,\sin^2\left(2\pi L\,\cos(\theta)/\lambda\right)}}{\displaystyle\int_0^{\infty} d\lambda \, d(\lambda)^2 \int_0^{\pi} d\theta \, \sin(\theta) \frac{\left[1 + R - 2\sqrt{R}\cos\left(4\pi z_0 \cos(\theta)\,\lambda\right)\right]}{(1-R)^2 + 4R\,\sin^2\left(2\pi L\,\cos(\theta)/\lambda\right)} \left(1 + \cos^2(\theta)\right)}$$

$$(29)$$

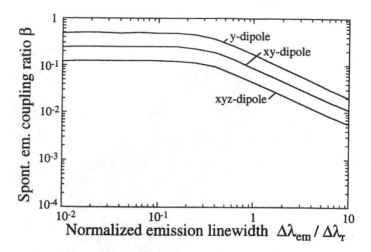

FIGURE 10 The spontaneous emission coupling ratio β vs. the normalized emission linewidth $\Delta\lambda_{em}/\Delta\lambda_{FWHM}$. Cavity parameters are $L = \lambda_{em}/2$, $R = 0.99$, and $z_0 = L/2$.

For a high reflectivity cavity, the denominator reduces to Equation 25. For convenience we have expressed the dipole moment coupling strength as a normalized function of wavelength instead of optical frequency. It is fairly easy to see that if $d^2 = d_0^2 \delta(\lambda_r - 2L)$, where δ denotes the delta function, $z_0 = L/2$, and $R \approx 1$, expression 29 simplifies into 28. The integrals in Equation 29 cannot be solved analytically in the general case, but numerical solutions are not difficult to obtain. In Figure 10 we have solved Equation 29 numerically for $R = 0.99$ assuming a Gaussian gain function d^2 of width (standard deviation) $\Delta\lambda_{em}$. As can be seen in the figure, the spontaneous emission coupling ratio β is roughly constant as long as the emission linewidth $\Delta\lambda_{em}$ is smaller than the cavity resonant linewidth $\Delta\lambda_{FWHM}$. However, if $\Delta\lambda_{em} > \Delta\lambda_{FWHM}$, β is smaller than its optimum value by a factor $\approx \Delta\lambda_{FWHM}/\Delta\lambda_{em}$.

For a cavity of length 3λ, and for an emission linewidth of 10 nm at 1 µm center wavelength, the maximum reflectivity to satisfy the condition $\Delta\lambda_{em} < \Delta\lambda_{FWHM}$ is only about 81%. Unfortunately such a low reflectivity will make it virtually impossible to compensate the 19% power loss per pass by gain provided by semiconductor quantum wells, since a typical quantum well gives only 0.1–1% power gain per pass. In reality no planar semiconductor microcavity laser reported so far has had a β value even near the optimum. The main reason has been that to achieve lasing, the mirror reflectivity has been too high to realize the optimum β (which is done when $\Delta\lambda_{FWHM} > \Delta\lambda_{em}$). However, there are additional reasons for the fact that reported values of β have been about an order of magnitude lower than the ideal value. This will be the topic in the following sections.

III. FUNDAMENTALS OF DIELECTRIC BRAGG MIRRORS

As mentioned in the introduction, it is not possible to make mirrors that have wavelength and reflection angle independent reflectivity as we assumed when deriving the results for the ideal planar cavity. The imperfections of the Bragg mirrors, or rather, the need to make realistic assumptions about the cavity mirrors is going to change the picture of the (ideal) planar cavity given in the last section. In this section we are going to discuss the main limitations of the Bragg mirrors such as the finite reflection angle and the mirror penetration depth.

A Bragg mirror consists (ideally) of a stack of quarter-wavelength thick slabs of dielectric with alternating high and low refractive index. At the Bragg vacuum wavelength λ_B, defined by

$$\lambda_B = 2(l_1 n_1 + l_2 n_2),\tag{30}$$

where l_i, n_i are the length and refractive index of slab i, the round-trip path length between the Fresnel reflections in successive dielectric interfaces is half a wavelength. However the reflection coefficient of a wave propagating in a slab with a higher refractive index, entering a slab with low refractive index, is positive, while the reflection coefficient of the opposite case, going from a lower refractive index to a higher, is negative. Hence the two reflections from subsequent interfaces add up in phase if we take the half wavelength optical path-length difference into account. In a thick stack of all these individual small reflections add up, making the total transmittivity of a nonabsorbing mirror decrease approximately exponentially with mirror thickness, or equivalently, with the number of slab layers.

The greatest advantage of Bragg mirrors is that they may have very high reflectivity. Epitaxial mirrors have been reported having $R > 99.9\%$.[5,38] Additional advantages are that the reflectivity can be tailor-made to suit the needs of a specific device, that they may have very low loss, and that they may be grown epitaxially. Furthermore, the conductivity of epitaxially grown mirrors may be high or low, depending on the doping. A metal mirror, in contrast, has a maximum reflectivity of about 95%, and even a moderately thick metal mirror has zero transmittivity. A metal mirror is also electrically conductive from DC into the ultraviolet region.

At the Bragg wavelength, the complex reflection coefficient of a Bragg mirror is always real. The sign of the coefficient can be chosen to be either positive or negative depending on the stacking order. If the stacking order (going from left to right) is HLHL..., where H (L) represents the higher (lower) index of refraction, then a wave incident from the left at the mirror normal

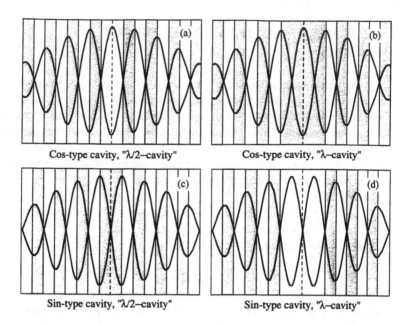

FIGURE 11 Schematic drawing of dielectric microcavities with cos-type, (a) and (b), and sin-type, (c) and (d), modes. The shaded areas represent the slabs with the higher refractive index. (From Weisbuch, C. and Burstein, E., *Confined Electrons and Photons: New Physics and Devices*, Plenum Press, London, 1995. With permission.)

direction has a negative reflection coefficient at the Bragg wavelength. Conversely, if the stacking order is LHLH..., the coefficient is positive. Hence, it is possible to make half-wavelength-long cavities with either the resonance node or the antinode at the cavity center, as shown in Figure 11. From the discussion above, it is clear that taking an ...HLHL mirror and an LHLH... mirror and putting them face to face (effectively forming a stack ...HLHLLHLH..., the resonator round-trip phase would be $0 + 0 = 0$, which is an even multiple of 2π. Therefore the cavity is resonant at the Bragg wavelength. The lowest cavity resonant mode is even (cos-type) with an antinode at the center of the cavity. The round-trip phase for an ...LHLH + HLHL... = ...LHLHHLHL... cavity is $\pi + \pi = 2\pi$ and this cavity is resonant, too. Its lowest-order resonant mode is odd (sin-type) and has a node at the cavity center. It should be noted that in order to form a cavity resonant at the Bragg wavelength, the dielectric stack cannot be perfectly periodic. In both cases an additional quarter-wavelength-thick slab is inserted into the otherwise periodic stack. In the terminology of solid-state physics this slab will form a "lattice defect" in our perfect one-dimensional dielectric "crystal". (The periodic refractive dielectric modulation in our stack corresponds to the periodic potential in a crystal.) In microcavity laser engineering language this extra slab is often referred to as a "spacer" (in this chapter a slightly different terminology is used), while in the quantum electronics community the slab is simply referred to as a "quarter wave phase shift". A cavity so formed is called a "quarter-wavelength-shifted distributed feedback" cavity, or, if the spacer is thicker (e.g., $3\lambda_B/4$) as in Figure 1(b) and (d), the cavity is often referred to as a "distributed Bragg reflector" cavity.[39,40] Recently, in the cavity QED community, the cavity has been called a "one-dimensional photonic bandgap" cavity, drawing an analogy with the three-dimensional photonic bandgap structures proposed.[41]

A localized "impurity" mode will hence form in the center of the "forbidden gap", the frequency (or wavelength) band where modes are prevented to propagate due to the periodic refractive dielectric modulation. The electric field of the "impurity mode" is largest at the impurity (i.e., right between the two mirrors) and the standing-wave envelope function is decaying exponentially on both sides. Actually the slab does not have to be exactly $\lambda_B/4$, $3\lambda_B/4$, $5\lambda_B/4,\ldots$ thick. However, if the spacer is just slightly longer (shorter) than $\lambda_B/4 + m\lambda_B/2$, the resonance will shift the resonance to the longer (shorter) wavelength side. The cavity quality factor (Q-factor) will drop in both cases.

For semiconductor microcavities, one of the "bulk materials" depicted in Figure 1 is air, which has a substantially lower index of refraction than the semiconductor wafer the structure is grown on. Upon leaving the cavity on the "air side", the field sees an additional (relatively large) reflection in the semiconductor-air interface. Hence, if the two Bragg mirrors are identical otherwise, the "air-side" mirror will have a different reflectivity than the semiconductor "substrate-side" side mirror. Usually the "air-side" Bragg mirror stack is designed so that this large reflection adds in-phase with the other reflections. In this case, the "air-side" mirror will have a higher reflectivity than the "substrate-side" mirror. Often this asymmetry is compensated for by making the "substrate-side" mirror thicker to achieve equal reflectivities as seen from the spacer looking outward.

In Figure 12, one minus the reflectivity at normal incidence of a GaAs/AlGaAs mirror has been plotted as a function of the mirror thickness. The refractive indices of GaAs and AlGaAs have been taken as 3.6 and 3.0 respectively. Since the mirror has been assumed to have some loss, the transmittivity $T \neq 1 - R$. It is seen that the absorption will saturate the reflectivity at some finite thickness. When this thickness is reached, the mirror transparency decreases without the corresponding increase in reflectivity, so knowing the mirror absorption is of considerable help for the microcavity engineer.

In Figures 13(a) and (b), the plane-wave reflectivity has been plotted as a function of the normalized wavelength λ_0/λ_B and the incidence angle for a S- and P-polarized wave, respectively. The mirror has been assumed to be 30 periods (60 layers) thick and fabricated from GaAs/AlGaAs. Furthermore, it has been assumed that the cavity is surrounded on both sides by bulk materials with a refractive index of 3.6. This models the "substrate-side" mirror reflectivity as seen from the cavity. From the plot it can be seen that near normal incidence the mirror has only high reflectivity within a finite wavelength range around the Bragg wavelength. To first order, the wavelength range is independent of the number of mirror layer pairs (this is not true for very thin and weakly reflecting Bragg mirrors). The width of the high reflectivity wavelength band at normal incidence is roughly given by [42]

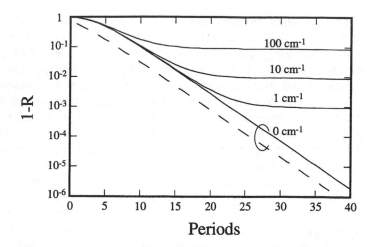

FIGURE 12   Plot of $1 - R$ vs. the number of mirror pairs for a dielectric mirror with slab refractive indices of 3.6 and 3.0. The absorption per unit length is the parameter. For an ideal mirror the $1 - R$ goes approximately as $\exp(-0.33M)$, where M is the number of dielectric layer pairs. The dashed line shows the corresponding reflectivity for the "air-side" mirror.

$$\Delta\lambda_{stopband} = \frac{2\lambda_B \Delta n}{\pi n_{eff}}, \tag{31}$$

where Δn is the refractive index difference between the dielectric layers, and n_{eff} is the effective refractive index of the mirror. For small relative refractive index differences n_{eff} can be replaced by the arithmetic mean of the refractive indices in the stack, while for large relative differences the geometric mean is more suitable. For our mirror, the stopband is predicted by Equation 31 to be $4\lambda_B(3.6-3.0)/(\pi[3.6+3.0]) \approx 0.116\lambda_B$. The FWHM of the computed reflectivity curve at $\theta = 0$ in Figure 13 is $0.128\lambda_B$, in good agreement with Equation 31.

From the figure it is also seen that the mirror is highly reflecting for waves near normal incidence and for waves that are evanescent in the $n = 3.0$ layers (for incidence angles greater than about 65 degrees). A Bragg mirror is a resonant device where the high reflectivity is achieved by the means of many small reflections that add up. Perfect addition (high reflectivity) can only take place at λ_B. At longer or shorter wavelengths (or rather, shorter or longer k_z-components) the interface reflections no longer add in phase and the reflectivity drops. In reality, the picture is a little more complicated, because the individual Fresnel reflection at each interface also depends on the incidence angle. However, to get a feel for the range of angles the mirror is highly reflecting, we assume that the drop in reflectivity at higher angle can be attributed solely to the phase mismatch due to k_z. For small deviations $\Delta\lambda$ and $\Delta\theta$ from the Bragg wavelength λ_B and normal incidence ($\theta = 0$), the z-component of the k-vector can be expanded as

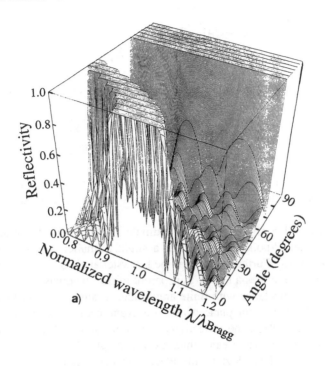

a)

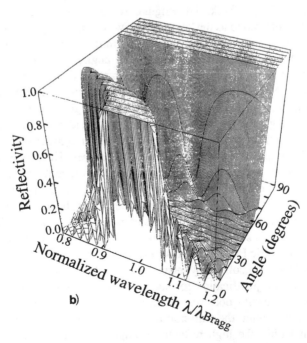

b)

FIGURE 13   The plane-wave reflectivity for an $n_1 = 3.6$, $n_2 = 3.0$ Bragg mirror, 30 periods thick, vs. normalized wavelength and incidence angle. In (a) the S-wave and in (b) the P-wave reflectivity is shown.

$$k_z = \frac{2\pi \cos(\theta)}{\lambda_B} \approx \frac{2\pi(1 - \Delta\lambda/\lambda_B)(1 - \Delta\theta^2/2)}{\lambda_B}.$$
(32)

Hence, the stopband angle can be estimated from (31) and (32) to be

$$\Delta\theta_{stopband} = \pm\sqrt{\frac{2\Delta\lambda_{stopband}}{\lambda_B}} = \pm 2\sqrt{\frac{\Delta n}{\pi n_{eff}}}.$$
(33)

Plugging in our numbers for a GaAs/AlGaAs Bragg mirror, it is found that the mirror is only highly reflecting for angles smaller than 0.48 radians, or 28 degrees.

Figure 12 confirms our rough estimate in (33). From the figure it may be seen that the P-polarized wave has a slightly smaller stop band than the S-polarized wave. Furthermore, near the Brewster angle $\theta_{Brewster} = \arctan(n_2/n_1) \approx 40°$, the P-wave sees a totally transparent mirror, whereas the S-wave sees an oscillating reflectance with finite amplitude. At angles greater than $\arcsin(n_2/n_1) \approx 50°$ the incident plane waves are evanescent in the AlGaAs ($n = 3.0$) layers. Nonetheless, the reflectivity does only approach unity (total internal reflection) for angles greater than $\arcsin(n_2/n_{eff}) \approx 65°$. For incidence angles between 50 and 65 degrees the electromagnetic field tunnels evanescently through the Bragg stack. The angular region between 28 to 65 degrees is sometimes referred to as the "open window" of the Bragg mirror. Hence, as far as the reflectivity is concerned, the Bragg mirror mimics an ideal mirror only within a finite range of wavelengths and angles. The restriction on usable wavelengths is irrelevant for our purposes since typical emission lines have linewidths of at most a few percent of the center wavelength. As just shown, Bragg mirrors can easily be designed to have high reflectivity over a wavelength range greater than 10% of the Bragg wavelength. Therefore, the entire emission spectrum will fit within the Bragg mirror high reflectivity wavelength range with ample margins for fabrication tolerances. The problem lies rather in the finite high reflectivity angular range. Unfortunately little solid angle is available around the normal direction, so actually the Bragg mirrors are transparent for most of the incident free space zero-point fluctuation modes and consequently most of the spontaneous emission will be radiated into unwanted leaky modes. To estimate the spontaneous emission coupling ratio β of a GaAs/AlGaAs dielectric Bragg mirror we can simply calculate the ratio between the solid angle we can control by means of the dielectric mirrors and divide by 4π steradians. The justification for such an estimate is that, as shown above, a planar mirror cavity does not modify the spontaneous emission rate by any appreciable amount, the emission is simply redistributed in space. Hence, the radiation within the high reflectivity angular region can be coerced into a narrow pencil-like beam in the forward direction, while the emission outside this region will essentially be unmodified. Hence, only a fraction

$$\frac{\int_0^{\Delta\theta_{stopband}} d\theta \, \sin(\theta)}{\int_0^{\pi/2} d\theta \, \sin(\theta)} = 1 - \cos(\Delta\theta_{stopband}) \approx 0.11$$

(34)

of the total solid angle can be controlled. In the calculation we have implicitly assumed that the emission linewidth is much narrower than the cavity resonant wavelength and that the atomic dipole moment is oriented along a fixed direction in the cavity plane (e.g., a y-dipole). If the emission linewidth $\Delta\lambda_{em}$ is larger than the cavity resonant wavelength $\Delta\lambda_{FWHM}$, the spontaneous emission coupling radio is decreased by the factor $\Delta\lambda_{FWHM}/\Delta\lambda_{em}$ from Equation 34 as explained in Section II. If the emitters have randomly oriented dipole moments in the cavity plane (xy-dipole), the spontaneous emission coupling ratio is only about 5% (half of (34) due to polarization degeneracy of the cavity mode). As we shall see in the next section, this is a surprisingly accurate estimate of the actual β of such a cavity.

Equations 33 and 34 tell us that the only way to increase the spontaneous emission coupling ratio is to increase the solid angle over which the mirror has high reflectivity. This, in turn, can only be accomplished by increasing the refractive index ratio $\Delta n/n_{eff}$. Unfortunately, if the mirror is to be grown epitaxially, lattice matching severely restricts the possible quaternary compositions possible to grow on a specific substrate. The AlGaAs system is one of the better systems in this respect; nonetheless, the maximum $\Delta n/n_{eff}$ is still only about 0.15. Some trials have been made with mirrors where every second layer has been etched away, leaving air in place of the low refractive index slabs.[43] Such mirrors have an extremely high refractive index difference and a potentially high β, but they are fragile, cumbersome to fabricate, and electrically insulating.

One more thing needs to be pointed out before we move on to a more exact calculation of the spontaneous emission pattern and the emission rate of a dielectric cavity. In the ideal mirror the reflection takes place at a well-defined plane. In a Bragg mirror, as we have seen, there is no such well-defined reflection plane, the reflections are distributed, and hence, some energy is stored in the mirrors. This inevitably lead to dispersion, so that the reflectance phase is wavelength dependent. In Figure 14 we have plotted the reflectance phase for two different mirror thicknesses. It can be seen that the dispersion close to the Bragg wavelength is independent of the number of mirror stacks. The reason is that close to the Bragg wavelength where the mirror is highly reflecting, the wave decays exponentially as a function of the distance from the mirror surface. Hence very little energy is stored in the slabs furthest from the surface, and adding a few periods will not significantly change the stored energy. Since the stored energy does not change with the mirror thickness the dispersion will also be essentially independent of the mirror thickness (but the mirror reflectivity will of course continue to increase with increasing mirror thickness).

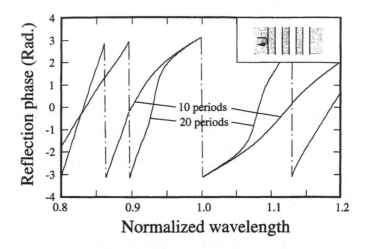

FIGURE 14   Computed Bragg mirror reflectivity phase as a function of the normalized wavelength λ_0/λ_B. The reference plane location is shown in the inset figure where the shaded slabs represent the $n = 3.6$ layers and the white slabs represent the $n = 3.0$ layers. (From Weisbuch, C. and Burstein, E., *Confined Electrons and Photons: New Physics and Devices,* Plenum Press, London, 1995. With permission.)

A simple equivalent model of the Bragg mirror is an ideal mirror sitting some distance l_{pen} away from the reference plane at which the mirror reflectance is calculated. A more detailed model can be found in References 44 and 45. In the following we will refer to l_{pen} as the Bragg mirror penetration depth. As a function of wavelength, the complex reflection coefficient will be $r \exp(-j2kl_{pen}) = r \exp(j\Phi)$ (where the electric field time dependence has been assumed to be $\exp(j\omega t)$. Differentiating the phase factor with respect to λ_0/λ_B gives

$$\frac{d\Phi}{d(\lambda_0/\lambda_B)} = \frac{4\pi n_{eff} l_{pen} \lambda_B}{\lambda_0^2},\tag{35}$$

where λ_0 is the vacuum wavelength. From the figure it can be seen that the slope of the reflection phase as a function of the normalized wavelength is approximately 6π. Using (35) at $\lambda_0 = \lambda_B$, we find that the penetration length of the wave into the Bragg mirror is roughly

$$l_{pen} = \frac{d\Phi}{d(\lambda_0/\lambda_B)}\frac{\lambda_B}{4\pi n_{eff}} \approx \frac{6\pi\lambda_B}{4\pi n_{eff}} = \frac{3\lambda_B}{2n_{eff}}.\tag{36}$$

We see that even for a GaAs/AlGaAs mirror, with a relatively high refractive index ratio, the penetration depth into each mirror is about 1.5 wavelengths. This means that effectively, a device with a quarter-wavelength-long spacer will work as an approximately three-wavelength-long cavity within the high reflectivity region. The most serious implication of the penetration depth is that the cavity mode spacing and the cavity resonance linewidth will be correspondingly narrower, placing more stringent requirements on the emission linewidth of the emitters to realize a high spontaneous emission coupling factor.

IV. DIELECTRIC CAVITY SPONTANEOUS EMISSION PATTERN

In Section II we developed the basic mathematics needed to calculate the spontaneous emission pattern and lifetime of a planar cavity. We will use the same tools in this section with minor additions. The first amendment needed in our theory is to incorporate the fact that dielectric cavities are not surrounded by "free space". At least on one side, epitaxially grown planar cavities have a substrate, an often absorbing dielectric with a high index of refraction. We will assume that the substrate is sufficiently thick and slightly absorbing as to consider it semi-infinite and therefore completely filling one of our half-spaces separated by the cavity, Figure 1. The bulk density of states is proportional to the refractive index, and therefore it is no longer possible to use the same density of states for the modes incident from the "substrate side" as from the "air side". Snell's law, which is the manifestation of the conservation of momentum in the cavity plane, dictates that modes in the left half-space (consisting of, e.g., air) propagating at an angle θ_l from the cavity normal will travel at an angle $\theta_r = \arcsin(n_L \sin(\theta_l)/n_R)$ in the right half-space (which we will assume is filled by the substrate), where n_L and n_R are the refractive indices of the left and right half-spaces. Assuming that the density of states per unit solid angle in the left half-space is isotropic so that $\zeta_l(\omega)$ is angularly independent, and using Snell's law, it is found that

$$\zeta_r(\omega, \theta_r) = \zeta_l(\omega) \frac{n_R^2 \cos(\theta_r)}{n_L^2 \cos(\theta_l)}, \tag{37}$$

where θ_l can be expressed in θ_r. The equation looks disturbing at first, since the right and the left half-space, for reasons of symmetry, should enter our calculation on equal footing. The symmetry is restored by looking at the zero-point fluctuation electric field. If these too, are supposed to be independent of the incidence angle in the left half-space, then the zero-point fluctuation electric field intensity for the modes incident from the right half-space must obey the relation[46]

$$E_{0r}^2(\theta_r) = \begin{cases} E_{0l}^2 \dfrac{n_L \cos(\theta_l)}{n_R \cos(\theta_r)} & \text{if } n_L \sin(\theta_l)/n_R \leq 1 \\ 0 & \text{if } n_L \sin(\theta_l)/n_R > 1 \end{cases} \tag{38}$$

to ensure that the zero-point fluctuations of the left and of the right half-spaces are in thermal equilibrium. Hence the product

$$\zeta_r E_{0r}^2 = n_{R\zeta} E_{0l}^2 / n_L \tag{39}$$

is independent of angle. In the introduction it was argued that no experiment can separate the influence of one of these factors from the other. Therefore, one of the factors can have any dependency on the angular coordinate as long as the other factor has the inverse dependency.

Equation 39 tells us that the decay rate for an atom is proportional to the refractive index of the medium in which it is sitting. This is actually an oversimplification due to the fact that we have only treated the macroscopic polarizability. A more careful analysis reveals that in addition, the so-called local field correction factor (the Clausius-Mossotti-Lorentz-Lorenz factor) also plays a role, and hence the decay rate is proportional to[46-48] $n \cdot 9n^4/(2n^2 + 1)^2 = 9n^5/(2n^2 + 1)^2$. For our needs, the local field correction factor is not important since it treats the influence of the local polarizability within atomic distances of the emitter. In semiconductor emitters, the excited atoms are part of a very large, high quality crystal lattice. Every semiconductor crystal thus defines its own local environment and therefore it is logical to incorporate the local field correction factor into the intrinsic lifetime of the semiconductor emitter. For example, in dye-molecule emitters, the dye solvent could modify even the local dielectric environment of the dye solution, and in this case the local field correction factor may play a role. In the following we will simply ignore the local field correction factor.

Another difference between the ideal mirror cavity and a dielectric cavity is that in the latter, bound planar waveguide modes may exist. If, for example, the spacer (cavity) has the highest index of refraction of all layers, including the substrate, bound modes propagating in the spacer plane will exist. The lowest-order mode in a dielectric planar waveguide has no cutoff, so independent of the spacer thickness, at least one mode will be present. In a real sample this mode will couple to radiation modes in unpumped, and therefore absorbing, regions of the planar cavity. In a lossless cavity the coupling will take place at the sample edges. In our model we have no edges *per definition,* and therefore these modes cannot couple to any of the modes in the orthogonal set we are expanding our field in. With the formalism presented here, such cavities cannot be treated in a self-consistent way. Since these modes do not couple to modes propagating in the cavity normal direction, and most planar microlasers are designed to be used as surface emitters, the structure is usually deliberately designed not to support bound modes. The simplest way to accomplish that is to make, e.g., the substrate have the highest index of refraction.

In such a structure, the waves propagating close to grazing incidence in the substrate will be evanescent in at least some of dielectric layers. Since the structure is periodic, we will see that evanescent wave resonant tunneling modes will form at angles slightly higher than the evanescent cut-off angle. These modes may have a substantial Q-value, and hence superficially "look" like bound modes. Since they do couple to radiation modes in the substrate, we will be able to treat them within a unified formalism.

An observation worth making relating to Snell's law is that only a very small portion of the emission cone from the active region can actually be seen from the "air side". If the dielectric cavity refractive index is taken to be 3.6, only the light-cone with a top half angle of 17 degrees (0.29 rad) will fill all the available 2π steradians on the "air side". The emission between 17 and 90 degrees will be totally internally reflected at the semiconductor-air interface and the emission between 90 and 180 degrees will (per definition) exit the device through the substrate. Hence only about 2% of all the solid angle can be monitored from the "air side". The unwanted radiation (remember that a typical dielectric Bragg mirror will have essentially zero reflectivity for incidence angles above 25 degrees) cannot be seen by any measurement apparatus on the "air side" without taking special measures to increase the output coupling. This was one of the oversights of the early planar microcavity experiments. Since only the cavity resonant mode could be seen by detectors on the "air side" of the cavity some overly optimistic spontaneous emission coupling ratios were reported.

To calculate the emission rate per unit solid angle we once more assume that the rate is proportional to the product of the mode density per solid angle and the zero-point fluctuation electric field intensity at the location of the emitters. Since a typical dielectric microcavity consists of perhaps 100 different layers, it is no longer possible to express the local field at the cavity center analytically as a function of the incident field, but we have to resort to numerical techniques. The standard technique is to use matrices to relate either the electric and magnetic fields on one side of a dielectric boundary to the same quantities on the other side,[49] or to relate the electric-field modes propagating in the right and in the left direction on one side of the interface to the corresponding modes on the other side.[46,50] Since all necessary field relations are linear, the two methods are equivalent, and it is largely a matter of convenience what matrices to use.

Our field expansion lends itself best to expressing the two propagating plane wave amplitudes at the left of some interface, or slab, as functions of the two propagating plane wave amplitudes on the right of the interface (or slab). In matrix notation we can write the equation for an S-polarized wave

$$\overline{E}_l^S = \overline{\overline{M}}^S \overline{E}_r^S \tag{40}$$

where $\overline{E}_l^S$ is the column vector

$$\overline{E}_l^S = \begin{pmatrix} E_{rl}^S \\ E_{ll}^S \end{pmatrix} \tag{41}$$

and $\overline{\overline{M}}^S$ is the S-wave transfer matrix. The indices rl and ll denote the plane waves on the left side traveling toward the right and the left respectively. The equations for the P-polarized wave follow trivially with the substitution $S \rightarrow P$ in the equations above. Only two transfer matrices are needed to calculate the dielectric planar microcavity. One is the matrix for a homogeneous slab of material which reads

$$\begin{pmatrix} \exp(jk_z l) & 0 \\ 0 & \exp(-jk_z l) \end{pmatrix}$$

(42)

for a slab of thickness l. The homogeneous slab matrix is independent of the polarization of the wave. The equation implicitly assumes a harmonic time dependence $\exp(j\omega t)$. The other is the matrix for a planar dielectric interface which read[46]

$$\frac{1}{2}\begin{pmatrix} 1+\dfrac{k_{rz}}{k_{lz}} & 1-\dfrac{k_{rz}}{k_{lz}} \\ 1-\dfrac{k_{rz}}{k_{lz}} & 1+\dfrac{k_{rz}}{k_{lz}} \end{pmatrix}, \frac{n_2}{2n_1}\begin{pmatrix} 1+\dfrac{n_L^2 k_{rz}}{n_R^2 k_{lz}} & 1-\dfrac{n_L^2 k_{rz}}{n_R^2 k_{lz}} \\ 1-\dfrac{n_L^2 k_{rz}}{n_R^2 k_{lz}} & 1+\dfrac{n_L^2 k_{rz}}{n_R^2 k_{lz}} \end{pmatrix}$$

(43)

for the S- and P-polarization respectively. Although not explicitly stated in these equations, all three matrices are angular and wavelength dependent since k_{lz} depends on both angle and wavelength. To calculate the electric field at the center of the dielectric cavity, we chain multiply all the matrices corresponding to the slabs and interfaces to the left of the emitters, and denote this product matrix $\overline{\overline{A}}^S$. Similarly, we denote the product matrix of all the slabs and interfaces to the right of the emitters, and call this matrix $\overline{\overline{B}}^S$. Finally we compute the product of these two matrices and call this matrix, representing the transfer matrix of the entire structure $\overline{\overline{AB}}^S$. Expressed in these matrices, the S-polarized electric field at the emitter position can be written[46]

$$E^S = \frac{B_{11}^S + B_{21}^S}{AB_{11}^S} E_{rl}^S + \left(B_{12}^S + B_{22}^S - \frac{AB_{12}^S(B_{11}^S + B_{21}^S)}{AB_{11}^S} \right) E_{rl}^S.$$

(44)

Exchanging the index P for S, the equation formally looks the same for the P-polarized wave. Using time reversal, it can be shown that the emission due to the zero-field fluctuations impinging from the left-hand side bulk will be emitted into the left side bulk. Hence, the emission rate per unit solid angle into the left side S-polarized modes for a y-dipole emitter with an infinitesimally narrow linewidth can be written

$$\gamma_{left}(\theta,\ \varphi) = \frac{3\Gamma_0 \cos^2(\varphi) n_L}{8\pi n_{QW}} \left| \frac{B_{11}^S(\theta_l) + B_{21}^S(\theta_l)}{AB_{11}^S(\theta_l)} \right|^2, \tag{45}$$

where n_{QW} is the refractive index of the quantum well. The equation for the emission into the right-hand side S-polarized modes is likewise

$$\gamma_{left}(\theta,\ \varphi) = \frac{3\Gamma_0 \cos^2(\varphi) n_R}{8\pi n_{QW}} \left| B_{12}^S(\theta_r) + B_{22}^S(\theta_r) - \frac{AB_{12}^S(\theta_r)\left(B_{11}^S(\theta_r)\right) + \left(B_{21}^S(\theta_r)\right)}{AB_{11}^S(\theta_r)} \right|^2. \tag{46}$$

The equation for the P-polarized wave follows trivially letting $S \to P$ and $\cos^2(\varphi) \to \sin^2(\varphi)\cos^2(\theta)$ in (45) and (46). If we left the whole structure simplify to a quantum-well slab embedded between two bulk half-spaces with the same refractive index as the quantum well itself, then all off-diagonal matrix coefficients in $\overline{\overline{A}}$, $\overline{\overline{B}}$, and $\overline{\overline{AB}}$ are identically zero, and the diagonal coefficients reduce to trivial phase-factors. Hence both (45) and (46) simplify to $3\Gamma_0 \cos^2(\varphi)/8\pi$. Compared to Equation 4, it is obvious that Equations 45 and 46 give the emission rate normalized to the case where the emitters are immersed in index matching bulk materials.

In Figure 15(a) we have plotted the emitted intensity as a function of angle for the S- and P-waves assuming that we have an xy-dipole in the cavity. The corresponding plot for the z-dipole emitter is not very interesting since the mirrors are transparent at high incidence angles where the coupling between the z-dipole and the P-polarized wave is strongest. The spacer (see Figure 1) is assumed to have zero thickness, and the quantum well(s) have been assumed to be much thinner than a wavelength, so the cavity is a half-a-wavelength-long cos-type cavity, Figure 11(a). We have made three plots in which the emission patterns into the "air side" and the "substrate side" are plotted separately. In (a) and (b) the Bragg mirror slab refractive indices are the same, but the number of mirror pairs differ, making the mirrors in (a) about 95% reflecting while the mirrors in (b) are 99.5% reflecting. In (c), the mirror reflectivity is roughly the same as in (b), but the low refractive index slabs have an index of only 2.0. The number of slab pairs on the "air side" of (a), (b), and (c), are always chosen to be smaller than the number on the "substrate side" to get equal right- and left-hand mirror reflectivity. In all three plots the emission linewidth has been assumed to be much narrower than the cavity resonance. It is seen that on the "air side", only the cavity resonant lobe can be seen, all the emission into the Bragg mirror open window is radiated into the substrate due to total internal reflection for angles greater than ≈ 16 degrees in the $n = 3.6$ material and ≈ 19.4 degrees in the $n = 3.0$ material. This should be compared to the high reflection region of the "substrate-side" mirror, which was computed in Section III to have a high reflectivity angular range of ≈ 28 degrees. The emission per unit

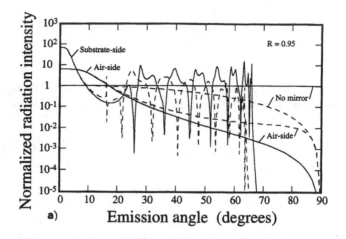

a)

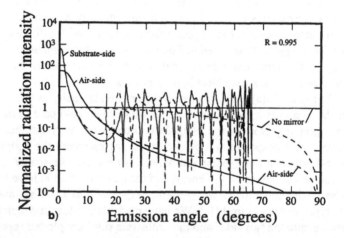

b)

FIGURE 15 Computed spontaneous emission per unit solid angle from a dielectric cavity with $n_L = 1$, $n_R = 3.6$, $n_1 = 3.6$, $n_2 = 3.0$, $n_S = 3.0$ and $n_{QW} = 3.6$. In (a) the plot for an 8/12 mirror period cavity (yielding $R_1 \approx R_2 \approx 0.95$) is shown, in (b) the corresponding plot for a cavity with 14/18 mirror pairs ($R_1 \approx R_2 \approx 0.995$) is shown. In (c), finally, the emission for a 5/6 mirror period cavity with $n_2 = 2.0$, yielding $R_1 \approx R_2 \approx 0.997$, is plotted. The solid lines are the emission with the S-polarization; the dashed lines represent the emission with P-polarization. The sudden jump in the P-wave emission intensity into the substrate at $\theta \approx 16$ degrees is the onset of total internal reflection.

solid angle in the cavity normal direction has an intensity about 12 times smaller on the "air side" than on the "substrate side". On the other hand, due to diffraction (Snell's law), the lobe is spread out over a solid angle about $n_R/n_L)^2 \approx 13$ times larger. The integrated emission within the cavity resonance lobe is therefore the same on the "air side" and the "substrate side". This should be the case since, seen from the emitter, the two mirrors have the same

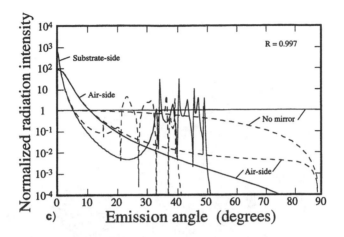

FIGURE 15 (continued)

reflectivities. The cavity resonance lobe FWHM can be calculated from (19). Taking the mirror penetration depth given by (36) into account, so that the effective cavity length in (a) is 3λ, $\Delta\theta_{FWHM} \approx 0.104$ radians or 5.9 degrees. The more exact transfer matrix calculation in Figure 15(a) yields the value 6.2 degrees. On the air side the angle should be about 3.6 times larger, or 22.3 degrees. The matrix calculation yields a value of 23 degrees, which tells us that (19) is quite accurate even for a nonideal dielectric mirror, provided we compensate for the mirror penetration depth.

In Figure 15(b) the emission pattern is plotted for the same parameters as in Figure 15(a), but with thicker and therefore more highly reflecting mirrors. It is seen that the main effect is to simultaneously increase the spontaneous emission rate per unit solid angle in the cavity normal direction and narrow the emission lobe. According to Equations 10 and 12, the emission intensity in the forward direction is enhanced by a factor $\approx 4/(1 - R)$ in a planar cavity, so the (b) cavity will emit ten times more power per solid angle in the forward direction than the (a) cavity. Equation 19 states that the emission lobe width is proportional to $\sqrt{1 - R}$. The (b) cavity will therefore have a $1/\sqrt{10}$ times narrower cavity resonance lobe than the (a) cavity. The cavity resonance lobe FWHM computed in (b) is 2.0 degrees. Hence, integrated over the full emission lobe solid angle, the cavity modeled in (b) will emit the same power into the cavity mode as the cavity modeled in (a). This confirms the assertion made in Section III, that what counts in a planar dielectric cavity is the refractive index difference of the cavity mirror slab pairs. The results also show that the dielectric cavity mimics the behavior of an ideal cavity within the Bragg mirror high reflectivity angular region given by (33). Outside this region the periodic mirror structure will only modulate the emission about its value in absence of mirrors. The modulation in the open window is "semiperiodic" with increasing emission angle, and the thicker the mirrors are, the shorter is the "period". This

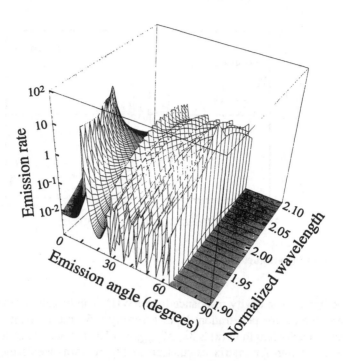

FIGURE 16   The spontaneous emission rate into the substrate per unit solid angle as a function of normalized emission wavelength λ_0/L and emission angle for an $R = 0.95$ dielectric cavity. The dipole moment is equally distributed in the x and y directions, and the emitter is located at $z_0 = L/2$. The dielectric cavity parameters are the same as in Figure 15(a).

is again a consequence of the relation between stored energy and dispersion. The more stored energy, the more rapid is the dispersion and the shorter is the "modulation period". What the mirrors do is merely to redistribute the zero-point fields within the open window. Therefore the total integrated emission is almost independent of the thickness (and hence reflectivity) of the Bragg mirrors. We will delve deeper into this issue in the next section.

For very high refractive index ratios, e.g., 3.5:1.0, the open windows of the S- and P-waves no longer overlap.[21] The P-wave open window is always centered on the Brewster angle and lies closer to normal incidence than the S-wave open window. Still, about the same amount of spontaneous emission is radiated into the two windows, since the P-window is wider than the S-window. This compensates for the fact that as a function of θ, less solid angle is available near the cavity normal than at higher angles.

In reality the situation is slightly more complicated than outlined above. The reason is the need to take realistic emission linewidths into account. In Figure 16, the emission per unit solid angle and unit wavelength into the substrate is plotted as a function of wavelength and angle. The emission line shape has been assumed to be Gaussian, but the results are rather insensitive

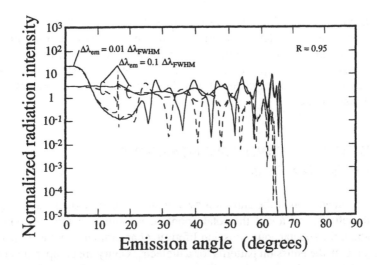

FIGURE 17 The wavelength integrated spontaneous emission rate per unit solid angle into the substrate as a function of angle. The cavity has the same parameters as in Figure 15(a). The gain linewidths have been assumed to be Gaussian with FWHM of $0.01\lambda_{em}$, and $0.1\lambda_{em}$.

to the exact line shape. The emission width $\Delta\lambda_{FWHM}$ in this plot has been assumed to be and $0.1\lambda_{em}$. It is instructive to compare this figure with Figure 9. The two cavities have the same reflectivities and the same emission linewidths. There are two main differences between the plots. First, in spite of having the Bragg mirrors mounted face to face, the dielectric cavity is effectively three wavelengths long, whereas the ideal cavity modeled in Figure 9 is only half a wavelength long. Consequently the dielectric cavity resonance lobe is narrower than the corresponding lobe of the ideal cavity. Second, it can be seen that only a fraction of the total spontaneous emission of the dielectric cavity is emitted into the cavity resonant mode. In addition to the emission along the "resonance ridge", there is substantial emission into the open window of the dielectric cavity. This emission is absent in the ideal cavity, and accounts for most of the discrepancy between the two models.

To compute the emission rate per unit solid angle for a finite linewidth emitter, the data in Figure 16 must be integrated over all wavelengths. This will model the far-field pattern that would be measured by a broadband detector such as a photodiode. The plot then changes in that the emission lobe appears wider, and all the fine details of the emission into the open window are averaged out. The wavelength averaged plot of the data in Figure 16 is shown in Figure 17. The radiation intensity in the forward direction decreases, for a large portion of the emission line is now at wavelengths outside the cavity resonance. On the other hand, the cavity resonances at slightly shorter wavelengths along the "resonance ridge" may couple strongly to the wide linewidth emitter. As the emission line becomes sufficiently broad, almost all the structure in the wavelength integrated far-field emission pattern will be washed away. The most important lesson to be learned from this plot is that unless angular and wavelength filtering is employed when measuring the emission pattern, the emission into the cavity mode may be severely overestimated. The emission in what appears to be a single cavity mode (lobe) is actually composed of emission of different wavelengths (and transverse modes) within the emission line.

V.   DIELECTRIC CAVITY SPONTANEOUS EMISSION LIFETIME

Integration of the emission over all solid angles and all wavelengths gives us information about the total spontaneous emission rate, or, equivalently, about the spontaneous emission lifetime of an emitter in a dielectric cavity. In general, the emission patterns for a dielectric cavity are computed numerically, and the ensuing emission patterns are so complex that the integration over angle and wavelength must also be computed numerically. In this section we will do so. We will stick to our definition of the cavity mode. Only the radiation within the FWHM of the emission lobe in the cavity normal direction belongs to the mode, and if this lobe has twofold polarization degeneracy, only the radiation into one of the polarizations should be included. We will assume that the quantum well can be modeled by an xy-dipole emitter. Since the result for this case is simply the superposition of the results of one x-dipole and one y-dipole, the spontaneous emission lifetime will be the same for all three cases. The spontaneous emission coupling ratio will be half as large for the xy-dipole than for the x- or y-dipole. The normalized lifetime we are going to compute and plot is the ratio between the lifetime of the emitter sitting in the cavity and the lifetime of the emitter (a thin, $\ll \lambda$, quantum well with a refractive index $n_{QW} = 3.6$), sitting between two semi-infinite bulk slabs also with the refractive index 3.6.

In Figure 18 we have plotted the normalized spontaneous emission lifetime and the spontaneous emission coupling ratio β for a dielectric cavity with the same parameters as that in Figure 15(a), as a function of the number of Bragg mirror pairs. The number of mirror pairs for the "air-side" mirror is plotted at the bottom, the number of "substrate-side" mirror pairs has in each case been chosen to be four pairs greater to give equal mirror reflectivities (cf. Figure 12). The mirror reflectivity for the 5/9 mirror pair cavity is about 0.82, while the reflectivity for the 15/19 pair cavity is approximately 0.995. It is seen that the emission lifetime and β both are insensitive to the Bragg mirror reflectivity. This we predicted in Section IV, where an increased mirror reflectivity merely redistributed the emission. The cavity mode narrows with increasing mirror reflectivity, but at the same time the emission intensity increases. The results are not exactly constant because of the finite numerical accuracy both in estimating the emission lobe FWHM, and the finite accuracy in the numerical integration routines (the relative accuracy goal in the integration routine was

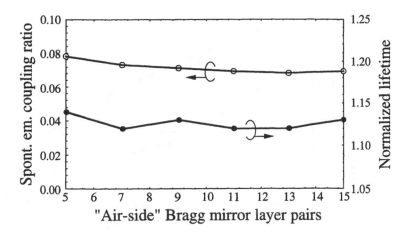

FIGURE 18   The spontaneous emission ratio β and the normalized lifetime as a function of the number of Bragg mirror slab pairs (and hence as a function of reflectivity). Neither quantity depends on the number of slab pairs.

set to 1% to ensure quick convergence). In the calculation of β we have included the radiation emitted from both mirrors. The effective β using only the radiation from one of the mirrors is half of that we compute, but increasing the number of "substrate-side" mirrors so that $R_1 \ll R_2$, the emission becomes effectively single-sided without changing the cavity Q significantly. In practice, of course, a single sided output is generally desired so the cavities should be (and are) designed to have unequal reflectivity.

In Figure 19 we have plotted β and the normalized spontaneous emission lifetime for a structure similar to that in Figure 15(a), but as a function of the refractive index of the n_2 slabs of the mirrors. The number of Bragg mirror pairs has been adjusted so that the reflectivities are approximately equal to 0.99 at each point. Hence, the cavity-resonant-mode linewidth remains constant. The number of Bragg mirror pairs go from 13/17 when $n_2 = 3.0$ to 2/3 when $n_2 = 1.0$. The emission linewidth has been assumed to be much smaller than the cavity linewidth. Should this not be the case, then, just as in the case of the ideal planar cavity, β drops, but the emission lifetime remains constant.[32] It is seen that the coupling constant increases from about 0.07 to about 0.25 when n_2 goes from 3.0 to 1.0. As expected, the β for an xy-dipole (a) is slightly higher than for an xyz-dipole (b). The difference is small, however, because the P-wave is blocked effectively by the Bragg mirrors at rather small angles from normal incidence, where the coupling to the z-dipole component is small. Therefore this dipole component contributes very little to the total emission, and consequently the normalized emission lifetime becomes longer for a given n_2 for the xyz-dipole than for the xy-dipole. The increase in spontaneous emission lifetime can be understood from two different points of view. One is that the emission lifetime for a very thin sample should be proportional to the refractive index of the surrounding bulk material. This is the essence of Equation 39. In

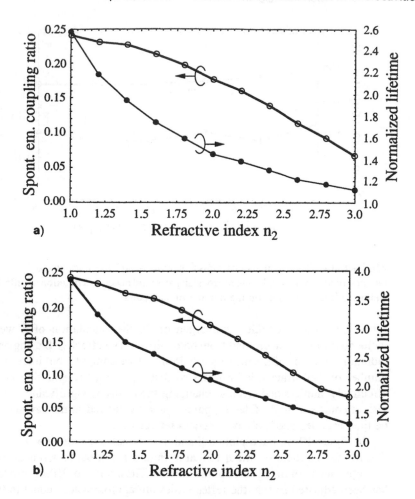

a)

b)

FIGURE 19 The spontaneous emission ratio β and the normalized lifetime as a function of the refractive index n_2. In (a) the dipole moment has been assumed to be distributed isotropically in the xy-plane, while in (b) the dipole moment is distributed isotropically over xyz-space.

our case, the two Bragg mirrors surrounding the quantum well have effective refractive indices somewhere between 3.3 and 1.0. The reference lifetime is computed for a bulk refractive index of 3.6. Therefore, it is reasonable to expect the spontaneous emission lifetime to increase by factors somewhere between $3.6/3.3 \approx 1.1$ and $3.6/1.0 = 3.6$. This is roughly what we get in Figure 19. The other point of view is to see the Bragg mirrors as a "shield", protecting the quantum well from the zero-point fluctuations impinging from the substrate. The low refractive index slabs in the Bragg mirrors reflect all waves above some critical angle so that the quantum well only "sees" the weak evanescent tail of these modes. Therefore essentially no spontaneous emission is emitted at angles close to $\theta = 90$ degrees.

The result in Figure 19 is quite generally true, independent of the number of Bragg mirror pairs as shown in Figure 18. The lifetime is determined mainly by the Bragg mirror refractive indices, not by the mirror reflectivity. For a mirror with an index ratio of 3.6:3.0, the calculated β of a y-dipole emitter is twice that calculated in Figure 19 (for an xy-dipole), or 0.136. In (34) above, using only the ratio of solid angle controllable by a Bragg mirror, we predicted a β of 0.11 in this case. This is a surprisingly accurate estimate of the actual β. Both the lifetime and the spontaneous emission coupling ratio also depend on the quantum well position and the cavity spacer length. These results above should be compared to the results in Reference 46. In that paper, the cavity mode was defined more generously to include all the emission in the cavity resonance lobe, not only the emission within the lobe FWHM. The radiation from both polarizations was also included in the "coupling efficiency". These two effects each contribute a factor of 2 in the estimation of β, and therefore the result for β in this chapter is a factor of 4 smaller than the "coupling efficiency" calculated in Reference 46. In Figure 7(a) and (b) of that reference one finds that initially the spontaneous emission lifetime increases with a decreasing n_2, but starts to decrease again for $n_2 < 1.5$. We believe this result is incorrect and that it is due to an erroneous numerical integration of the cavity resonant mode. In that calculation the number of slab pairs was kept constant, effectively making the cavity Q-factor astronomical when $n_2 \rightarrow 1$. This makes the cavity resonant lobe very narrow and difficult to integrate over. We believe that the more recent results in this chapter are the correct ones.

The bottom line of these calculations is that it is quite difficult to modify the spontaneous emission lifetime by any appreciable amount by a planar dielectric cavity. The alteration in lifetime will be of the order of unity, and most of the effect can be attributed to the change of local dielectric environment, not to any cavity effect. This agrees with the conclusions in Chapter by Brorson, where it is shown that not only longitudinal, but also transverse confinement is needed to get large enhancements of the spontaneous emission rate.

It is also difficult to realize spontaneous emission coupling ratios greater than a few percent with planar dielectric cavities, at least with epitaxial semiconductor devices. In practice it is difficult to find emitters with sufficiently narrow linewidths even to obtain these modest ratios. A typical device, even while cooled to 4 K, has a coupling ratio one order of magnitude smaller[32] than the optimum calculated in Figure 19.

VI. DIELECTRIC CAVITY STIMULATED EMISSION

In regular lasers with three-dimensional confinement of the optical mode, the spatial cavity modes are essentially independent of both the pumping and the cavity mirror reflectivity; i.e., the cavity near- and far-field mode patterns are identical below and above threshold and also for a range of mirror reflectivities (say, for all $R \geq 0.9$). On the other hand, the laser temporal modes do change greatly both with pumping and with mirror reflectivity. Below threshold, the coherence length of the cavity mode(s) is independent of the pumping but is proportional to $(1 - R)^{-1}$. Consequently the mode linewidth $\propto 1 - R$. Above threshold, the lasing mode coherence length is essentially proportional to the pumping, and hence the laser linewidth is inversely proportional to the pumping.

This is not the case for the planar cavity. In the sections above we have assumed that the planar cavity extended to infinity in the xy-plane so that there are no boundaries in this plane. Nonetheless, Equation 19 shows that the spontaneous emission is emitted into a relatively well-defined lobe which can be associated with a finite area. The explanation for this somewhat unexpected behavior is mirror dissipation. Emitters separated by a sufficient distance in the plane cannot "communicate" with each other using radiation propagating close to the cavity normal because after a finite number of reflections between the mirrors (and therefore finite propagation distance), the photons have left the cavity. However, the spatial and temporal modes of a planar cavity are intimately connected as manifested by Equations 18 and 19. Above threshold, these relations imply that as the spectral linewidth of the planar cavity lasing mode gets narrower with increasing pumping, so will the emission lobe. While there (in principle) exists no limit to the minimum spectral width of the emission line, there does exist a limit for the emission lobe width (or divergence). Every sample has a finite lateral size, and this sets the fundamental limit to the divergence. In reality, a typical sample is only pumped over a very small area, typically a spot about 20 to 50 μm across. This sets the practical limit to the minimum divergence.

The expression for the emission lobe above the threshold can be written

$$\Delta\theta_{FWHM,a} \approx \sqrt{\frac{2\lambda_{r0}n_{cav}^2 n_{sp}}{\pi c_0 p \tau_p}}, \tag{47}$$

where $n_{sp} \equiv N_{upper}/(N_{upper} - N_{lower})$ is the laser population inversion factor, and N_{upper} (N_{lower}) is the population of the upper (lower) lasing level. The mean cavity photon number is denoted p. We sill not go through the derivation of the expression (47) here; the interested reader is referred to Reference 32. Using the relation $P = h\nu p/\tau_p$, where P is the emitted power from the laser, Equation 47 can be re-expressed as

$$\Delta\theta_{FWHM,a} \approx \Delta\theta_{FWHM} \sqrt{\frac{\pi h \nu c_0 n_{sp}}{2\lambda_{r0}n_c^2 P}} \tag{48}$$

where $\Delta\theta_{FWHM}$ is the "cold-cavity" lobe angle given by (19). The two expressions above are only valid as long as the effective mode radius above threshold, given by

$$\Delta r_a \approx \sqrt{\frac{\lambda_{r0} c_0 p \tau_{sp}}{\pi n_{cav} n_{sp}}} \tag{49}$$

is smaller than the pump spot size. For higher pump powers the lobe divergence is given only by the pump radius and the wavelength. In this regime, by pumping with an oblong spot, it should be possible to get far-field emission patterns that do not have rotational symmetry.

The narrowing of the emission lobe has so far only been predicted theoretically; no experimental observations have been reported. While the phenomenon in principle should be present, various cavity imperfections such as cavity taper and gain filamentation may render the effect difficult to observe.

VII. CONCLUSIONS AND OUTLOOKS

In this chapter we have tried to give a relatively complete picture of the possibilities of spontaneous emission engineering in planar dielectric cavities. We started out by considering a highly idealized model, where the planar mirrors were assumed to be "ideal" in the sense that the mirror reflectivity was independent of both wavelength and reflection angle. It was shown that even in this highly idealized case, the spontaneous emission could never be enhanced by great amounts. On the other hand, spontaneous emission inhibition could be realized quite effectively in very short cavities. The effect of wavelength size cavities was essentially to coax the emission into a single mode, provided the dipole moment of the emitters has a specific orientation. For emitters with the dipole moment distributed isotropically in the mirror plane, the emission is shared between two polarization degenerate modes. It was also shown that as soon the emitter spectral linewidth exceeds the cold-cavity linewidth, the spontaneous emission coupling ratio starts to drop.

In the subsequent section it was shown that dielectric Bragg mirrors actually have quite different characteristics than an ideal mirror. Specifically, Bragg mirrors are only highly reflecting within a limited wavelength band around the Bragg wavelength, and, more important, within a limited angular range close to the mirror normal. For most angles and wavelengths the mirrors are more or less transparent. The greater the refractive index ratio between the slab pairs in the Bragg mirrors, the greater the high reflection bands are. In addition to these effects, the mirrors have a spatially distributed reflectivity, meaning that the reflected wave penetrates a finite distance into the Bragg mirror. The consequence is that the "cavity" will effectively be a bit longer than simply the distance between the Bragg mirrors (the spacer length). The shortest "cavity" realizable in, e.g., lattice-matched AlGaAs on GaAs is about three wavelengths long.

The effect of the nonideal characteristics of the Bragg mirrors was found to be that, in most cases, the lion's share of the spontaneous emission was emitted into leaky modes and not into the cavity resonant mode. For a cavity made from lattice-matched AlGaAs on GaAs the best β achievable was shown to be about 0.07 for a cavity with the emitter dipole moment oriented isotropically in the mirror plane. For an emitter with a unidirectional dipole moment in the cavity plane (e.g., a y-dipole) the coupling ratio is twice as high. Again, the ratio will drop if the emission spectral linewidth is wider than the cold-cavity resonant mode linewidth. While the β's calculated above are about one order of magnitude smaller than the optimum value of unity, they are several orders of magnitude higher than the coupling ratio of conventional semiconductor lasers. Therefore, even the planar cavity devices hold some promise for device applications.

It was also shown that the spontaneous emission lifetime was increased rather than decreased in a dielectric cavity, even when the cavity was optimized for spontaneous emission enhancement. This is in contrast to the ideal planar cavity case and the result is rather disappointing. A rapid decay would have meant a more efficient and potentially faster device.

The findings in this chapter, in combination with the material in Chapters 5 and 7, clearly show that two- and three-dimensional confinement pronounces the cavity QED effects more than one-dimensional confinement does. It is also clear from these chapters that some characteristics, e.g., the decrease of β when the emission linewidth exceeds the cold-cavity linewidth, are quite general. The only exception are the photonic bandgap cavities first proposed by Eli Yablonovitch, but a discussion of these cavities lies outside the scope of this paper. The interested reader is referred to Reference 51 for a review of their detailed properties and present status. The photonic bandgap cavities are quite insensitive to the emitter linewidth, since in these devices a truly "three-dimensional Bragg mirror" is realized and not simply a planar one. The limitations of one-dimensional confinement must be seen as the trade-off for the relatively simple fabrication technology needed to make these structures.

We believe that the future for microcavity devices lies very much in the hands of the semiconductor microfabrication engineers. The microcavity laser field is driving the state of the art in semiconductor fabrication, and at present, experimental advances are restrained by the limitations in fabrication technology. This is good rather than bad news, because it means that for the near future, the progress in the microcavity QED field is not hindered by any fundamental limits, only practical ones. With the present rate of progress in microfabrication this means that we may expect quite a bit of development and experimental demonstration of new, interesting, and potentially useful devices in the not too distant future.

ACKNOWLEDGMENTS

The material in this chapter has been collected over several years during which time the authors have worked in different constellations with many different co-workers. Most of these have helped shape our present understanding of how spontaneous emission can be modified in the presence of cavities. We are grateful to Drs. Susumu Machida, Henrich Heitmann, Franklin Matinaga, Ricardo Horowicz, and Kazuhiro Igeta, who, at various periods, worked at NTT Basic Research Labs in Tokyo, Japan. We also thank Dr. Anders Karlsson, Prof. Lars Thylén, and Prof. Olle Nilsson at the Royal Institute of Technology, Stockholm, Sweden. Lastly, we should like to acknowledge discussions with some of our international colleges in the field, notably with Dr. Hiroyuki Yokoyama, Prof. Stuart Brorson, Prof. Francesco DeMartini, Dr. Jonathan Dowling, Prof. Eli Yablonovitch, Prof. Seng-Tiong Ho, Dr. Richart Slusher, and Dr. Xiaomei Wang.

REFERENCES

1. Hall, R. N., Fenner, G. E., Kingsley, J. D., Soltys, T. J., and Carlson, R. O., Coherent light emission from GaAs junctions, *Phys. Rev. Lett.*, **9**, 366, 1962.
2. Nathan, M. I., Dumke, W. P., Burns, G., Dill, Jr., F. H., and Lasher, G. J., Stimulated emission of radiation from GaAs p-n junction, *Appl. Phys. Lett.*, **1**, 62, 1962.
3. Holonyak, N. and Bevacqua, S. F., Coherent (visible) light emission from $Ga(As_{1-x}P_x)$ junctions, *Appl. Phys. Lett.*, **1**, 82, 1962.
4. Quist, T. M., Rediker, R. H., Keyes, R. J., Krag, W. E., Lax, B., McWorther, A. L., and Ziegler, H. J., Semiconductor maser of GaAs, *Appl. Phys. Lett.*, **1**, 91, 1962.
5. Jewell, J. L., Harbison, J. P., Scherer, A., Lee, Y. H., and Florez, L. T., Vertical-cavity surface-emitting lasers: Design, grown, fabrication, characterization, *IEEE J. Quantum Electron.*, **27**, 1332, 1991.
6. Jewell, J. L., Harbison, J. P., and Scherer, A., Microlasers, *Scientific American*, **265**, 56, November, 1991.
7. Yamamoto, Y., Machida, S., and Björk, G., Micro-cavity semiconductor lasers with controlled spontaneous emission, *Opt. Quantum Electron.*, **24**, S215, 1992.
8. Yokoyama, H., Nishi, K., Anan, T., Nambu, Y., Brorson, S. D., Ippen, E. P., and Suzuki, M., Controlling spontaneous emission and thresholdless laser oscillation with optical microcavities, *Opt. Quantum Electron.*, **24**, S245, 1992.
9. Jewell, J. L., Olbright, G. R., Bryan, R. P., and Scherer, A., Surface-emitting lasers break the resistance barrier, *Photonics Spectra*, **26**, 126, November, 1992.
10. Yamamoto, Y. and Slusher, R. E., Optical processes in microcavities, *Phys. Today*, **46**, 66, June, 1993.
11. Purcell, E. M., Spontaneous emission probabilities at radio frequencies, *Phys. Rev.*, **69**, 681, 1946.
12. Kastler, A., Atomes à l'intérieur d'un interféromètre Perot-Fabry, *Appl. Opt.*, **1**, 17, 1962.
13. Barton, G., Quantum electrodynamics of spinless particles between conducting plates, *Proc. R. Soc. London Ser. A*, **320**, 251, 1970.
14. Stehle, P., Atomic radiation in a cavity, *Phys. Rev. A*, **2**, 102, 1970.

15. Milonni, P. W. and Knight, P. L., Spontaneous emission between mirrors, *Opt. Commun.*, **9**, 119, 1973.

16. Philpott, M. R., Fluorescence from molecules between mirrors, *Chem. Phys. Lett.*, **19**, 435, 1973.

17. Agarwal, G. S., Finite boundary effects in quantum electrodynamics, *in: Quantum Electrodynamics and Quantum Optics*, Ed. Barut, A. O., Plenum, New York, 1984.

18. Cook, R. J. and Milonni, P. W., Quantum theory of an atom near partially reflecting walls, *Phys. Rev. A*, **35**, 5081, 1987.

19. Haroche, S., Spontaneous emission in confined space, *in: Lecture Notes in Physics, Vol. 282 — Fundamentals of Quantum Optics II*, Ed. Ehlotzky, F., Springer-Verlag, Berlin, 1987.

20. Brorson, S. D., Yokoyama, H., and Ippen, E., Spontaneous emission rate alteration in optical waveguide structures, *IEEE J. Quantum Electron.*, **26**, 1492, 1990.

21. Yamamoto, Y., Machida, S., Igeta, K., and Björk, G., Controlled spontaneous emission in microcavity semiconductor lasers, *in: Coherence, Amplification, and Quantum Effects in Semiconductor Lasers*, Ed. Yamamoto, Y., John Wiley & Sons, New York, 1991.

22. Dowling, J. P., Scully, M. O., and DeMartini, F., Radiation pattern of a classical dipole in a cavity, *Opt. Commun.*, **82**, 415, 1991.

23. Dowling, J. P., Spontaneous emission in cavities: How much more classical can you get? *in: Foundations of Physics, Vol. 23*, Plenum, New York, 1993, 895.

24. Milonni, P., *The Quantum Vacuum*, Academic Press, New York, 1994.

25. Hinds, E. A., Cavity quantum electrodynamics, *in: Advances in Atomic Molecular and Optical Physics, Volume 28*, Eds. Bates, D. and Bederson, B., Academic Press, New York, 1991, 237.

26. Morin, S. E., Wu, Q., and Mossberg, T. W., Cavity quantum electrodynamics at optical frequencies, *Opt. Photon. News*, **3**, 8, August, 1992.

27. Haroche, S., Cavity quantum electrodynamics, *in: Fundamental Systems in Quantum Optics*, Eds. Dalibard, J., Raimond, J. M., and Zinn-Justin, J., Elsevier Science Publishers B.V., Amsterdam, 1992.

28. Haroche, S. and Raimond, J. M., Cavity quantum electrodynamics, *Scientific American*, **268**, 26, April, 1993.

29. Meystre, P. and Sargent III, M., *Elements of Quantum Optics*, Springer-Verlag, Berlin, 1991.

30. DeMartini, F., Marrocco, M., and Murra, D., Transverse correlations in the active microscopic cavity, *Phys. Rev. Lett.*, **65**, 1853, 1990.

31. Ujihara, K., Spontaneous emission and the concept of effective area in a very short cavity with plane-parallel dielectric mirrors, *Jpn. J. Appl. Phys.*, **30**, L901, 1991.

32. Björk, G., Heitmann, H., and Yamamoto, Y., Spontaneous emission coupling factor and mode characteristics of planar dielectric microcavity lasers, *Phys. Rev. A*, **47**, 4451, 1993.

33. Yamamoto, Y., Machida, S., Horikoshi, Y., Igeta, K., and Björk, G., Enhanced and inhibited spontaneous emission of free excitons in GaAs quantum wells in a microcavity, *Opt. Commun.*, **80**, 337, 1991.

34. Ochi, N., Shiotani, T., Yamanishi, M., Honda, Y., and Suemune, I., Controllable enhancement of excitonic spontaneous emission by quantum confined stark effect in GaAs quantum wells embedded in quantum microcavities, *Appl. Phys. Lett.*, **58**, 2735, (1991).

35. Baba, T., Hamano, T., Koyama, F., and Iga, K., Spontaneous emission factor of a microcavity DBR surface emitting laser, *IEEE J. Quantum Electron.*, **27**, 1347, 1991.

36. Yokoyama, H. and Brorson, S. D., Rate equation analysis of microcavity lasers, *J. Appl. Phys.*, **66**, 4801, 1989.

37. Björk, G. and Yamamoto, Y., Analysis of semiconductor microcavity lasers using rate equations, *IEEE J. Quantum Electron.*, **27**, 2386, 1991.

38. Houdré, R., Stanley, R. P., Oesterle, U., Ilegems, M., and Weisbuch, C., Room temperature exciton-photon Rabi-splitting in a semiconductor microcavity, *J. Phys.*, **3**, 51, 1993, supplement JP II, no. 10.

39. Kogelnik, H. and Shank, C. V., Coupled wave theory of distributed feedback lasers, *J. Appl. Phys.*, **43**, 2327, 1972.

40. Haus, H. and Shank, C. V., Antisymmetric taper of distributed feedback lasers, *IEEE J. Quantum Electron.*, **QE-12**, 532, 1976.

41. Yablonovitch, E., Inhibited spontaneous emission in solid-state physics and electronics, *Phys. Rev. Lett.*, **58**, 2059, 1987.

42. Yariv, A., *Quantum Electronics, 3rd edition,* John Wiley & Sons, New York, 1989, chapter 22.5.

43. Ho, S. T., McCall, S. L., Slusher, R. E., Pfeiffer, L. N., West, K. W., Levi, A. F. J., Blonder, G. E., and Jewell, J. L., High index contrast mirrors for optical microcavities, *Appl. Phys. Lett.*, **57**, 1387, 1990.

44. Babic, D. I. and Corzine, S. W., Analytic expressions for the reflecting delay, penetration depth, and absorbance of quarter-wave dielectric mirrors, *IEEE J. Quantum Electron.*, **28**, 514, 1992.

45. Babic, D. I., Chung, Y., Dagli, N., and Bowers, J. E., Modal reflection of quarter-wave mirrors in vertical-cavity lasers, *IEEE J. Quantum Electron.*, **29**, 1950, 1993.

46. Björk, G., Yamamoto, Y., Machida, S., and Igeta, K., Modification of spontaneous emission rate in planar dielectric microcavity structures, *Phys. Rev. A*, **44**, 669, 1991.

47. Von Hippel, A. R., *Dielectrics and Waves,* John Wiley & Sons, New York, 1954.

48. Glauber, R. J. and Lewenstein, M., Quantum optics of dielectric media, *Phys. Rev. A*, 43, 467, 1991.

49. Born, M. and Wolf, E., *Principles of Optics,* Pergamon, Oxford, 1980, p. 66.

50. Björk, G. and Nilsson, O., A new exact and efficient numerical matrix theory of complicated laser structures: Properties of asymmetric phase-shifted DFB lasers, *IEEE J. Lightwave Technol.*, LT-5, 140, 1987.

51. Yablonovitch, E., Special issue about photonic bandgaps, *J. Opt. Soc. Am. A*, 10, 1993.

7 Spontaneous Emission in Microcavity Surface Emitting Lasers

Toshihiko Baba and Kenichi Iga

TABLE OF CONTENTS

I. INTRODUCTION

A vertical cavity surface emitting laser (VCSEL or SEL)[1,2] is a novel type of semiconductor laser that has a structure and lasing characteristics fundamentally different from those of conventional edge emitting lasers with horizontal stripe cavity. After the first achievement of room temperature continuous wave operation in 1988,[3] SELs have been attracting much attention and a lot of devices have been demonstrated using various III–V compounds.[4-8]

0-8493-3786-0/95/$0.00+$.50

The laser cavity of SELs is constructed normal to the substrate plane by stacking multilayer films including active layer and distributed Bragg reflectors (DBRs) made of dielectric materials[9] and/or semiconductors.[10] The cavity length ranges from several microns to submicrons and much shorter than that of edge emitting lasers. The optical gain length is also very short; it corresponds to the active layer thickness and is around 10 nm for a quantum-well active layer in extreme cases. In spite of this condition, lasing oscillation is possible owing to the reflectors exhibiting ultrahigh reflectivity over 99%. Thus, apart from the size limit of edge emitting lasers, SELs can be very compact and their threshold current can possibly be reduced to μA order.[11] In such short cavities, the longitudinal mode separation is accordingly very wide ($\geq$ 100 nm) and stable single longitudinal mode operation is easily obtained. When the lateral cavity size is less than 10 μm, fundamental transverse mode operation is also obtained with circular output beam compatible to single mode fibers.[12,13] In addition, many SEL chips can be monolithically integrated to form a two-dimensional laser array,[14-16] which is an essential component required for parallel optical signal processing, optical interconnects, optical computing, and so on.

Meanwhile, the concept of spontaneous emission control in microcavities[17,18] is becoming of great interest for presenting ultimate and novel performance of SELs. In 1982, this concept was discussed in the range of optical frequency for the first time, based on a closed cavity model.[19] The so-called *thresholdless lasing operation* in an ideal cavity was predicted from a rate equation analysis.[20] In 1987, the thresholdless operation was first observed by employing a dielectric planar microcavity with emission from a dye solution.[21] Afterward, this topic has been discussed theoretically and experimentally in the field of semiconductor lasers and especially microcavity SELs. Some dynamic properties of thresholdless lasers were analyzed and the disappearance of relaxation oscillation was predicted.[22] Noise characteristics in thresholdless lasers were calculated.[23,24] The spontaneous emission control was quantitatively estimated using planar and three-dimensional (3-D) cavity models.[25,26] Electronic confinement in low-dimensional quantum structures was discussed simultaneously with the spontaneous emission control.[27] Experimentally, the enhancement and inhibition of spontaneous emission were observed by tuning/detuning cavity resonance to the emission from the quantum-well active region, using planar cavities at low temperature,[28] at room temperature,[29] and using a post-shape 3-D cavity at low temperature.[30] A high speed on/off switching of spontaneous emission faster than the lifetime limit was demonstrated by modulating cavity resonance.[31]

In this chapter, the authors discuss some spontaneous emission characteristics in microcavity SELs including the topics above. The main issue is whether actual SELs can be thresholdless lasers or not. The next section discusses how the spontaneous emission is related to optical modes in microcavities. Then, the general expression for spontaneous emission in a 3-D SEL model is obtained by solving the classical wave equation. Modified distribution of modes in the SEL is calculated and displayed over the optical wave vector space (**k**-space). The spontaneous emission factor, which is essential to the estimation of spontaneous emission control, is calculated. The spontaneous emission control is discussed with electron quantum confinements. The change of spontaneous emission lifetime is estimated. With these results, the possibility of thresholdless operation is investigated by solving the rate equations.

II. SPONTANEOUS EMISSION IN A MICROCAVITY

In this section, the relation between cavity modes (allowed light) and the emission spectrum from material of active region (emitted light) is discussed in order to examine the behavior of spontaneous emission. This is considered in the **k**-space, as illustrated in Figure 1. Here, the cavity modes are schematically drawn by dots and the emission spectrum in free space is a 1/8 spherical shell. The shell radius corresponds to the emission angular frequency ω divided by the velocity of light c in the active region. The angular frequency ω is related to $k = |\mathbf{k}|$ as $\omega/c = k = \sqrt{k_x^2 + k_y^2 + k_z^2} = 2\pi n_a/\lambda$, where n_a is the refractive index of active region and λ is the vacuum wavelength. In the cavity, the emission couples to the modes inside the spectrum shell. For an open cavity much larger than wavelength, such as an edge emitting laser, the mode field in the cavity well couples to outer space, and many modes distribute almost uniformly inside the spectrum. On the other hand, if the cavity is about as large as the wavelength and almost closed by boundaries, the modes in the spectrum are limited and their distribution is strongly influenced by the boundary conditions. If only a single mode is allowed in such a microcavity, emission does not occur except into this mode. This is the ultimate condition of spontaneous emission control and in the absence of nonradiative decay corresponds to thresholdless operation.

Thus, the coupling efficiency of spontaneous emission to the lasing mode is important for the evaluation of the spontaneous emission control. The efficiency is called the *spontaneous emission factor* or sometimes *spontaneous emission coupling efficiency,* and denoted as m_1/m,[32] β,[33] C,[34] α,[35] etc. In the following, the authors use C for this factor to avoid confusion with the α parameter in noise theory[36] and the propagation constant β of guided waves.[37] It is defined as

$$C \equiv \frac{(\text{Radiation energy coupled to a lasing mode})}{(\text{Total radiation energy})} \qquad (1)$$

From this definition, one can see that C must be less than or equal to 1. It is easily expected from the above discussion that a larger C factor will be obtained by size reduction of the closed cavity. From rate equation analysis, the thresholdless operation is derived for $C = 1$.[20,22,28]

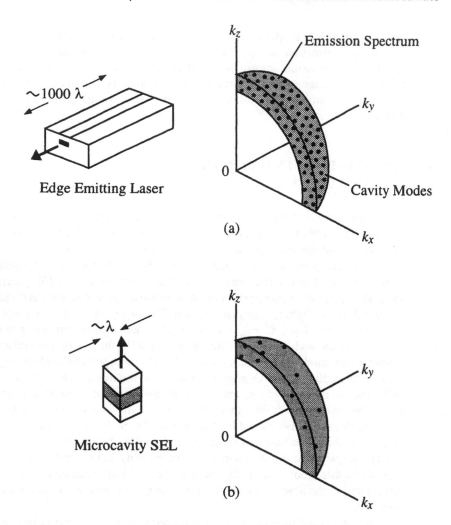

FIGURE 1   Distribution of cavity modes over wave vector space (a) in a conventional large cavity edge emitting laser, and (b) in a microcavity SEL.

As the closed cavity, one may imagine a cavity whose surfaces, except an aperture for light output, are surrounded by the combination of dielectric film and metals. However, the mirror loss at the metal is large since the reflectivity of actual metal is no higher than 98% in the optical frequency (and typically less than 80% after some thermal process). The photonic band structure proposed in 1987[38,39] is an approach that aims at the realization of an ideal closed cavity. This is a kind of 3-D periodic structure of dielectric media or semiconductor materials, which inhibits all modes or allows only one mode, depending on its design. However, a serious surface recombination problem has disturbed the observation of such effects in such structures made of highly processed semiconductors. Therefore, the planar vertical cavity of SELs, as shown in Figure 2, has often been employed in theories and experiments.

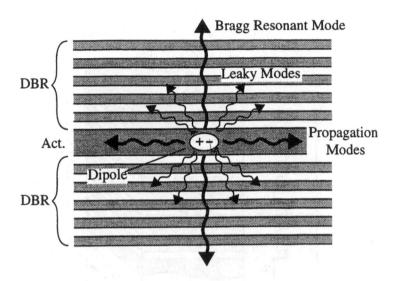

FIGURE 2   Planar cavity constructed by a pair of DBR mirrors. (From Baba et al., *IEEE J. Quantum Electron.*, 27, 1347, 1991. With permission © 1991 IEEE.)

This cavity is constructed by a pair of DBRs with a spacer inserted between them and an active layer which is located inside the spacer. The thickness of the spacer is designed to a multiple of half the emission wavelength in order to maintain Bragg resonant modes. When sufficient multilayers are stacked as DBRs, their reflectivity can be very high (0.99 to 0.999), and the field of the resonant mode is well separated from outer space. Here, it should be noticed that, since this type of cavity is open in lateral directions and DBRs have dispersive features against incident angle of light, there exist not only the resonant modes but also leaky modes and propagation modes, as illustrated in Figure 2. In some experiments, the influence of these unwanted modes was eliminated by utilizing an excitonic emission in quantum wells at low temperature, which exhibits a very narrow spectrum of less than 1 nm. When such a narrow spectrum is tuned to the resonant mode, the spontaneous emission is remarkably enhanced by the high contrast of the clearly defined resonant mode with the ambiguously expanding unwanted modes. However, it is supposed that the spontaneous emission energy coupled into these modes is unexpectedly large when the spectral width is as broad as 20 to 50 nm, typically observed in semiconductors at room temperature. If one discusses the spontaneous emission control in SELs with such a cavity, all of the modes should be fully taken into account.

In addition, if one discusses thresholdless operation by solving rate equations, one must clearly define the lasing mode by considering a 3-D cavity which is realized by etching or partly exciting the planar cavity. Often the symbol β has been used to denote the coupling efficiency of spontaneous emission into a group of Bragg resonant modes in the planar cavity. In this chapter, however, the *C* factor was defined for a fully defined lasing mode in the 3-D cavity.

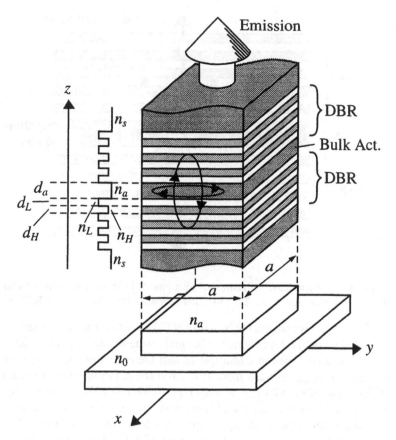

FIGURE 3   Analysis model of a 3-D microcavity SEL with bulk active region. (From Baba et al., *IEEE J. Quantum Electron.*, 27, 1347, 1991. With permission © 1991 IEEE.)

III. EXPRESSION OF RADIATION ENERGY

In this section, the radiation energy of spontaneous emission from a dipole located in microcavity SEL is expressed in the form of energy density which is a function of **k**. The wavevector **k** indicates orientation and energy of light. The energy density distribution corresponds to the radiation pattern of various emission spectrum.

Figure 3 shows an analysis model which simulates some index-guiding-type devices fabricated by etching processes.[4–6] In this model, a GaAs bulk active region, AlAs/AlGaAs semiconductor DBRs, and GaAs substrate and cap layer are assumed. Of course, this is generally applicable to other material systems. Although the air, polyimide, and AlGaAs buried layer are imagined as the surrounding region, air was assumed for this region to simplify the following analysis. In the following, a and d_a are used for the width and thickness of the active region, respectively, and n_H, n_L, and n_0 for indexes of

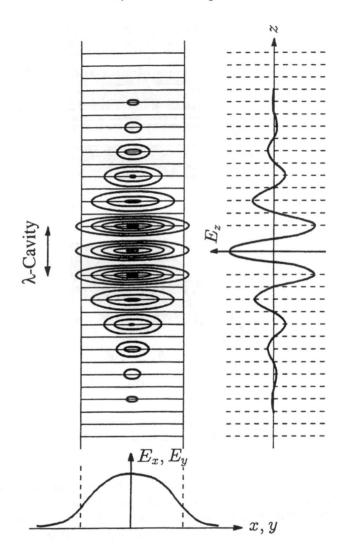

FIGURE 4   Schematic field profile of the lasing mode in a modeled microcavity SEL. (From Baba et al., *IEEE J. Quantum Electron.*, 28, 1310, 1992. With permission © 1992 IEEE.)

high and low index layers of DBR and the surrounding region, respectively. The thickness of each layer in the DBRs is designed to be a quarter of the targeted Bragg wavelength λ_B divided by each index. The schematic field profile of the lasing mode in the model is illustrated in Figure 4 when $d_a = \lambda_B / n_a$.

The analysis starts from the following classical wave equation with respect to the electric field vector **E**:

$$\nabla^2 \mathbf{E} - 2\kappa \left(\varepsilon_0 \mu_0 n^2 \right) \frac{\partial \mathbf{E}}{\partial t} - \varepsilon_0 \mu_0 n^2 \frac{\partial^2 \mathbf{E}}{\partial t^2} = \mu_0 \frac{\partial^2 \mathbf{P}}{\partial t^2} \qquad (2)$$

244

Spontaneous Emission and Laser Oscillation in Microcavities

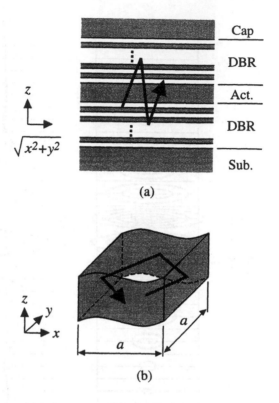

(a)

(b)

FIGURE 5 Separated cavities from a 3-D microcavity: (a) vertical cavity, and (b) lateral cavity. (From Baba et al., *IEEE J. Quantum Electron.*, 27, 1347, 1991. With permission © 1991 IEEE.)

where κ is the decay constant of the mode energy by absorption loss, n is the refractive index, and ε_0 and μ_0 are the permittivity and the permeability in vacuum, respectively. $\mathbf{P}$ is the polarization vector characterized by an electric dipole located at point $\mathbf{r}_e = (x_e, y_e, z_e)$ and represented as

$$\mathbf{P} = \mathbf{P}_0 e^{(-1/\tau_s + j\omega_e)t} \delta(\mathbf{r} - \mathbf{r}_e)$$

(3)

where $\mathbf{P}_0$ is the initial dipole vector with the amplitude P_0, τ_s is the spontaneous emission lifetime, ω_e is the angular frequency of dipole oscillation, δ is the Dirac delta function, and $\mathbf{r} = (x, y, z)$.

Since it is difficult to solve $\mathbf{E}$ directly in the 3-D model, the 3-D cavity is separated into two cavities, as shown in Figure 5. A cavity mode is treated as the product of a longitudinal mode and a transverse mode. Since the two end layers (GaAs substrate and cap layer) have the largest index in the model, any light suffers radiation loss into these end layers. Thus the longitudinal mode is treated as a continuous mode specified by the continuous eigenvalue.

On the other hand, the transverse modes are discrete owing to the total reflection at semiconductor-air boundaries unless the k-vector of the modes is oriented almost normal to the boundaries. In addition, the index step between the semiconductor and the air surrounding region is very large, so the lateral light confinement is very strong and for most modes. Thus the transverse modes are determined by the cavity width a and almost independent of the respective refractive indexes. For the separated cavities, Equation 2 can be reduced to scalar wave equation with respect to electric field amplitude E expressed as

$$E_{\nu\mu}\left(\mathbf{r},\, k_z,\, t\right) = \int_{-\infty}^{\infty} dk_z \cdot S_{\nu\mu}\left(k_z,\, t\right) f_\nu(x) g_\mu(y) h\left(k_z,\, z\right) e^{j\omega t}$$

$$(4)$$

where $\nu\mu$ is the order of transverse modes, $S_{\nu\mu}$ is a function slowly varying with time, and f_ν, g_μ, and h are the spatial distribution functions. In the separated cavities, f is given by $f_\nu = \cos k_\nu x$ or $\sin k_\nu x$ in the active region as solutions of the scalar wave equation, when the center of the active region is chosen as the origin of coordinate system. g_μ and h are given similarly with discrete wave number k_μ and continuous k_z, respectively.

Let us substitute Equations 3 and 4 into Equation 2 rewritten in the scalar description. After multiplying both sides by $f_{\nu'}(x) g_{\mu'}(y) h(k_z', z)$ ($f_{\nu'} g_{\mu'} h'$ for short), integration over all the real space and the k-space gives

$$\frac{\partial^2 S_{\nu\mu}}{\partial t^2} + 2\kappa \frac{\partial S_{\nu\mu}}{\partial t} + \omega^2 S_{\nu\mu} = \frac{\omega_e^2}{\varepsilon_0} P_0 e^{(-1/\tau_s + j\omega_e)t} \cdot \frac{(fgh)_{\mathbf{r}=\mathbf{r}_e}}{\displaystyle\iiint d^3\mathbf{r} \cdot \sum_{\nu'\mu'} \int_{-\infty}^{\infty} dk_z \cdot n^2 f_\nu f_{\nu'} g_\mu g_{\mu'} h h'}$$

$$(5)$$

The general solution of Equation 5 is given by

$$S_{\nu\mu} = \frac{\omega_e P_0}{j 2\pi\varepsilon_0} \cdot \frac{e^{(-\kappa + j\omega)t} - e^{(-1/\tau_s + j\omega_e)t}}{(1/\tau_s - \kappa) + j(\omega - \omega_e)} \cdot \frac{(fgh)_{\mathbf{r}=\mathbf{r}_e}}{A^2 \displaystyle\iint_{-\infty}^{\infty} dxdy \cdot n_s^2 f_\nu^2 g_\mu^2}$$

$$(6)$$

where A is the amplitude of h in the substrate and cap layer, and the following formula has been used for continuous modes:[40]

$$\int_{-\infty}^{\infty} n^2 h h' dz = \pi n_s^2 A^2 \delta\left(k_z - k_z'\right)$$

$$(7)$$

From Equation 6, the energy density $W_{\nu\mu}(k_z)$ is given by

$$W_{\nu\mu}(k_z) = \frac{1}{2}\int_0^\infty dt \cdot 2\kappa \int\int\int d^3r \cdot \sum_{\nu'\mu'}\int_{-\infty}^\infty dk_z' \cdot \varepsilon_0 n^2 E_{\nu\mu}(r, k_z, t) E_{\nu'\mu'}^*(r, k_z', t)$$

$$= \frac{P_0^2}{8\varepsilon_0\pi} \cdot \frac{\tau_s(1/\tau_s + \kappa)\omega_e^2}{(1/\tau_s + \kappa)^2 + (\omega - \omega_e)^2} \cdot \frac{(f_\nu^2 g_\mu^2 h^2)_{r=r_e}}{A^2 \int\int_{-\infty}^\infty dxdy \cdot n_s^2 f^2 g^2} \qquad (8)$$

Next, $W_{\nu\mu}(k_z)$ is averaged with respect to the direction, position, and emission spectrum of dipoles.[34] The average with respect to direction is given by replacing P_0^2 with $\overline{P_0^2}$, where $\overline{P_0^2} = P_0^2/3$ for bulk active region. The average with position r_e is given by integrating $W_{\nu\mu}(k_z)$ in the active region and dividing by the volume V ($= a^2 d_a$) of the active region. The average with spectrum is given by multiplying $W_{\nu\mu}(k_z)$ by an emission spectrum function $F(k_e)$ normalized so that its integration with k_e becomes 1, where $k_e = n_a\omega_e\sqrt{\varepsilon_0\mu_0}$. Thus, the average energy density is given by

$$\overline{W}_{\nu\mu}(k_z) = \int_0^\infty dk_e \int\int\int_{\text{(active region)}} d^3r_e \cdot \frac{F(k_e)W_{\nu\mu}(k_z)}{V}.$$

$$= \int_0^\infty dk_e \frac{\overline{\tau}_s \overline{P_0^2}}{16\varepsilon_0 a^2} k^2 F(k_e) \left[\frac{(1/\overline{\tau}_s + \kappa)\sqrt{\varepsilon_0\mu_0}\, n_a}{\pi\left\{(1/\overline{\tau}_s + \kappa)^2 \varepsilon_0\mu_0 n_a^2 + (k - k_e)^2\right\}} \right] \times$$

$$\sum_{\substack{\cos \\ \sin}} \frac{1}{A^2}\left(1 \pm \frac{\sin k_z d_a}{k_z d_a}\right)\left(\int\int_{-a/2}^{a/2} dxdy \cdot n_a^2 f_\mu^2 g_\mu^2 \Bigg/ \int\int_{-\infty}^\infty dxdy \cdot n_s^2 f_\nu^2 g_\mu^2\right) \qquad (9)$$

where $\overline{\tau}_s$ is the average lifetime. Now, let us assume that the decay by absorption loss and relaxation is much slower than by mirror loss, i.e., $(1/\tau_s + \kappa)\sqrt{\varepsilon_0\mu_0} \ll k - k_e$. Then the term in the square brackets becomes equivalent to Dirac delta function $\delta(k - k_e)$ and $\overline{W}_{\nu\mu}(k_z)$ is reduced to

$$\overline{W}_{\nu\mu}(k_z) = \frac{\overline{\tau}_s \overline{P_0^2}}{8\varepsilon_0}(4\pi k^2)F(k)D_{xy}^{(\nu\mu)}D_z \qquad (10)$$

where $k^2 = k_\nu^2 + k_\mu^2 + k_z^2$. $D_{xy}^{(\nu\mu)}$ is the transverse mode density in the unit area of active region. D_z is the longitudinal mode density in the unit thickness of active region and unit interval of k_z. They are given separately for s-waves and p-waves by

$$D_{xy}^{(\nu\mu)} = \frac{1}{4a^2} \cdot \left(\int\limits_{-a/2}^{a/2}\!\!\int dxdy \cdot n_a^2 f_\nu^2 g_\mu^2 \middle/ \int\limits_{-\infty}^{\infty}\!\!\int dxdy \cdot n_s^2 f_\nu^2 g_\mu^2 \right) \approx \frac{1}{4a^2} \qquad (11)$$

$$D_z = \sum_{\substack{\cos \\ \sin}} \frac{\xi_r}{2\pi A^2} \qquad (12)$$

where the s-wave polarization is parallel to the substrate plane and the p-wave polarization lies inside the incident plane. The energy density $\overline{W}_{\nu\mu}(k_z)$ is given by the sum of the average energies for these two waves.

The factor ξ_r is called the relative confinement factor[41] and indicates how efficiently the mode field overlaps the active region. For the s-wave, it is given by

$$\xi_r^{(s)} = \begin{cases} 1 + \dfrac{\sin k_z d_a}{k_z d_a} & (\cos-\text{type}) \\[2ex] 1 - \dfrac{\sin k_z d_a}{k_z d_a} & (\sin-\text{type}) \end{cases} \qquad (13)$$

On the other hand, since the electric field of a p-wave has parallel and normal components to the substrate plane, ξ_r is given by the sum of these two components, as follows:

$$\xi_r^{(p)} = \begin{cases} \dfrac{k_h^2}{k^2}\left[1 + \dfrac{\sin k_z d_a}{k_z d_a}\right] + \dfrac{k_z^2}{k^2}\left[1 - \dfrac{\sin k_z d_a}{k_z d_a}\right] & (\cos-\text{type}) \\[2ex] \dfrac{k_h^2}{k^2}\left[1 - \dfrac{\sin k_z d_a}{k_z d_a}\right] + \dfrac{k_z^2}{k^2}\left[1 + \dfrac{\sin k_z d_a}{k_z d_a}\right] & (\sin-\text{type}) \end{cases} \qquad (14)$$

where $k_h^2 = k_\nu^2 + k_\mu^2$. In bulk GaAs SELs, d_a is comparable to or larger than the wavelength and $\xi_r \sim 1$.

Here let us consider the amplitude A in Equation 12. A is the function of k_z and k, i.e., a function of k_z and k_h, and is numerically calculated by an operating characteristic matrix[42,43] including indexes and thicknesses of DBR to the field in active region. When Φ is the field parallel to the substrate plane and Ψ is inside the normal plane, A is given by

$$A^2 = \Phi^2 + \left(\frac{\zeta_s \Psi}{k_s}\right)^2 \qquad (15)$$

$$\begin{bmatrix} \Phi \\ \Psi \end{bmatrix} = \prod_{i=1}^{L} M_i \begin{cases} \begin{bmatrix} \cos k_z d_a/2 \\ (-k_z/\zeta_a)\sin k_z d_a/2 \end{bmatrix} & (\cos-\text{type}) \\[20pt] \begin{bmatrix} \sin k_z d_a/2 \\ (k_z/\zeta_a)\cos k_z d_a/2 \end{bmatrix} & (\sin-\text{type}) \end{cases} \tag{16}$$

where

$$k_s^2 = \left(\frac{n_s}{n_a}\right)^2 k^2 - k_h^2, \quad k_i^2 = \left(\frac{n_i}{n_a}\right)^2 k^2 - k_h^2 \tag{17}$$

$$M_i = \begin{cases} \begin{bmatrix} \cos|k_i|d_i & (\zeta_i/|k_i|)\sin|k_i|d_i \\ (-|k_i|/\zeta_i)\sin|k_i|d_i & \cos|k_i|d_i \end{bmatrix} & (k_i^2 > 0) \\[20pt] \begin{bmatrix} \cosh|k_i|d_i & (\zeta_i/|k_i|)\sinh|k_i|d_i \\ (|k_i|/\zeta_i)\sinh|k_i|d_i & \cosh|k_i|d_i \end{bmatrix} & (k_i^2 < 0) \end{cases} \tag{18}$$

$$\zeta_i = \begin{cases} 1 & (s-\text{wave}) \\ n_i^2 & (p-\text{wave}) \end{cases} \tag{19}$$

Subscripts a, s, and i denote parameters of active region, substrate, and ith layer in DBR, respectively. L is the total number of multilayers in DBR.

IV. MODES IN MICROCAVITY SELS

In this section, mode density calculated for the modeled SEL is presented. After the density of continuous longitudinal mode D_z and the distribution of discrete transverse modes $D_{xy}^{(\nu\mu)}$ are shown separately, the 3-D mode density is shown by taking product of them. The fundamental mode is defined and the mode dispersion is discussed briefly.

In the numerical calculation, $\lambda_B = 0.88$ μm, $n_a = 3.6$, $n_H = 3.6$, $n_L = 2.95$, $n_0 = 1.0$, and $L = 31$ (15.5 pairs) are assumed for the model, where a certain aluminum composition is supposed for the AlGaAs layer in DBRs. The mode density around the Bragg wave number $k_B = 2\pi n_a/\lambda_B = 7.14 \times 10^4$ cm^{-1} will be mainly discussed.

A. LONGITUDINAL MODE DENSITY D_z

Figures 6 and 7 show D_z in the k_z-k_h plane. They were calculated for a λ-cavity ($d_a = \lambda_B/n_a = 0.244$ µm) and a $\lambda/2$-cavity ($d_a = \lambda_B/2n_a = 0.122$ µm), respectively, and are displayed in a shaded drawing in four levels; the larger D_z is, the darker the drawing is.

For the s-wave, the curve of the Bragg resonant mode indicated by A is seen in Figure 6(a), which satisfies $k_z \approx k_B$. Along the curve, there are white regions indicating the stopband of the DBRs. From the detailed evaluation of this figure, it is found that (1) spontaneous emission for k_B is enhanced toward the z-direction, (2) emission for $k < k_B$ ($\lambda > \lambda_B$) toward the z direction is suppressed, (3) emission for $k > k_B$ ($\lambda < \lambda_B$) is enhanced toward the oblique direction, and (4) by adopting high index contrast materials for DBRs, the stop band is expanded and the maximum D_z of the resonant mode is increased, resulting in the more remarkable enhancement of emission. Figure 6(a) includes not only the Bragg resonant mode but also other various modes. The halftone areas spread around the stop band show weakly resonant leaky modes. The boundary indicated by **B** shows the critical condition of total internal reflection at interfaces between the active layer and the DBRs. The curve indicated by **C** shows the dispersion of the propagation mode. It is seen by comparing Figures 6 and 7 that the number of propagation modes increases as d_a increases.

For the p-wave, the Brewster angle breaks the curve of the Bragg resonant mode off, as indicated by **D** in Figure 6(b). However, D_z is almost similar to that for s-wave at $k \sim k_B$.

B. TRANSVERSE MODES $D_{xy}^{(\nu\mu)}$

Light is expanded into discrete transverse modes if the incident angle θ is larger than the critical angle of total reflection at semiconductor-air boundaries, i.e., $\theta > 16$ to $17°$. If not, light radiates to the air and should be treated as continuous modes. As seen in Figures 6 and 7, the longitudinal mode is almost absent around k_B when $\theta < 20$ to $21°$. This means that only discrete transverse modes are necessary for taking cross points of longitudinal and transverse modes. In addition, since the incident angle may have the transverse component, as illustrated in Figure 5(b), the 3-D incident angle is much larger than the critical angle. Thus, strong lateral light confinement can be assumed and $D_{xy}^{(\nu\mu)}$ becomes almost independent of the indexes. $D_{xy}^{(\nu\mu)}$ is approximated by $1/4a^2$.

Figure 8 displays the distribution of transverse modes when $a = 2\lambda_B/n_a = 0.49$ µm. In **k** space, the discrete mode region and the radiation mode region are separated by the critical plane of total reflection. The dispersion curve closest to the k_z-axis indicates the fundamental transverse mode $\nu\mu = 00$. The

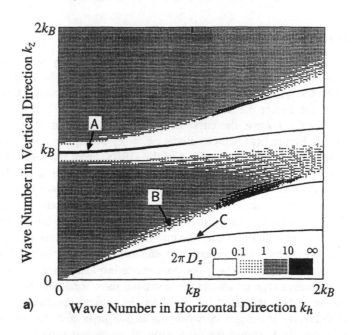

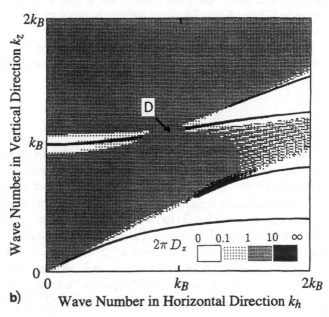

FIGURE 6 Distribution of longitudinal mode density D_z for a λ-cavity model: (a) s-wave, and (b) p-wave. (From Baba et al., *IEEE J. Quantum Electron.*, 27, 1347, 1991. With permission © 1991 IEEE.)

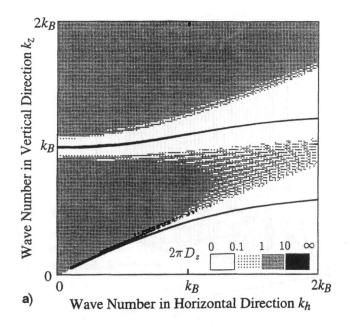

a)

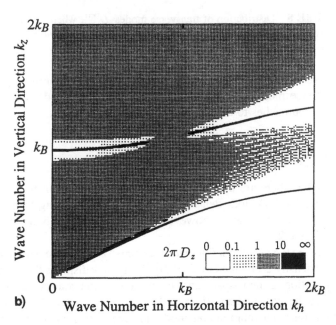

b)

FIGURE 7 Distribution of longitudinal mode density D_z for a $\lambda/2$-cavity: (a) s-wave, and (b) p-wave. (From Baba et al., *IEEE J. Quantum Electron.*, 27, 1347, 1991. With permission © 1991 IEEE.)

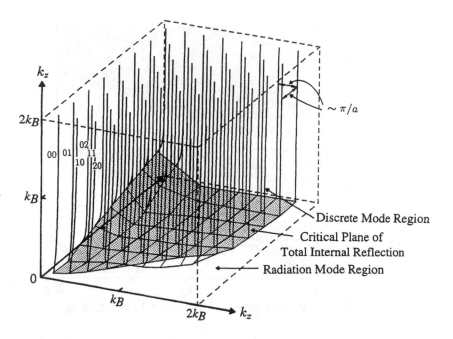

FIGURE 8 Distribution of transverse modes $D_{xy}^{(\nu\mu)}$, where the lateral size a of the cavity is $2\lambda_B/n_a = 0.49$ µm. (From Baba et al., *IEEE J. Quantum Electron.*, 27, 1347, 1991. With permission © 1991 IEEE.)

interval between neighboring modes is nearly π/a. Therefore, as a^2 increases, transverse modes distribute more densely and the mode density in the k_x-k_y plane approaches to that in the free plane, i.e., $(1/2\pi)^2$.

C. 3-D CAVITY MODES $D(\kappa)$

When there are no lateral confinement structures in the modelled cavity ($a = \infty$), the 3-D mode density is obtained by multiplying D_z in Figures 6 or 7 by $(1/2\pi)^2$. In this cavity, however, no modes are defined clearly and drawn as dots in the **k**-space.

When the lateral confinement is assumed as in Figure 3 (a is finite), the 3-D mode density is obtained by $D_z \times D_{xy}^{(\nu\mu)}$. Figure 9 displays the s-wave 3-D mode distribution in the **k**-space and its projection after rotating around the k_z-axis. Here, a λ-cavity and $a = 2\lambda_B/n_a = 0.49$ µm are assumed. The cross point of the Bragg resonant longitudinal mode and the fundamental transverse mode is defined as the fundamental mode of the cavity. The mode energy of the fundamental mode concentrates in a small region of k_z around the eigenwave vector $\mathbf{k}_0$. This extent is determined by the resonant quality of the cavity and thus by the mirror loss. The emission spectrum function $F(k)$ around k_B is drawn by a 1/8 spherical shell. As understood from Equation 10, the radiation energy couples to the modes inside $F(k)$. Coupled energy to the fundamental

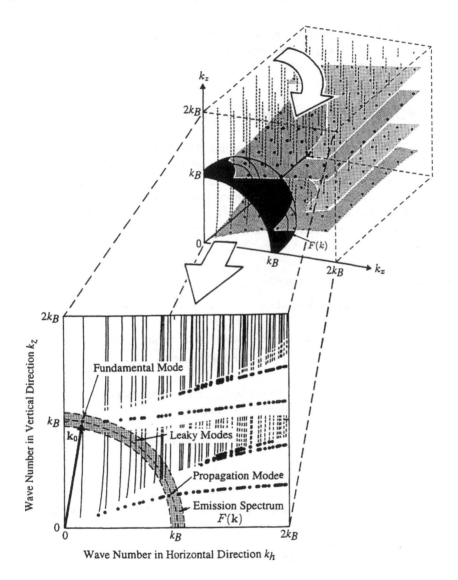

FIGURE 9 Distribution of mode density in a 3-D cavity $D(\mathbf{k})$ over the k-space and its projection after rotating around the k_z-axis: $a = 2\lambda_B/n_a = 0.49$ μm. (From Baba et al., *IEEE J. Quantum Electron.*, 27, 1347, 1991. With permission © 1991 IEEE.)

mode is obtained by integrating $\overline{W}_{00}(k_z)$ around $\mathbf{k}_0$. It is seen from the projection in Figure 9 that some amount of light couples to leaky modes and several propagation modes.

Last, let us mention the detuning of resonance caused by the dispersion of the fundamental transverse mode. The resonant wave number $k_0 = |\mathbf{k}_0|$ is slightly larger than k_B due to the transverse mode dispersion. The normalized wave number shift $\delta = (k_B - k_0)/k_B$ is calculated to be 0.017 and 0.027 for

$a = 1.2$ μm and 0.49 μm, respectively. If the cavity is buried by a polyimide or AlGaAs, δ will decrease owing to the light penetration from the active region to the surrounding region.

V.   SPONTANEOUS EMISSION FACTOR C

In this section, the spontaneous emission factor C is calculated to evaluate the amount of spontaneous emission control in SELs and the possibility of thresholdless operation. First, the C factor is estimated by an approximate formula. Next, computer calculation based on the expression of radiation energy obtained in Section III is shown. The difference between the two results, which is evident at $C > 0.1$, is discussed.

A.   ESTIMATION FROM A FORMULA

The C factor is defined as

$$C \equiv \frac{(\text{Radiation energy coupled to a lasing mode})}{(\text{Total radiation energy})} \tag{20}$$

If the cavity modes distribute uniformly so that same radiation energy couples to every mode, the definition is rewritten as

$C \equiv (\text{Total number of modes in emission spectrum})^{-1}$

$= [(\text{Mode density in free space}) \times$

$(\text{Effective mode volume}) \times (\text{Emission spectral width in k space})]^{-1}$

$$= \left[\left(\frac{1}{2\pi}\right)^3 \times \left(\frac{n_{eq}^3 V}{\xi}\right) \times \left(4\pi k_e^2 \cdot \pi \Delta k_e\right) \right]^{-1} \tag{21}$$

where ξ is the confinement factor, n_{eq} is the equivalent index of mode, and k_e and Δk_e are the peak wave number and the full width at half maximum (FWHM) of the emission spectrum, respectively. Converting k_e and Δk_e by λ_e and $\Delta\lambda_e$, Equation 21 is rewritten as

$$C = \frac{\lambda_e^4}{4\pi^2 V_m \Delta\lambda_e} \tag{22}$$

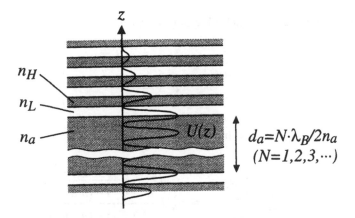

FIGURE 10 Schematic energy profile of the Bragg resonant mode in a vertical cavity. (From Baba et al., *IEEE J. Quantum Electron.*, 27, 1347, 1991. With permission © 1991 IEEE.)

where V_m is the effective mode volume given by

$$V_m = n_{eq}^3 \, V/\xi \tag{23}$$

Let us estimate the C factor in the modeled SEL by using Equations 22 and 23. For this, n_{eq} and ξ must be given for the fundamental mode. Since the strong lateral confinement is assumed in the model, n_{eq} and ξ are calculated by considering the field penetration from active region to DBRs. When d_a is a multiple of the half wavelength $\lambda_B/2n_a$, the field profile of the Bragg resonant mode in the z-direction is schematically described in Figure 10. The normalized mode energy U_a confined inside the active region is given by

$$U_a = n_a d_a /2 \tag{24}$$

Now the DBRs are assumed to be sufficiently reflecting so that the radiation to the substrate and the cap layer is small. Then the energy U_o outside the active region is given by

$$U_o \simeq 2 \times \left[\frac{1}{2} n_L \frac{\lambda_B}{4n_L} + \sum_{i=1}^{\infty} \left(\frac{1}{2} n_H \frac{\lambda_B}{4n_H} + \frac{1}{2} n_L \frac{\lambda_B}{4n_L} \right) \left(\frac{n_L}{n_H} \right)^{2i} \right]$$

$$= \frac{\lambda_B}{4} \cdot \frac{1 + \left(n_L/n_H \right)^2}{1 - \left(n_L/n_H \right)^2} \tag{25}$$

From Equations 24 and 25, ξ is given by

$$\xi = \frac{U_a}{U_a + U_o} \simeq \frac{d_a}{d_a + \frac{\lambda_B}{2n_a} \cdot \frac{1+(n_L/n_H)^2}{1-(n_L/n_H)^2}}$$

(26)

Similarly, n_{eq} is given by

$$n_{eq}^2 = \frac{\int_{-\infty}^{\infty} n^2 U dz}{\int_{-\infty}^{\infty} U dz} \simeq n_H^2 \frac{d_a + \frac{\lambda_B}{2n_a} \cdot \frac{2(n_L/n_H)^2}{1-(n_L/n_H)^2}}{d_a + \frac{\lambda_B}{2n_a} \cdot \frac{1+(n_L/n_H)^2}{1-(n_L/n_H)^2}}$$

(27)

If the cavity is buried by a polyimide or AlGaAs, the field also penetrates into the surrounding region. For this case, ξ and n_{eq} are obtained by regarding the cavity as a hollow waveguide having a core of refractive index given by Equation 27 and cladding of index n_0. The index n_{eq} and lateral confinement factor are calculated by using some waveguide analysis, e.g., Marcatili's method.[45] The final value for ξ is given by the product of the vertical confinement factor of Equation 26 and the lateral confinement factor.

Figure 11 shows the calculated C factor as a function of width a, where air, polyimide, and AlGaAs are assumed as the surrounding region. The C factor simply increases with the size reduction of cavity and becomes 0.015 for $a = 1$ μm. This value is very large compared with typical values of 10^{-5} to 10^{-6} for edge emitting lasers. When the cavity is buried by AlGaAs, this increase saturates below $C < 0.1$ because of the decrease of the lateral confinement for $a < 1$ μm. On the other hand, C exceeds 0.1 when the cavity is etched or buried by a polyimide and $a < 0.4$ μm. This size seems possible to realize experimentally.[4]

B. COMPUTER SIMULATION

Equation 22 indicates that a very small V_m and very narrow $\Delta\lambda_e$ provide a large C factor. However, it is required from the definition that C is not larger than 1. This implies that, for such V_m and $\Delta\lambda_e$, the assumption of the uniform mode distribution is no longer valid. When the mode distribution in a microcavity is too different from that in free space, the C factor should be evaluated by comparing the energy w_0 coupled to a linear polarized fundamental mode with total radiation energy w_t. In a λ-cavity, the Bragg resonant mode has the cos-type field. From Equations 10 and 11, C is given by

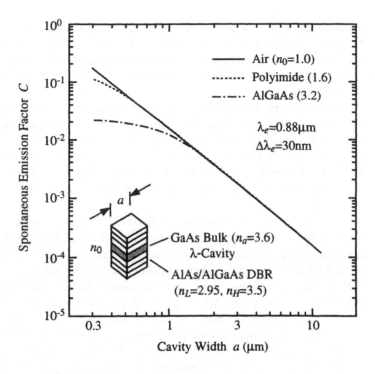

FIGURE 11 Spontaneous emission factor C of the fundamental mode vs. the lateral size a of a cavity, which is estimated by using Equation 22.

$$C = \frac{w_0}{w_t} = \frac{p \int dk_z \cdot \overline{W}_{00}(k_z)}{\sum_{\nu\mu} \int dk_z \cdot \overline{W}_{\nu\mu}(k_z)}$$

$$\approx \frac{p \sum_{s,p} \int dk_z \cdot \left[\frac{(\Delta k_e/2)\xi_r/A^2}{(\Delta k_e/2)^2 + (k - k_e)^2} \right]_{\substack{\cos \\ k_x,k_y = \pi/a}}}{\sum_{\nu\mu} \sum_{s,p} \sum_{\substack{\cos \\ \sin}} \int dk_z \cdot \left[\frac{(\Delta k_e/2)\xi_r/A^2}{(\Delta k_e/2)^2 + (k - k_e)^2} \right]_{\substack{k_x = \nu\pi/a \\ k_y = \mu\pi/a}}}$$

(28)

where the following normalized Lorentzian function was assumed for the spectrum $F(k)$ to simplify the analysis:

$$F(k) \equiv \left(\frac{\Delta k_e}{2} \right) \Big/ \pi \left[\left(\frac{\Delta k_e}{2} \right)^2 + (k - k_e)^2 \right]$$

(29)

In Equation 28, p is the polarization factor given by the spontaneous emission ratio of one polarization of the fundamental mode. For a symmetric cross section of the cavity and the bulk active region, two polarizations are degenerate at the wavevector $\mathbf{k}$, and p is 0.5 for any $\mathbf{k}$. Thus the upper limit of C is fixed to 0.5.

In the numerical calculation, the region inside the spectrum function $F(k)$ was divided into 20 sections. The integration and the summation of the denominator in Equation 28 were performed in each section and adding by using a supercomputer ETA-10. Also those of the numerator were performed along the dispersion curve of fundamental transverse mode $\nu\mu = 00$ around $\mathbf{k}_0$ within the extent in which the term in the square bracket is larger than 10^{-6} times its maximum.

Figures 12(a) and (b) show the obtained C factor as a function of spectral width, where Δk_e is converted to $\Delta\lambda_e$ by the relation $\Delta\lambda_e/\lambda_e = \Delta k_e/k_e$, and a λ-cavity and an air surrounding region are assumed. In Figure 12(a), it is also assumed that the emission is always tuned to the cavity resonance, i.e., $\lambda_e = \lambda_0$. Solid curves and dashed lines show calculated results and estimation by Equation 22, respectively. When $\Delta\lambda_e > 10$ nm, the calculated results almost agree with Equation 22. However, the results reasonably saturate when C approaches to the upper limit 0.5 with reductions of a and $\Delta\lambda_e$. This difference between the result and Equation 22 seems to be caused by the change of total radiation energy w_t, which is identical to the change of spontaneous emission lifetime. When $a = 0.49$ μm, C is estimated to be 0.12, 0.20, and 0.27 for $\Delta\lambda_e = 20$, 10, and 5 nm, respectively. Such narrow $\Delta\lambda_e$ may be obtained at low temperature. Further enhancement of C is possible by introducing an oblong cross section of the cavity, because the wave vector $\mathbf{k}_0$ including two degenerate polarizations is split into two wave vectors, and only one polarized mode can be chosen by tuning this mode to k_e.

In this figure, the experimental estimation for fabricated SELs with a bulk GaAs active region is also shown by an open circle. It was 3×10^{-5} (45) and much smaller than the theoretical expectation. This seems to be caused by the very thick (5.5 μm) and wide (7 μm) cavity fabricated in the experiment. As seen from Equation 22, reduction of the cavity thickness will provide a larger C. However, it should be noticed that when the active layer is thinner than λ_B/n_a, C becomes dependent on the position of active region against the antinode of the mode field. This effect will be discussed in the next section.

From Figure 12(a), the dependence of the C factor on mirror reflectivity R can be discussed by using the relation between $\Delta\lambda_e$ and the resonant bandwidth $\Delta\lambda_0$. In the modeled λ-cavity, $\Delta\lambda_0$ is estimated to be 0.14 nm. As seen in this figure, C for $a = 3.7$ μm saturates far below the upper limit 0.5 when $\Delta\lambda_e < \Delta\lambda_0$. This is because the energy coupled to the fundamental mode no longer increases when the emission spectrum is completely included into the resonant bandwidth. Conversely speaking, the results of Figure 12(a) do not depend on the resonant bandwidth and hence on R when $\Delta\lambda_e > \Delta\lambda_0$. The dependence on absorption loss in cavity is discussed similarly. With no excitation, the typical absorption coefficient in bulk GaAs is 1,000 to 2,000 cm^{-1} at $\lambda = 0.88$ μm.[44]

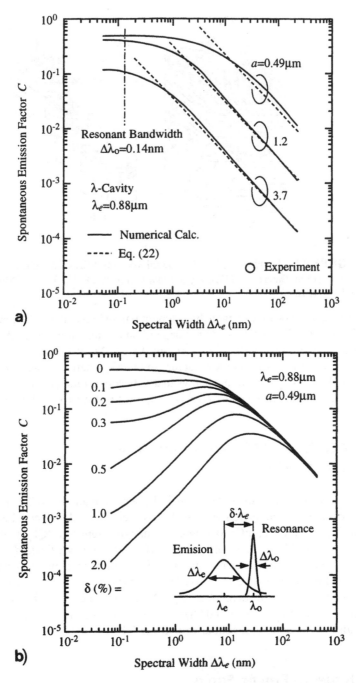

FIGURE 12   Computed C factor of the fundamental mode vs. the spectral width $\Delta\lambda_e$. (a) The peak emission wavelength λ_e is tuned to the resonant wavelength λ_0, and (b) λ_e is detuned from λ_0 by the normalized wavelength shift $\delta = (\lambda_0 - \lambda_e)/\lambda_e$. The experimental data in Figure 12(a) is from Reference 45. (From Baba et al., *IEEE J. Quantum Electron.*, 27, 1347, 1991. With permission © 1991 IEEE.)

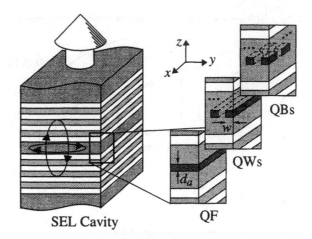

FIGURE 13 Analysis model of a 3-D microcavity SEL with a quantum-well active region. (From Baba et al., *IEEE J. Quantum Electron.*, 28, 1310, 1992. With permission © 1992 IEEE.)

This absorption may expand the resonant bandwidth to 0.7 to 1.0 nm. However, the results of Figure 12(a) do not depend on the absorption when $\Delta\lambda_e > 1.0$ nm. The increase of C will saturate below 0.5 for $\Delta\lambda_e < 1.0$ nm.

  Figure 12(b) shows the C factor versus $\Delta\lambda_e$ when λ_e is detuned from λ_0 by $\delta = (\lambda_0 - \lambda_e)/\lambda_e$. This figure shows that C becomes rather small when $\delta > 0.5\%$ and $\Delta\lambda_e < 10$ nm. To realize $C > 0.1$, λ_0 must be tuned to λ_e within 0.5% error.

VI. EFFECTS OF ELECTRON QUANTUM CONFINEMENT

  In this section, the C factor in SELs with a quantum-well active region is calculated and compared with the bulk case. The analysis model is illustrated in Figure 13. Here, InGaAs single-strained quantum film (QF), multiple quantum wires (QWs) and quantum boxes (QBs) with $n_a = 3.6$ and $d_a = 80$ nm are considered for the active region. The QF is sandwiched by GaAs barrier layers ($n_c = 3.6$) to construct a λ-cavity. The thickness d_c of the barriers is determined so that it satisfies the condition $n_a d_a + 2 n_c d_c = \lambda_B$. The QWs are arranged toward the x-direction. For QWs and QBs, GaAs barrier region is also assumed, which satisfies the condition similar to that for QF.

A. OUTLINE OF EXPECTED EFFECTS

  The effects of quantum wells are outlined as follows:

  1. The mode field overlaps the thin quantum wells more or less efficiently depending on their relative position.

2. The average polarization $\overline{P_0^2}$ is anisotropic due to the restricted orientation of dipoles in quantum wells.
3. The spectral width $\Delta\lambda_e$ is narrower than in bulk due to the quantized energy levels.

The detail of each effect is discussed in the following.

Although the quantum wells are inserted into the cavity, the mode volume V_m does not change from that for bulk as long as the cavity structure is the same. However, the mode density is changed by the change of the relative confinement factor ξ_r, as can be seen in Equation 12. The factor ξ_r given by Equations 13 and 14 expresses how efficiently the field overlaps the active region. The relative positions of the mode field and active region are illustrated in Figure 14. As discussed in Section III, ξ_r is almost 1 and this effect is leveled when d_a is large enough, as in bulk active region. For QF, ξ_r changes from nearly 2 to nearly 0 depending on whether the QF is located near an antinode or a node of the field. For the model of Figure 13, ξ_r for the cos-type field is nearly 2. Thus the energy w_0 of the fundamental mode becomes almost twice as that for bulk case. On the other hand, the total energy w_t does not change so much with the change of ξ_r because the increased energy for the cos-type field is compensated with decreased energy for the sin-type field. Similarly the fundamental mode is enhanced when a lot of QWs or QBs are arranged uniformly inside the xy plane. If a single QW is located at the center of the cavity, ξ_r increases in both the y- and z-directions and becomes 4. For a single QB, ξ_r maximally becomes 8. When one calculates the mode volume V_m in quantum wells by using Equation 23, one should use the approximate formula for ξ, which is rewritten from Equation 26 to

$$\xi \simeq \frac{\xi_r d_a}{d_a + 2d_c + \dfrac{\lambda_B}{2n_a} \cdot \dfrac{1 + \left(n_L/n_H\right)^2}{1 - \left(n_L/n_H\right)^2}} \tag{30}$$

where $n_a \simeq n_c$ is assumed.

The anisotropic emission is caused by the restriction for dipole orientation. In QF, dipoles are likely to be oriented along the xy plane. Since the fundamental mode has the wave vector almost toward the z-axis and the electric field vector inside the plane, the relative emission for this mode is stronger than the average emission for other modes by a factor of 1.5, when assuming the electron wave in a QF at the subband edge condition. However, since both x- and y-polarizations are equally allowed for this mode, p in Equation 28 is still 0.5. In QWs, the dipole orientation is further restricted to the x-direction. The emission for the fundamental mode is given by averaging the lateral component of slightly inclined dipoles in QWs. This inclination is determined by the aspect ratio of QW cross section, i.e., $r = w/d_a$. When the electron wave at the lowest energy level is strongly confined in the QWs, the emission toward the z-direction is changed from that in bulk by a factor of $0.75(1 + 2r^2)/(1 + r^2)$.

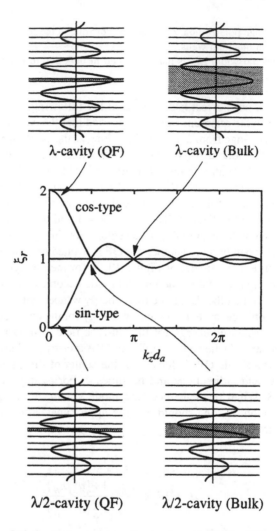

FIGURE 14 Relation between the cavity structure, the field profile of Bragg resonant mode and the relative confinement factor ξ_r. (From Baba et al., *IEEE J. Quantum Electron.*, 28, 1310, 1992. With permission © IEEE.)

In addition, the emission for x-polarization is stronger than for y-polarization and $p = (1 + r^2)/(1 + 2r^2)$. In total, the emission for the x-polarized fundamental mode is 1.5 times stronger than for other modes. In the case of QBs, the dipole orientation is determined by their three side lengths. The emission for the fundamental mode is increased or decreased in the range of 0.75 to 1.5 times depending on the orientation.

The spectral width $\Delta\lambda_e$ depends on the material and structure of active region and temperature. Typically it is 30 nm for bulk GaAs at room temperature.[49] By comparing the theoretical gain in bulk GaAs and in strained InGaAs QF,[50] $\Delta\lambda_e$ for the QF is estimated to be 15 nm. Ideally, $\Delta\lambda_e$ of several nanometers and subnanometers are expected for QW and QB, respectively.

TABLE 1
Quantum Effects for the C Factor in Various Regions

Act.	Bulk	QF $r=\infty$	QW $r=1$	$r=0.3$	QB
ξ_r	1	2	2~4	2~4	2~8
P_0	$P_0{}^{bulk}$	$\times 1.5$	$\times 1.5$	$\times 1.5$	$\times 1.5$
$\Delta\lambda_e$ (nm)	30	15	5	5	1
C	C_{bulk}	$\times 6$	$\times 18 \sim$	$\times 18 \sim$	$\times 90 \sim$
C_{max}	0.5	0.5	0.67	0.91	>0.5

From Baba et al., *IEEE J. Quantum Electron.*, 28, 1310, 1992. With permission © 1992 IEEE.

Table 1 summarizes the quantum effects for changes in the C factor. The C factor in QF is nearly 6 times larger than in bulk when it is much smaller than the upper limit 0.5. This comes from the increase of ξ_r (2×), anisotropic emission (1.5×), and narrow spectral width (2×). In the multiple QWs case, C will be multiplied over 18 times by the further reduction of $\Delta\lambda_e$. In the single QW case, C will be enhanced more than 30 times. It is a significant feature of the QW that, if the cross section of the QW is vertically long ($r \to 0$), the upper limit of C determined by $p = (1 + r^2)/(1 + 2r^2)$ approaches 1. In QB, p is smaller than 1 since dipoles in QB cannot absolutely orient in one direction. However, it may be much easier to make C approach its upper limit using QBs than using other quantum wells due to their extremely narrow emission spectral width.

B.   CALCULATION OF MODES AND THE C FACTOR

The mode density distribution and the C factor in quantum-well SELs are calculated using Equations 10 to 19. It is found from basic theories[47,48] and some modifications,[26] that the average polarization P_0^2 in quantum wells is a function of the orientation of the wavevector **k**. This was taken into account in the calculation, while the strained effect in the quantum wells was ignored. As in the bulk case, the longitudinal mode density D_z for the QF is displayed in Figures 15 and 16. Here a λ-cavity is assumed, and the cos-type field and the sin-type one are shown individually for the s-wave, and the p-wave. For the s-wave, the cos-type field has a large mode density all over the k-space and

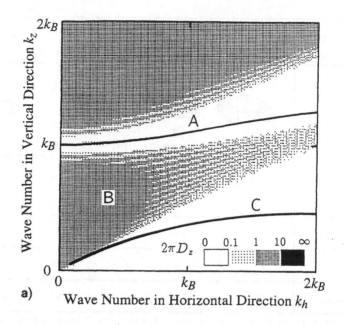

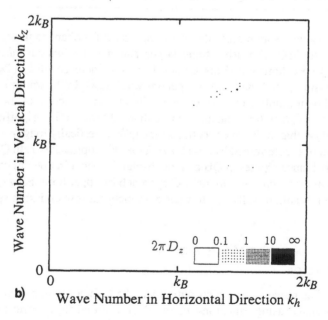

FIGURE 15 Distribution of the longitudinal mode density D_z of the s-wave in a QF model: (a) cos-type field, and (b) sin-type field. (From Baba et al., *IEEE J. Quantum Electron.*, 28, 1310, 1992. With permission © 1992 IEEE.)

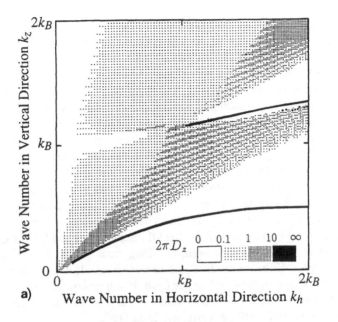

a)

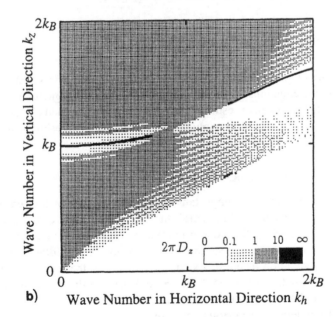

b)

FIGURE 16 Distribution of the longitudinal mode density D_z of p-wave in a QF active region: (a) cos-type field, and (b) sin-type field. (From Baba et al., *IEEE J. Quantum Electron.*, 28, 1310, 1992. With permission © 1992 IEEE.)

maintains the Bragg resonant modes (one of them is indicated by symbol **A**), leaky modes (symbol **B**), and propagation modes (symbol **C**). In contrast to this, the sin-type field has quite a small mode density and maintains no such modes. Obviously, these are caused by the difference of ξ_r between the cos-type and sin-type. For both cos-type and sin-type fields of p-waves, D_z has some variation in values owing to the complicated expressions of ξ_r in Equation 14.

Figure 17(a) shows the calculated C factor for bulk, QF, and QW active regions. As the spectral width $\Delta\lambda_e$, values summarized in Table 1 were used. In addition, $\lambda_0 = \lambda_e$ was assumed. The result for bulk almost agrees with the estimation in Figure 11 except for the point that the increase of C slightly saturates below 0.5 values of C in QFs and QWs exceed 0.01 even with $a = 3$ μm. By analogy to the bulk case, this value will not be changed even if the cavity is buried by AlGaAs or polyimide. When a is reduced to $2\lambda/n_a = 0.54$ μm, $C = 0.12$ for the QF, and $C = 0.4$ and 0.6 for QWs of square cross section ($r = 1$) and vertically long cross section ($r = 0.3$), respectively.

Figure 17(b) shows the C factor versus $\Delta\lambda_e$. If $\Delta\lambda_e$ is much narrower than 10 nm, C approaches each upper limit. If subnanometer $\Delta\lambda_e$ is available at low temperature and applicable to some actual devices, the extreme value $C > 0.9$ can be realized with the vertically long QWs.

VII. LASING CHARACTERISTICS

In this section, the lasing characteristics of microcavity SELs are examined from rate equations. First, the change of average spontaneous emission lifetime $\bar{\tau}_s$ in a microcavity is estimated. Next, using the obtained C factor and $\bar{\tau}_s$, rate equations are solved to evaluate the possibility of thresholdless operation.

A. SPONTANEOUS EMISSION LIFETIME

The lifetime $\bar{\tau}_s$ is important to evaluate the spontaneous emission control in microcavity SELs. In general, the threshold current of SELs is reduced by the increase of $\bar{\tau}_s$. Such condition will be obtained if cavity modes except the lasing mode are restricted. Once SELs exhibit thresholdless operation, a very high modulation response speed is expected by the reduction of $\bar{\tau}_s$. Such a condition will be obtained if the mode density of the lasing mode is enhanced and simultaneously other modes are restricted.

In general, spontaneous emission lifetime $\bar{\tau}_s$ is given by

$$\frac{N}{\bar{\tau}_s} = \int_0^\infty T(\omega)D(\omega)d\omega \qquad (31)$$

where N is the injected carrier density, $T(\omega)$ is the spontaneous emission probability in free space, and $D(\omega)$ is the mode density for the angular frequency ω. As described in the previous section, it is convenient for estimating

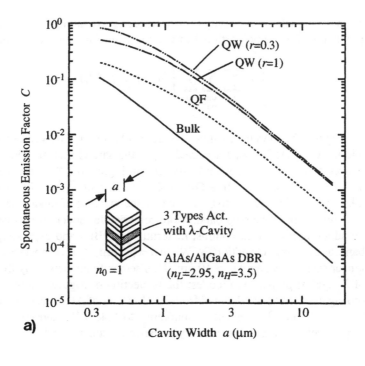

a)

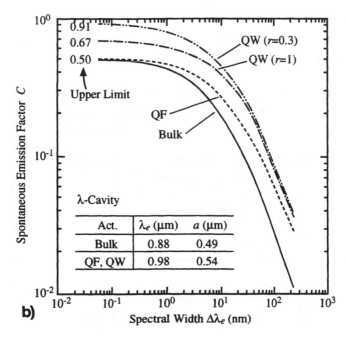

b)

FIGURE 17 Computed C factor of the x-polarized fundamental mode (a) vs. lateral size a and (b) vs. spectral width $\Delta\lambda_e$. (From Baba et al., *IEEE J. Quantum Electron.*, 28, 1310, 1992. With permission © 1992 IEEE.)

$\bar{\tau}_s$ in the modeled microcavity to convert ω to wave numbers k_v, k_μ, and k_z. The probability $T(\omega)$ is proportional to the spectrum function $F(k)$. Thus,

$$\frac{1}{\bar{\tau}_s} \propto \sum_{s,p} \sum_{\substack{\cos \\ \sin}} \sum_{v\mu} \int dk_z \cdot F(k) D_{xy}^{(v\mu)} D_z$$

(32)

Figure 18 shows the relative change of lifetime $\bar{\tau}_s / \bar{\tau}_{so}$ versus the spectral width $\Delta\lambda_e$, where $\lambda_e = \lambda_0$ is assumed. $\bar{\tau}_{so}$ is the natural lifetime when the active region is buried by a large bulk. As seen in Figure 18, $\bar{\tau}_s$ is reduced even in a planar cavity ($a = \infty$). In a QF, 40% reduction of $\bar{\tau}_s$ will be observed for $\Delta\lambda_e = 15$ nm. This seems to be caused by the increase of mode energy around k_B collected from the stop band. More remarkable reduction will be obtained by adopting high index contrast materials for DBRs and expanding the stop band, as discussed in Section IV. When $a = 0.5$ μm, $\bar{\tau}_s$ is comparable to or even longer than $\bar{\tau}_{so}$ when $\Delta\lambda_e > 10$ nm. This seems to be caused by the detuning of λ_e against propagation modes; the projection of Figure 9 shows that three propagation modes are detuned from λ_e. However, if $\Delta\lambda_e$ is reduced to 5, 1, and 0.5 nm, $\bar{\tau}_s$ in QF decreases remarkably to 0.4, 0.15, and 0.08 times $\bar{\tau}_{so}$, respectively, owing to the strongly enhanced resonant mode.

B.   RATE EQUATION ANALYSIS

For carrier density N and photon density of the lasing mode S, the rate equations are

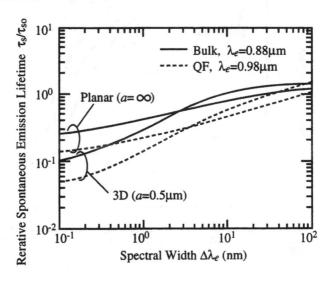

FIGURE 18   Relative change of spontaneous emission lifetime $\bar{\tau}_s / \bar{\tau}_{so}$ in microcavity SELs vs. spectral width $\Delta\lambda_e$.

$$\frac{dN}{dt} = \frac{I}{eV} - \xi GS(N - N_0) - \frac{N}{\overline{\tau}_s}$$

(33)

$$\frac{dS}{dt} = \xi GS(N - N_0) + \frac{CN}{\overline{\tau}_s} - \frac{S}{\tau_p}$$

(34)

where I is the injection current, e is the electron charge, G is the differential gain coefficient, N_0 is the carrier density for transparency, and τ_p is the photon lifetime in the cavity. Static solutions of these equations for N and S are derived by $dN/dt = 0$ and $dS/dt = 0$.

Let us show some numerical results for quantum-well SELs. For an 80-nm-thick $In_{0.2}Ga_{0.8}As$ QF emitting at a wavelength of 0.98 μm, $G \simeq 1.1 \times 10^{-5}$ cm^3/s and $N_0 = 1.5 \times 10^{18}$ cm^{-3} (49). Now considering the surrounding air region, the confinement factor ξ is estimated from Equation 30 to be 0.015. If the internal absorption loss is 10 cm^{-1} and the reflectivity of DBR is 0.998, τ_p is calculated to be 4×10^{-12} s. The light output power converted from the photon number of the lasing mode SV/ξ is calculated and plotted versus current I, as shown in Figure 19(a). Here, C factors from Figure 17(a) and $\overline{\tau}_s$ in Figure 18 with $\overline{\tau}_{so} = 1.5 \times 10^{-9}$ s at $\Delta\lambda_e = 15$ nm were used. The threshold current, which is indicated by the abrupt increase of light power, is reduced with reduction of the width a. Simultaneously the abruptness becomes ambiguous with the increase of C factor. However, C is still less than 0.3 even with $a = 0.3$ μm, and the curve cannot be absolutely straight. Figure 17(b) shows the characteristics for a QW SEL, where $G = 3.3 \times 10^{-5}$ cm^3/s, $N_0 = 1.2 \times 10^{18}$ cm^{-3}, $\xi = 0.008$, and $r = 0.3$ are assumed. As well as the threshold reduction, it is found that an almost straight line of quasi-thresholdless operation is obtained with an $a = 0.3$ μm cavity.

VIII. SUMMARY

In this chapter, the authors discussed spontaneous emission control and the possibility of thresholdless operation in actual microcavity SELs.

Based on the classical wave equation, we derived the general expression of spontaneous emission energy. From the numerical simulation of the mode energy distribution over all k-space, it was confirmed that a large amount of mode energy is concentrated around the wavevector of Bragg resonant mode in microcavity SELs. However, since the emission spectrum from semiconductors is typically several 10 nm in width, the spontaneous emission energy of ambiguous leaky modes and lateral propagation modes inside the spectrum is large so that it cannot be ignored to evaluate the net effect of the enhanced Bragg resonant mode.

The spontaneous emission factor C for the Bragg resonant mode in microcavity SELs is as large as 10^3 to 10^5 times that in conventional horizontal

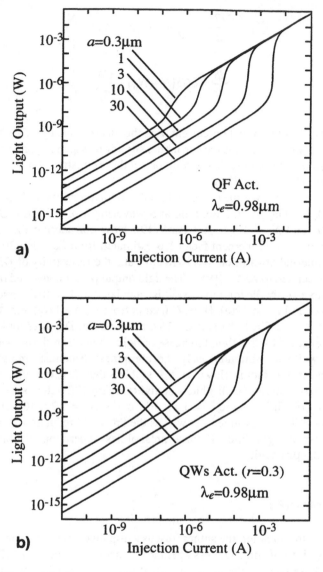

a)

b)

FIGURE 19 Light output vs. current characteristics for (a) QF and (b) QWs SELs. Calculated C factor and lifetime $\bar{\tau}_s$ are taken into account. (From Baba et al., *IEEE J. Quantum Electron.*, 28, 1310, 1992. With permission © 1992 IEEE.)

cavities. It is not so dependent on the vertical cavity quality (or reflectivity of DBRs) but is almost inversely proportional to the cavity lateral size. The C factor becomes larger than 0.1 in a submicron SEL with bulk active region. When one employs a quantum-well active region efficiently overlapping the antinode of the field of Bragg resonant mode, the C factor will increase from that for the bulk case by a factor of over 6 (at least 3 even with the broadening of spectral width by nonradiative recombination). Especially, a submicron SEL with quantum wires possibly has a C factor of over 0.9 by the polarization selectivity of quantum wires.

The spontaneous lifetime will also be altered in microcavity SELs. However, it is limited to within 0.5 to 2 times that in a conventional cavity. This small change is due to the unexpectedly large mode energy coupled to the leaky modes and propagation modes.

With the estimated C factors and lifetimes, the lasing characteristics were examined from the rate equations. The almost thresholdless operation will be obtained in a submicron SEL with quantum wires. To realize such an SEL experimentally, the reduction of nonradiative recombination at the microcavity surface will be the important issue. In addition, investigation of the absolutely closed cavity structure will provide more remarkable effects of microcavities.

REFERENCES

1. Soda, H., Iga, K., Kitahara, C. and Suematsu, Y., GaInAsP/InP surface emitting injection lasers, *Jpn. J. Appl. Phys.,* 18, 2329, 1979.
2. Iga, K., Koyama, F. and Kinoshita, S., Surface emitting semiconductor lasers, *IEEE J. Quantum Electron.,* 24, 1845, 1988.
3. Koyama, F., Kinoshita, S. and Iga, K., Room-temperature continuous wave lasing characteristics of GaAs vertical cavity surface-emitting laser, *Appl. Phys. Lett.,* 55, 221, 1989.
4. Jewell, J. L., Harbison, J. P., Scherer, A., Lee, Y. H. and Florez, L. T., Vertical cavity surface-emitting lasers: design, growth fabrication, characterization, *IEEE J. Quantum Electron.,* 27, 1332, 1991.
5. Geels, R. S., Corzine, S. W. and Coldren, L. A., InGaAs vertical-cavity surface-emitting lasers, *IEEE J. Quantum Electron.,* 27, 1359, 1991.
6. Numai, T., Kurihara, K., Ogura, I., Kosaka, H., Sugimoto, M. and Kasahara, K., High electronic-optical conversion efficiency in a vertical-to-surface transmission electro-photonic device with a vertical cavity, *IEEE Photon. Technol. Lett.,* 5, 136, 1993.
7. Baba, T., Yogo, Y., Suzuki, K., Koyama, F. and Iga, K., First room temperature cw operation of GaInAsP/InP surface emitting laser, *IEICE Trans. Electron.,* E76-C, 1423, 1993.
8. Lott, J. A. and Schneider Jun, R. P., Electrically injected visible (639–661 nm) vertical cavity surface emitting laser, *Electron. Lett.,* 29, 830, 1993.
9. Kinoshita, S., Sakaguchi, T., Odagawa, T. and Iga, K., GaAlAs/GaAs surface emitting laser with high reflective TiO_2/SiO_2 multilayer Bragg reflector, *6th Conference on Lasers and Electro-Optics,* FO4, 1986.
10. Sakaguchi, T., Koyama, F. and Iga, K., Vertical cavity surface-emitting laser with an AlGaAs/AlAs Bragg reflector, *Electron. Lett.,* 24, 929, 1988.
11. Tamanuki, T., Koyama, F. and Iga, K., Estimation of threshold current of microcavity surface emitting laser with cylindrical waveguide, *Jpn. J. Appl. Phys.,* 30, L593, 1991.
12. Baba, T., Yogo, Y., Suzuki, K., Koyama, F. and Iga, K., Near room temperature continuous wave lasing characteristics of GaInAsP/InP surface emitting laser, *Electron. Lett.,* 29, 913, 1993.

13. Tai, K., Hasnain, G., Wynn, J. D., Fischer, R. J., Wang, Y. H., Weir, B., Gamelin, J. and Cho, A. Y., 90% coupling of top surface emitting GaAs/ AlGaAs quantum-well laser output into 8 μm diameter core silica fiber, *Electron. Lett.*, 26, 1628, 1990.

14. Uchiyama, S. and Iga, K., Two-dimensional array of GaInAsP/InP surface-emitting laser, *Electron. Lett.*, 21, 162, 1985.

15. Iga, K., Koyama, F. and Kinoshita, A., Surface emitting semiconductor laser array: its advantage and future, *J. Vac. Sci. Technol. A*, 7, 842, 1989.

16. Orenstein, M., von Lehmen, A. C., Chang-Hasnain, C., Stoffel, N. G., Harbison, J. P. and Florez, L. T., Matrix addressable vertical cavity surface emitting laser array, *Electron. Lett.*, 27, 437, 1991.

17. Purcell, E. M., Spontaneous emission probabilities at radio frequencies, *Phys. Rev.*, 69, 681, 1946.

18. Drexhage, K. H., Interaction of light with monomolecular dye layers, *Progress in Optics*, ed. E. Wolf, North-Holland, New York, 12, 165, 1974.

19. Kobayashi, T., Segawa, T., Morimoto, Y. and Sueta, T., Novel-type lasers, emitting devices, and functional optical devices by controlling spontaneous emission, *46th Fall Meeting of Japan Society of Applied Physics*, 29a-B-6, 1982 (in Japanese).

20. Kobayashi, T., Morimoto, Y. and Sueta, T., Closed micro-cavity laser, *National Topical Meeting on Radiation Science*, RS85–06, 1985 (in Japanese).

21. De Martini, F. and Jacobovitz, G. R., Anomalous spontaneous-stimulated-decay phase transition and zero-threshold laser action in a microscopic cavity, *Phys. Rev. Lett.*, 60, 1711, 1988.

22. Yokoyama, H. and Brorson, S. D., Rate equation analysis of microcavity lasers, *J. Appl. Phys.*, 66, 4801, 1989.

23. Agarwal, G. P. and Gray, G. R., Intensity and phase noise in microcavity surface-emitting semiconductor lasers, *Appl. Phys. Lett.*, 59, 399, 1991.

24. Björk, G., Karlsson, A. and Yamamoto, Y., On the linewidth of micro-cavity lasers, *Appl. Phys. Lett.*, 60, 304, 1992.

25. Baba, T., Hamano, T., Koyama, F. and Iga, K., Spontaneous emission factor of a microcavity DBR surface emitting laser, *IEEE J. Quantum Electron.*, 27, 1347, 1991.

26. Baba, T., Hamano, T., Koyama, F. and Iga, K., Spontaneous emission factor of a microcavity DBR surface emitting laser (II) — effect of electron quantum confinements., *IEEE J. Quantum Electron.*, 28, 1310, 1992.

27. Yamanishi, M. and Yamamoto, Y., An ultimately low-threshold semiconductor laser with separate quantum confinements of single field mode and single electron-hole pair, *Jpn. J. Appl. Phys.*, 30, L60, 1991.

28. Yamamoto, Y., Machida, S., Igeta, K. and Björk, G., Controlled spontaneous emission in microcavity semiconductor lasers, in *Coherence, Amplification and Quantum Effects in Semiconductor Lasers*, John Wiley & Sons, New York, 1991.

29. Yokoyama, H., Nishi, K., Anan, T., Yamada, H., Brorson, S. D. and Ippen, I. P., Enhanced spontaneous emission from GaAs quantum wells with monolithic optical microcavities, *Appl. Phys. Lett.*, 57, 2814, 1990.

30. Tezuka, T., Nunoue, S., Yoshida, H. and Noda, T., Spontaneous emission enhancement in pillar-type microcavities, *Jpn. J. Appl. Phys.*, 32, L54, 1993.

31. Yamanishi, M., Yamamoto, Y. and Shiotani, T., A novel modulation scheme in semiconductor light emitters with quantum microcavities: high speed intensity modulation by switching of coupling efficiency of spontaneous emission, *IEEE Photon. Technol. Lett.*, 3, 888, 1991.

32. Vilms, J., Wandinger, L. and Klohn, K. L., Optimization of the gallium arsenide injection laser for maximum cw power input, *IEEE J. Quantum Electron.*, QE-2, 80, 1966.

33. Villotte, J. P. and Garault, Y., Dynamic behaviour of semiconductor lasers, *Electron. Lett.*, 11, 206, 1975.

34. Suematsu, Y. and Furuya, K., Theoretical spontaneous emission factor of injection lasers, *Trans. IEICE Jpn.*, 60, 467, 1977.

35. Petermann, K., *IEEE J. Quantum Electron.*, QE-15, 566, 1979.

36. Henry, C. H., Theory of linewidth of semiconductor lasers, *IEEE J. Quantum Electron.*, QE-18, 259, 1982.

37. Tamir, T., Ed., *Guided-Wave Optoelectronics*, Springer-Verlag, Berlin, Germany, 1988, chap. 2.

38. Yablonovitch, E., Inhibited spontaneous emission in solid-state physics and electronics, *Phys. Rev. Lett.*, 58, 2059, 1987.

39. John, S., Strong localization of photons in certain disordered dielectric superlattices, *Phys. Rev. Lett.*, 58, 2486, 1987.

40. Suematsu, Y. and Furuya, K., Quasi-guided modes and related radiation losses in optical dielectric waveguides with external higher index surroundings, *IEEE Trans. Microwave Theory Tech.*, vol. MTT-23, 170, 1975.

41. Corzine, S. W., Geels, R. S., Scott, J. W., Yan, R. H. and Coldren, L. A., Design of Fabry-Perot surface-emitting lasers with a periodic Gain Structure, *IEEE J. Quantum Electron.*, 25, 1513, 1989.

42. Suematsu, Y. and Furuya, K., Propagation mode and scattering loss of a two-dimensional dielectric waveguide with gradual distribution of refractive index, *IEEE Trans. Microwave Theory Tech.*, MTT-20, 524, 1972.

43. Knittl, Z., Optics of thin films, in *An Optical Multilayer Theory*, Wiley-Interscience, New York, 1976.

44. *Handbook of Optical Constants of Solids*, Academic Press, Florida, 1985.

45. Marcatili, E. A. J., Dielectric rectangular waveguide and directional coupler for integrated optics, *Bell Syst. Tech. J.*, 48, 2071, 1969.

46. Koyama, F., Morito, K. and Iga, K., Intensity noise and polarization stability of GaAlAs-GaAs surface emitting lasers, *IEEE J. Quantum Electron.*, 27, 1410, 1991.

47. Asada, M., Kameyama, A. and Suematsu, Y., Gain and intervalence band absorption in quantum-well lasers, *IEEE J. Quantum Electron.*, QE-20, 745, 1984.

48. Asada, M., Miyamoto, Y. and Suematsu, Y., Theoretical gain of quantum-well wire lasers, *Jpn. J. Appl. Phys.*, 24, L95, 1985.

49. Yamada, M., Ishiguro, H. and Nagato, H., Estimation of the intra-band relaxation time in undoped AlGaAs injection lasers, *Jpn. J. Appl. Phys.*, 19, 135, 1980.

50. Corzine, S. W., Yan, R. H. and Coldren, L. A., Theoretical gain in strained InGaAs/AlGaAs quantum wells including valence-band mixing effects, *Appl. Phys. Lett.*, 57, 2835, 1990.

8 Spontaneous and Stimulated Emission in the Microcavity Laser

Hiroyuki Yokoyama

TABLE OF CONTENTS

0-8493-3786-0/95/$0.00+$.50
© 1995 by CRC Press, Inc.

I. INTRODUCTION

In the last decade, marked progress has been achieved in research aimed at controlling spontaneous emission by using wavelength sized cavities (hereafter we shall call this kind of cavity a microcavity). In microcavity research, much work has been specifically directed toward studying the fundamental physics of the interaction of matter with vacuum field fluctuations.[1-3] Originally, many experiments were carried out in the microwave region by using Rydberg state atomic beams.[4-8] More recently, attention has turned to the optical regime in experiments in which atomic beams,[9-11] organic dyes,[12-15] and semiconductors[16-19] are used. In addition to being attractive for studying the fundamental physics of the interaction between materials with vacuum field fluctuations, controlling spontaneous emission is also desirable for light-emitting-device applications.[20-27] In a conventional laser, only a small portion of the spontaneous emission couples into a single state of the electromagnetic field controlled by the laser cavity (that is, the cavity resonant mode formed by the cavity mirrors); the rest is lost to free space modes (that is, it radiates out the side of the laser). This is one of the essential mechanisms behind the occurrence of laser oscillation "threshold" behavior; intense stimulated emission output can be obtained only above a threshold input power that can overcome the spontaneous emission loss to free space modes. If a large amount of spontaneously emitted photons are confined in a cavity whose dimensions are on the order of a single wavelength, loss to free space modes is much decreased and the threshold input power can also be much reduced.

Even though quantum electrodynamics (QED) analyses give many insights for optical phenomena in the microcavity, the essential physics of light emission in microcavity is described by a rather simple formalism. In this article, discussions are mainly developed within the framework of Fermi's golden rule. We wish stress the basics of controlling the spontaneous emission and the laser oscillation of optical microcavities rather than advanced cavity QED approaches. In this manner, from the laser device physics point of view, we can extract an important but simple rule from a huge physical background; that is, "make the cavity small, then spontaneous emission is controlled and the laser oscillation threshold is reduced." This means that the scaling law is still adaptable for a microcavity device of practical size. In other words, the necessary operating power can be decreased linearly depending on the device volume. This chapter describes this essential feature of optical microcavities as well as its physical basis. As an application of the physical principle of the microcavity, a simplified discussion of device design methodology of the microcavity semiconductor laser is also presented.

In Section II, the basics of photon emission in a microcavity are described. Attention is focused on microcavity induced changes in the rate and the spatial distribution of spontaneous emission. Section III describes the operating principle and some operating properties of microcavity lasers. We show that controlling spontaneous emission plays an essential role in the operation of microcavity lasers. In Section IV, discussions concerning the microcavity laser are extended to the semiconductor laser device. It is shown that the strong absorption of a semiconductor drastically alters the light emission properties of the microcavity. Section V describes the prospects for device applications of optical microcavity physics. We show several necessary conditions for realizing novel kinds of light emitting devices which have wavelength sized cavities.

II. PHOTON EMISSION IN MICROCAVITIES

This section describes the basics of controlling spontaneous emission by a cavity. Instead of treating advanced topics of quantum electrodynamics in a cavity (cavity QED), we outline the microcavity induced changes in the rate and the spatial distribution of spontaneous emission. This treatment is based on the mode density modification in Fermi's golden rule, assuming a rather weak interaction between light and materials. However, the validity of the golden rule is also examined in comparison with a strong perturbation formalism using density matrix equations, which can describe oscillatory spontaneous emission, i.e., the so-called vacuum Rabi flopping.

A. FERMI GOLDEN RULE FORMULA AND EMISSION RATE ALTERATION

In the weak interaction limit, the transition rate of a system state is expressed by a first-order perturbation calculation based on Schrödinger's equation. This is well known as Fermi's golden rule. For an optical transition, the expression for the *photon emission rate* of a homogeneously broadened two-level atom in a cavity (or in free space) is given by[28]

$$R_{emi} = \frac{2\pi}{\hbar^2} \int_0^\infty |\langle f|H|i\rangle|^2 \rho(\omega)\, \mathscr{L}(\omega)d\omega. \tag{1}$$

Here, the symbols in the equation are H, the atom-field interaction hamiltonian; $|i\rangle$, the initial state of the system; $|f\rangle$, the final state of the system; $\rho(\omega)$, the density of photon state (mode density); and $L(\omega)$, the final state distribution function of the atomic transition, with the normalization condition

$$\int_0^\infty \mathcal{L}(\omega)d\omega = 1. \tag{2}$$

If the ω dependence of ρ (ω) is much more gentle than that of L (ω), as is schematically shown in Figure 1(a), (1) is reduced to

$$R_{emi} = \frac{2\pi}{\hbar^2}\left|\langle f|H|i\rangle\right|^2 \rho(\omega). \tag{3}$$

Considering the electric dipole transition under the interaction with *quasi-single-mode light*, and noting that $a^\dagger|s\rangle = \sqrt{s+1}\ |s+1\rangle$, the matrix element of (3) can be written as

$$\left|\langle f|H|i\rangle\right| = \omega|u^*|\left|\langle s+1, \Psi_l|(\mathbf{d}\cdot\mathbf{e})a^+|s, \Psi_u\rangle\right|$$

$$= \omega\sqrt{\frac{\hbar}{2\varepsilon_0\omega V}}d\sqrt{s+1}, \tag{4}$$

with

$$d = \sqrt{\frac{1}{3}}\left|\langle \Psi_l|\mathbf{d}\cdot\mathbf{e}|\Psi_u\rangle\right|, \tag{5}$$

and the symbols are V, the effective volume of the cavity (mode volume); $|u^*|$, the normalized spatial distribution function of the quantized electric field; $\mathbf{d}$, the electric dipole vector; $\mathbf{e}$, the unit vector of the electric field; $a^\dagger$, the photon creation operator; $\Psi_{u(l)}$, upper (lower) level wave function of the atom; s, the initial number of photons inside the cavity; ε_0, the dielectric permeability of free space; and $\sqrt{1/3}$, the statistical averaging factor for the electric dipole orientation. Thus, Equation 3 is represented as

$$R_{emi} = \left(\frac{d}{\hbar}\right)^2 \frac{\hbar\omega}{2\varepsilon_0 V}(s+1)\ 2\pi\rho(\omega). \tag{6}$$

The component proportional to s of $(s+1)$ denotes "stimulated emission", while the component proportional to 1 represents "spontaneous emission." Note that the interaction Hamiltonian usually also includes the photon annihilation operator a, which works as $a|s\rangle = \sqrt{s}\ |s-1\rangle$, and describes the (stimulated) photon absorption process. However, in the present discussion, the absorption process is ignored in order to focus on the photon emission process. Actually, in a quasi-four-level approximation as described in Section III, absorption processes are not important. The significant influence of the absorption will be discussed later for describing the operation of a semiconductor microcavity laser.

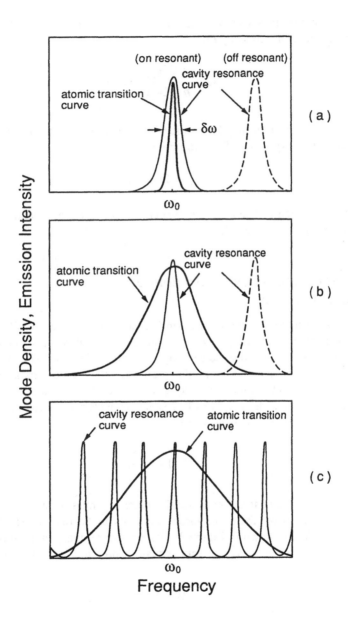

FIGURE 1 Schematic representation of several situations showing the relation between the cavity resonance curve and the transition spectrum of the material. (a) The single-cavity-mode resonance width is broader than the transition width of the material. (b) The material's-transition width is broader than the single-cavity-mode width, but there is still, at most, only one cavity-mode within the transition width. (c) The cavity is large, and there are many cavity resonant modes within the transition width of the material.

We here introduce, for the next section, a notation for the quantum mechanical Rabi frequency Ω:

$$\Omega = \frac{d}{\hbar} \sqrt{\frac{\hbar\omega}{2\varepsilon_0 V}} \sqrt{s+1}.$$

$$(7)$$

The Rabi frequency Ω describes the strength of the atom-field interaction. Using this expression, Equation 6 becomes

$$R_{emi} = \Omega^2 2\pi\rho(\omega)$$

$$(8)$$

If $s = 0$, then Ω becomes the "vacuum Rabi frequency" and R_{emi} describes "spontaneous emission".

B.   EFFECTIVE MODE DENSITY AND EMISSION RATE ALTERATION

Within the framework of Fermi's golden rule, one can easily describe the enhancement or the suppression of spontaneous emission by considering the alteration of the mode density by a cavity.[29] The emission rate given by Equation 6 depends on the mode volume and thus can be characterized by the volume normalized mode density $g(\omega) = \rho(\omega)/V$. $g(\omega)$ is here defined as the "effective mode density". For free space, considering a virtual box cavity, the mode density is calculated to be[30]

$$g_f(\omega) = \frac{\omega^2}{\pi^2 c^3}.$$

$$(9)$$

Regarding the relations between the material emission width and the cavity resonance curve, there are three situations to be distinguished. These are schematically shown in Figure 1. For a resonant quasi single-mode cavity in the case of Figure 1(a), the approximate expression for $g(\omega)$ is given by

$$g_c(\omega) = \frac{Q}{\omega V}.$$

$$(10)$$

Here, Q and V respectively denote the quality factor and the mode volume of the cavity. (10) means that *one mode exists within a frequency width of* $\delta\omega = \omega/Q$, and thus the volume normalized mode density is given by $1/(\delta\omega V_{cav})$. Therefore, the mode density ratio between free space and the cavity is given by

$$F \equiv \frac{g_c(\omega)}{g_f(\omega)} = \frac{\pi^2 c^3 Q}{\omega^3 V} = \frac{\lambda^3 Q}{8\pi V}.$$

$$(11)$$

If V is $(\lambda/2)^3$, F becomes $\sim Q$ and the photon emission rate is enhanced by a factor of $\sim Q$. This is "enhanced spontaneous emission".[29] On the other hand, if the cavity is off-resonant (dotted line curve in Figure 1a), the photon emission rate is decreased by a factor of $\sim 1/Q$, instead of being enhanced. This is "inhibited spontaneous emission".[29]

When the ω dependence of $\rho\,(\omega)$ is sharper than (or comparable to) that of $L\,(\omega)$, as shown in Figure 1(b), the emission rate change is evaluated by integrating (1) instead of (3). Note that, as shown in Figure 1(c), if there are many cavity resonant modes inside the spectral curve of $L\,(\omega)$, and the integrated value of $\rho\,(\omega)$ within the spectral curve of $L\,(\omega)$ is similar to that of free space, then the photon emission rate is not modified (i.e., the integration of (1) is not changed by the presence of the cavity). This is the general situation for conventional laser cavities.

Ideally, if the effective mode density is calculated for a certain cavity, one can obtain how much enhancement or suppression is induced in the spontaneous emission rate. For a few cavity configurations with complete reflectors, the direct mode density calculation can give closed form expressions of the spontaneous emission rate change.[23] However, for cavities with incomplete metallic reflectors or dielectric layered reflectors, the emission rate change is to be evaluated by the full integration of spatial emission distribution patterns as described in Chapter 5 by Brorson. In the latter case, the effective mode density is obtained as a final result of spontaneous emission rate calculation. Several examples of spontaneous emission rate calculations are shown in the different chapters in this book.

C. RABI FLOPPING IN THE CAVITY

Most of the operation properties of a typical microcavity laser are described by rate equation analyses, as is shown in the rest of this chapter. In this treatment, Fermi's golden rule works very well. However, on some occasions when there is a strong light-matter interaction, the golden rule description breaks down. In order to describe the strong interaction between an atom and the cavity field, a strong perturbation formula is necessary. The formula should also give a result similar to the golden rule calculation in the weak interaction limit. This approach will thereby also indicate the validity of the golden rule. Here, we assume that the only relaxation process experienced by the electric dipole is due to the atomic transition. For the simplest case with a single excited atom and a single cavity mode, the following quantum mechanical density matrix equations of motion can be used.[31]

$$\dot{\rho}_{11} = i\Omega\left(\rho_{12} - \rho_{21}\right) \tag{12}$$

$$\dot{\rho}_{22} = -\gamma\rho_{22} - i\Omega\left(\rho_{12} - \rho_{21}\right) \tag{13}$$

$$\dot{\rho}_{33} = -\gamma\rho_{22} \tag{14}$$

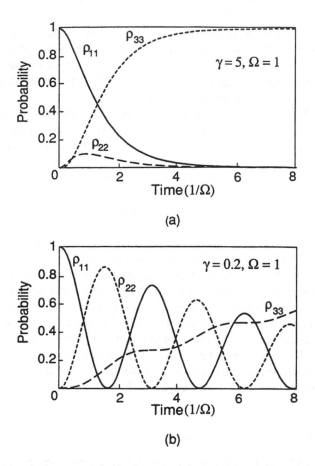

FIGURE 2 Numerical solutions of Equations 12 to 15, showing the photon emission of a single two-level atom in a cavity. (a) Strong damping regime: $\Omega = 1$, $\gamma = 5$. (b) Weak damping oscillatory regime: $\Omega = 1$, $\gamma = 0.2$.

$$\left(\dot{\rho}_{12} - \dot{\rho}_{21}\right) = 2i\Omega\left(\rho_{11} - \rho_{22}\right) - \frac{\gamma}{2}\left(\rho_{12} - \rho_{21}\right) \tag{15}$$

with variables Ω, the vacuum Rabi frequency (in the present case); ρ_{11}, the probability for an upper-level atom without photon (i.e., the probability for the state $|\Psi_u > |s = 0 >$); ρ_{22}, the probability for a lower-level atom with one photon (i.e., the probability for the state $|\Psi_1 > |s = 1 >$); ρ_{33}, the probability for a lower-level atom without a photon (or the probability for finding the emitted photon outside the cavity; i.e., for the state $|\Psi_l > |s = 0 >$); and $\gamma = \omega/Q$, the rate of photon damping in the cavity. Note that ρ_{22} is the probability of the photon existing inside the cavity, and the (statistically averaged) emitted power detectable outside the cavity is proportional to $\gamma \rho_{22}$.

A couple of numerical results are shown in Figure 2. For $\Omega \ll \gamma$, the excited atom shows a quasi-exponential decay like conventional spontaneous emission while for $\Omega \gg \gamma$, the oscillatory behavior of so-called vacuum Rabi flopping occurs. This means the reversible oscillatory behavior takes place even in spontaneous emission.[32] Figure 3 indicates the emission spectra obtained by Fourier transform of the time-domain behavior. The spectrum of the decaying Rabi flopping shows the two split peaks, which is called vacuum Rabi splitting.

Here, we derive approximate expressions of the above differential equations under a strong damping condition of $\gamma \gg \Omega$. Assuming that $d\rho_{nm}/dt \ll \gamma \rho_{nm}$, we obtain the following expression from Equation 15.

$$\rho_{11} - \rho_{22} = \frac{\gamma(\rho_{12} - \rho_{21})}{4i\Omega}. \tag{16}$$

Thus, choosing $\rho_{11}(t = 0) = 1$, and taking into account the resultant condition $\rho_{22} \ll \rho_{11}$, Equations 12 and 13 are reduced to

$$\dot{\rho}_{11} = -4\Omega^2 \frac{\rho_{11}}{\gamma} \tag{17}$$

$$\dot{\rho}_{22} = -\gamma\rho_{22} + 4\Omega^2 \frac{\rho_{11}}{\gamma}. \tag{18}$$

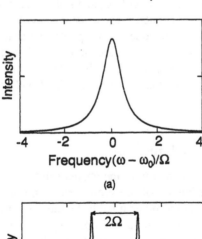

(a)

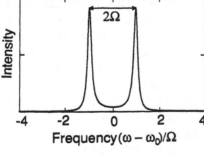

(b)

FIGURE 3 Fourier power spectra corresponding to Figure 2. (a) Strong damping regime: $\Omega = 1$, $\gamma = 5$. (b) Weak damping oscillatory regime: $\Omega = 1$, $\gamma = 0.2$.

Equation 17 indicates that ρ_{11} decays exponentially with a rate of $4\Omega^2/\gamma$. Therefore, the spontaneous emission rate modified by the cavity is

$$A_c = \frac{4\Omega_c^2}{\gamma} \equiv 4\left(\frac{d}{\hbar}\right)^2 \frac{\hbar\omega}{2\varepsilon_0 V} \frac{Q}{\omega},\qquad(19)$$

while the spontaneous emission rate in free space is

$$A_f = \Omega_f^2 2\pi\rho(\omega) \equiv \left(\frac{d}{\hbar}\right)^2 \frac{\hbar\omega^3}{\varepsilon_0 \pi c^3},\qquad(20)$$

where Ω_c and Ω_f respectively represent the vacuum Rabi frequency in the cavity and in free space. Taking the ratio of Equations 19 and 20, we obtain

$$F \equiv \frac{A_c}{A_f} = \frac{\lambda^3 Q}{4\pi^2 V}.\qquad(21)$$

This is approximately equal to Equation 11, which is derived from the Fermi's golden rule argument. It should be noted that Equation 18 is the photon rate equation with a single excited atom inside the cavity, and now the cavity effect is doubly included in this, once in the photon emission rate, and again in the photon damping rate. From the laser point of view, this can be an essential feature of the microcavity, which is different from the case of a large conventional cavity. This point is also described in Section III as related to the analysis of laser operation.

D. RADIATION PATTERN ALTERATION

Next, we discuss, instead of the changes in temporal behavior, the change in the spatial distribution of the spontaneous emission intensity caused by a microcavity. This is the other notable feature of a microcavity which has a wavelength sized geometry. As a simple example, we choose a planar cavity structure because the drastic changes are easily shown schematically for this case. However, we here assume a planar cavity filled with an absorptive medium in which light emitting dipoles are embedded. Discussing such a cavity is also useful for later sections.

A finite transmission loss of the reflectors is introduced in the present analysis. Then, the spatial radiation pattern of the spontaneous emission is calculated by using the reciprocity principle,[33] i.e., seeing how the atomic absorption of radiation is modified by the cavity. Equivalently, this can be considered from the quantum mechanical point of view as being due to the change in the zero-point fluctuation amplitude inside the cavity.

The field intensity inside the cavity is obtained by summing bidirectional multireflected waves and averaging over the medium's thickness (eliminating the standing wave effect).[34] The ratio of the intensity inside to the intensity outside, which describes the enhancement (or suppression) of an incident plane wave electromagnetic field in a FP cavity, becomes

$$\frac{I}{I_0} = \frac{(1-R)\{1-\exp(-\alpha L/\cos\theta_i)\}\{1+R\exp(-\alpha L/\cos\theta_i)\}}{(\alpha L/\cos\theta_i)\{1+R^2\exp(-2\alpha L/\cos\theta_i)-2R\exp(-\alpha L/\cos\theta_i)\cos(2\eta\omega L\cos\theta_i/c)\}}.$$

$$(22)$$

Here the symbols are I_0, the incident field intensity; I, the field intensity seen by the dipole inside the cavity; R, the power reflectivity for each reflector; θ_i, the incident angle in the x–z plane (inside the cavity); L, the cavity length; α, the power absorption coefficient of the medium; and η, the refractive index of the medium.

Figure 4 shows radiation patterns in the x–z plane for an x-axis oriented electric dipole distributed homogeneously inside the present planar cavity. The curves depicted in Figures 4(b) to (d) show spatial radiation distributions for slightly different cavity lengths around $L \sim \lambda/2$ assuming a reflectivity $R = 0.95$. The distance between a point on the curve and the origin corresponds to the emission intensity. The emission intensity distribution in free space is also shown in Figure 4(a) for comparison; the free space emission intensity in the z-axis direction is normalized to be unity. In Figure 4, the absorbance inside the cavity is assumed to be rather small but not negligible. For a monochromatic dipole having an emission wavelength equal to the cut-off frequency $\omega_c = \pi c/L$, the emission intensity around the cavity axis direction is enhanced by a factor of nearly $(1 + R) / (1 - R) \approx 40$ as shown in Figure 4(b), and the emission is greatly suppressed in the other directions. When the cavity length L is slightly larger than $\sim\lambda/2$, as is shown in Figure 4(c), the emission cone spreads off the cavity axis, while the emission intensity is decreased in all directions if L is smaller than $\sim\lambda/2$ (Figure 4(d)). The case of Figure 4(d) shows "inhibited spontaneous emission" below the cut-off frequency.[29]

The reader is reminded that considering such condensed materials as semiconductors or dyes, the emission will be broadband rather than monochromatic; this results in a broadening of the directionality of the spontaneous emission. This situation corresponds, in the spectral domain, to Figure 1(b). On the other hand, such a spatial distribution as shown in Figure 4(b) corresponds to the spectral distribution of Figure 1(a).

If we assume a pair of $R = 1$ reflectors and no absorption loss, the emission cone does not have a spread in the radiation angle for a monochromatic dipole (for example, the emission couples to just one single plane wave mode for the $L = \lambda/2$ cavity). Thus, it should be noted that a nonzero spread in the radiation angle of the emitted power in the present case is due to both the less than unity reflectivity and the absorption. This angular distribution reflects the mode size of a single cavity (quasi) mode; the beam divergence angle $\delta\theta$ is related to the

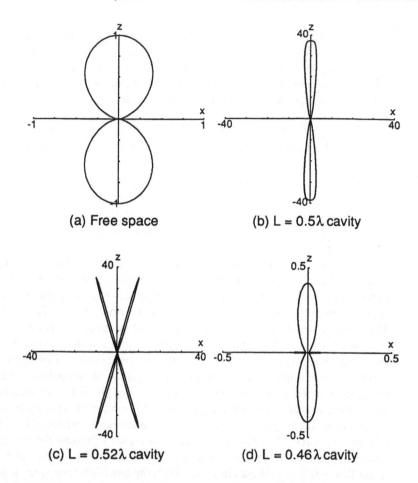

(a) Free space   (b) L = 0.5λ cavity

(c) L = 0.52λ cavity   (d) L = 0.46λ cavity

FIGURE 4 Emission intensity distribution patterns of dipoles directed along the x-axis; the views in the x-z plane. (a) Free space emission; (b) to (d) emission from the dipoles homogeneously distributed inside the planar cavity with slightly different mirror distances near λ/2. It is assumed that the absorbance is rather small ($\alpha L = 0.1$). (From Yokoyama et al., *Opt. Quantum Electron.*, **24**, S252, 1992. With permission from Chapman & Hall.)

mode diameter D by $\delta\theta \sim \lambda/D$. A theoretical base for the passive cavity mode of a planar microcavity was given by Ujihara.[35]

The influence of strong absorption is shown in Figure 5 as an example. This emission pattern is calculated by assuming a rather strong absorption while the other parameters are the same as those for Figure 4(b). We find a strongly absorptive resonant cavity only weakens the emission intensity instead of increasing it into the cavity axis. This fact significantly affects the device design concepts, which issue is described in Sections IV and V.

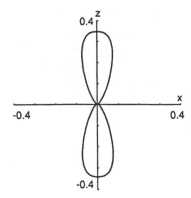

FIGURE 5 Emission intensity distribution patterns of dipoles directed along the x-axis inside the $\lambda/2$ planar cavity under a large absorbance of $\alpha L = 1$. The view is in the x-z plane.

III. MICROCAVITY LASERS

Controlling spontaneous emission is also strongly related with the stimulated emission processes. In this section, we analyze the laser oscillation properties of optical microcavities within the framework of Fermi's golden rule.

A. RATE EQUATIONS FOR A CLOSED CAVITY SINGLE MODE LASER

First, we consider the situation when two-level atoms are located inside a closed microcavity. Although our interest has been focused on spontaneous emission in Section II, emission rate alteration is also expected in stimulated emission as shown in Equations 6 to 11. The overall photon emission rate R_{emi} for an atom is re-expressed using Equation 6 as

$$R_{emi} = A_c \, (s + 1), \tag{23}$$

where A_c is the cavity modified spontaneous emission rate. Here, for simplicity, we assume that the lower laser state is rapidly depopulated, and there are no nonradiative processes, no inversion saturation effects, and no collective spontaneous emission effects of n atoms. (These conditions could be almost satisfied in a laser system consisting of such condensed materials as semiconductors and dyes.) Then, the rate equation for the number of photons inside the cavity can be written as[17,22,27]

$$\dot{s} = A_c(s+1)n - \gamma s. \tag{24}$$

The first term in the right-hand side represents the total photon emission rate, and the second term is the photon escape rate. Note in this equation that *the cavity effect doubly appears (i.e., in both the terms of the right-hand side).* As shown in Equation 18, a similar form is obtained starting from a set of density matrix equations for single atom emission under the strong damping condition (golden rule regime). The rate equation for the population inversion is expressed as a combination of the pumping rate p and the photon emission rate given in Equation 24 as

$$\dot{n} = p - A_c(s+1)n. \tag{25}$$

The static solution of these coupled rate equations is simple but noteworthy:

$$s = \frac{p}{\gamma}, \text{ and } n = \frac{p\gamma}{A_c(p+\gamma)}. \tag{26}$$

We see that the light output increases linearly with increased pumping for any pumping rate. In other words, this device works as a "thresholdless laser",[17,22,27,36] as long as we focus our attention on the output versus input characteristics. This occurs because all photons are emitted into the single cavity mode. Note, however, that as the pumping increases, n approaches a fixed value γ/A_c, which is the threshold inversion value. This means that the light output gradually changes from spontaneous emission to stimulated emission in nature. We will discuss this point again later. This thresholdless laser operation is different from that of the "one atom maser (laser)".[6] In that case, the operation criterion is much more strict, that is, the gain by single atom population inversion overcomes an extremely low cavity loss.

It should be noted that enhanced spontaneous emission ($A_c > A_f$), is not the necessary condition for the disappearance of an apparent threshold. The essential point is that only one (quasi) mode exists within the emission width of the laser material. However, an increase in the spontaneous emission rate has some advantages from the device point of view. For example, the dynamic response of the device to modulation will be improved as a result of the increased spontaneous emission rate.

B. RATE EQUATIONS FOR AN OPEN CAVITY SINGLE MODE LASER

So far, we have considered the case of a completely closed cavity resonator. Now we would like to generalize to the case of an open resonator. We assure that the number of cavity resonant modes is still one, but now other mode exist which correspond to photons leaving the open cavity. We assume that the spontaneous emission into the cavity mode can still be enhanced, but the free space modes have the free space spontaneous emission rate. This corresponds to the case discussed by Heinzen et al. for spontaneous emission

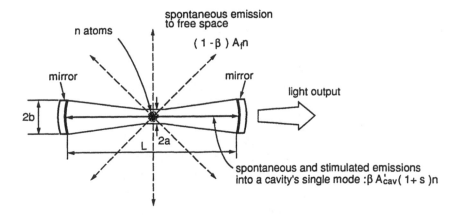

FIGURE 6   Schematic representation of photon emission into free space and into a cavity mode for a confocal FP cavity laser system. Spontaneous emission coupled to many free space modes occurs at the free space rate. On the other hand, within the solid angle of the single (degenerate) cavity mode, enhanced spontaneous emission takes place, as well as stimulated emission, depending on pumping power.

in a confocal or a concentric FP cavity.[10] The physical setup is shown in Figure 6. We take the fractional ratio of the solid angle subtended by the cavity mode to the free space modes to be β. For a confocal cavity, the relations among the value of β, the mode volume V, the cavity mode solid angle $\Delta\Omega$, the mirror radius b, the central beam diameter a, and the cavity length L are given by $a \sim \lambda L/2\pi b$, $V \sim 2\pi a^2 L = \lambda^2 L^3/(2\pi b^2)$, $\Delta\Omega = 8\pi b^2/L^2$, and $\beta = \Delta\Omega/4\pi = 2b^2/L^2 = \lambda^2 L/\pi V$.[10] Thus, β is proportional to the inverse of the mode volume V; this means that a smaller beam cross section results in larger β. If the relation between the cavity mode and the material's emission width is such as shown in Figure 1(a), the spontaneous emission rate within the cavity mode volume is given by $A_{oc} \sim A_f/(1-R)$. βA_{oc} gives the same expression with A_c as in Equation 19. It should be noted that this argument is precise only when the solid angle $\Delta\Omega$ is sufficiently small. However, the present picture is very convenient in order to understand the open microcavity system intuitively. Thus, we here assume that the same discussion is applicable to a large solid angle cavity geometry.

We now extend the discussion for laser oscillation. Taking into account the contribution of stimulated emission into the cavity mode, the rate equations can be represented as[17,22,27]

$$\dot{n} = p - (1-\beta)A_f n - \beta A_{oc}(s+1)n, \qquad (27)$$

$$\dot{s} = \beta A_{oc}(s+1)n - \gamma s. \qquad (28)$$

Here, s is now the number of photons in the cavity mode. In the right-hand side of the inversion rate equation, the second term represents the spontaneous

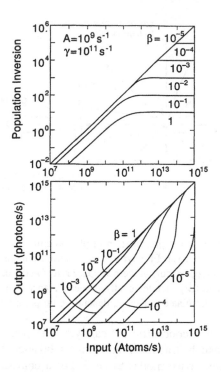

FIGURE 7　Calculated light output S_{out} and population inversion n vs. pumping p of microcavity four-level-like lasers on a logarithmic scale. $A_f = 10^9$ s^{-1}, $A_{oc} = 10\ A_f$, $\gamma = 10^{12}$ s^{-1}. (From Yokoyama et al., *Opt. Quantum Electron.*, **24**, S265, 1992. With permission from Chapman & Hall.)

emission rate for free space modes, and the third term means the overall photon emission rate in the cavity. Note that stimulated emission is taken into account only for the cavity mode.

C.　INPUT-OUTPUT PROPERTIES

Figure 7 shows steady-state solutions of Equations 27 and 28, with logarithmic scales. β is the parameter in this calculation. When the β value is very small, clear thresholds are observed in the input-output curves. (In conventional laser devices, β ranges from 10^{-10} to 10^{-5}.) In the mode point of view, the excited atoms are mostly coupled with free space modes in an open cavity of small β, even though there is only one cavity mode within the emission bandwidth. That is to say that most of the spontaneous emission escapes from the open side of a conventional laser cavity. In that situation, the cavity mode photon number can only increase rapidly above the "threshold" due to stimulated emission. Thus, the phase transition of photons in the cavity mode occurs at the threshold. It is seen that the threshold becomes unclear as β increases, and it disappears from the input-output curve at $\beta = 1$. This is because all the emitted photons couple into the single cavity resonance mode. However, if we pay attention to the behavior of the population inversion, the difference between the spontaneous emission dominant region and the laser oscillation region is recognized in the figure even for $\beta = 1$. With increasing pumping, n linearly increases within the spontaneous emission dominant region. On the other hand, in the laser oscillation region, n is clamped at the lasing threshold level. Thus, in that sense, a lasing threshold still exists although it does not appear in the input-output curves.

Note that both the threshold population inversion n_{th} ($= \gamma / \beta A_{oc}$) and the threshold pumping rate (i.e., input power) decrease with increasing β. This is related to the decrease in the mode volume. For a confocal or concentric cavity, in which the laser medium is located at the focal point, increasing β corresponds to an increase in the mirror size and thus a decrease in the beam diameter (keeping the cavity length constant).

When $\beta \sim 1$, the threshold pumping rate stays at $p_{th} = \gamma$ independent of A, although $n_{th} = \gamma / A$ decreases with increasing A. This is because more input power is necessary to give the same n_{th} value under an increased spontaneous decay rate A.

D.   RATIO AMOUNT OF SPONTANEOUS EMISSION COUPLED INTO THE CAVITY MODE

In Equations 27 and 28, there are two kinds of spontaneous emission rates; one is the free space rate and the other is the rate modified by a cavity. Introducing the total spontaneous emission rate A_T, the ratio amount of spontaneous emission coupled to the cavity mode (hereafter this is called the C factor) is defined as

$$C = \frac{\beta A_{oc}}{(1-\beta)A_f + \beta A_{oc}} \equiv \frac{\beta A_{oc}}{A_T}.$$

If we assume $A_T \approx A_f$, and express $A_T = A$, Equations 27 and 28 can be rewritten as

$$\dot{n} = p - (1-C)An - CA(s+1)n, \tag{29}$$

$$\dot{s} = CA(s+1)n - \gamma s. \tag{30}$$

These equations are convenient for analysis of the laser oscillation properties of arbitrary microcavity structures, ignoring the change in the total spontaneous emission rate. It should be noted here, in order to avoid confusion, that Baba and Iga, in Chapter 7, use the same notation C for the ratio amount of spontaneous emission captured in a single cavity mode, but Björk and Yamamoto, in Chapter 6, use the notation β for the same quantity. In this present chapter, the notation β is used for the fraction of the solid angle subtended by a single cavity mode instead of the fraction amount.

The ratio amount C of spontaneous emission coupled to the single cavity mode can also be mathematically expressed via Fermi's golden rule as follows. We now consider the photon emission process of a two-level atom put in a cavity whose size is reasonably larger than the wavelength of emitted light. In this situation, returning to the fundamental expression of the golden rule,[30] the photon emission rate of an excited atom inside the cavity is written as

$$R_{emi} = \sum_{\mathbf{k}} \frac{2\pi}{\hbar^2} \left| \langle f|H|i \rangle \right|_{\mathbf{k}}^2 \delta(\omega_0 - \omega_{\mathbf{k}}).$$

(31)

Here, the matrix element is now expressed as

$$\left| \langle f|H|i \rangle \right| = \omega_{\mathbf{k}} \left| u_{\mathbf{k}}^* \right| \left| \langle s_{\mathbf{k}}+1, \Psi_l | (\mathbf{d} \cdot \mathbf{e}_{\mathbf{k}}) a_{\mathbf{k}}^\dagger | s_{\mathbf{k}}, \Psi_u \rangle \right|$$

$$= \omega_{\mathbf{k}} \sqrt{\frac{\hbar}{2\varepsilon_0 \omega_{\mathbf{k}} V}} d_{\mathbf{k}} \sqrt{s_{\mathbf{k}}+1},$$

(32)

with

$$d_{\mathbf{k}} = \left| \langle \Psi_l | \mathbf{d} \cdot \mathbf{e}_{\mathbf{k}} | \Psi_u \rangle \right|.$$

(33)

In Equations 31 to 33, the subscript $\mathbf{k}$ represents the mode having wavevector $\mathbf{k}$, ω_0 is the angular frequency corresponding to the energy difference between the two levels of the atom, $\omega_{\mathbf{k}}$ is the angular frequency of the photon in $\mathbf{k}$ mode, and the other symbols have the same meanings as those in Equation 4 with the subscript representing the $\mathbf{k}$ mode. Taking into account the statistical average of the electric dipole orientation as in Equation 5, Equation 31 can be rewritten as

$$R_{emi} = \sum_{\mathbf{k}} \left(\frac{d}{\hbar}\right)^2 \frac{\hbar\omega_{\mathbf{k}}}{2\varepsilon_0 V}(s_{\mathbf{k}}+1)\delta(\omega_0 - \omega_{\mathbf{k}}).$$

(34)

This equation represents the summation of stimulated and spontaneous emission components for every $\mathbf{k}$ mode.

Now suppose that there are m cavity modes satisfying $|\omega_{\mathbf{k}}| \approx \omega_0$. In this case, Equation 34 can be expressed as

$$R_{emi} = \sum_{i}^{m} \frac{\omega_0 d^2}{2\hbar\varepsilon_0 cV}(s_i+1)\delta(k_0 - |\mathbf{k}_i|),$$

(35)

with $|k_i| = \omega_k/c$ for the ith mode. If there are no photons initially in any modes, we have $s_i = 0$, and the sum in Equation 35 gives the spontaneous emission rate A as

$$A = \sum_i^m \frac{\omega_0 d^2}{2\hbar\varepsilon_0 cV}\delta\left(k_0 - |\mathbf{k}_i|\right).$$

(36)

Representing the spontaneous emission rate in the ith mode as A_i and $A_i/A = C_i$, Equation 36 can be re-expressed as

$$A = \sum_i^m A_i = \sum_i^m C_i A.$$

(37)

Therefore, the total photon emission rate in the ith mode, including the stimulated emission, can be represented as

$$R_i = C_i A (s_i + 1).$$

(38)

If the spontaneous emission is equally divided into m modes, C_i becomes $1/m$. Assuming that, for whatever reason, the laser oscillation occurs in only one particular mode, we can denote the fractional amount of spontaneous emission coupled into this mode as C, and the other total fraction as $(1 - C)$. Thus, it is possible to obtain the expression for the spontaneous emission coupling ratio, i.e., the C factor, without considering the local increase (or decrease) in the mode density.

Now, a useful explicit expression for the C factor is discussed. We consider a certain cavity structure, for example, a box cavity filled by a material having a refractive index of η, but having dimensions which are still much larger than the wavelength of interest. In this situation, the C factor is the inverse of the number of cavity resonant modes m included in the material's emission bandwidth $\Delta\omega$. In order to obtain the number of modes m, we can utilize an expression of the volume normalized mode density. The mode density $g(\omega)$ is given by

$$g(\omega) = \frac{\omega^2 \eta^3}{\pi^2 c^3}.$$

(39)

This expression is based on Equation 9 but multiplied by η^3 because the cavity is now filled with a material having a refractive index of η. Supposing that the mode volume V is the same as the cavity volume (i.e., the mode is well confined in the box), then the number of cavity resonant modes m within the bandwidth $\Delta\omega$ (FWHM) is approximately given by

$$m \approx g(\omega)V \Delta\omega.$$

(40)

Therefore, using the relation $\Delta\omega = 2\pi\,\Delta\lambda c\,/\lambda^2$, the C factor is expressed as

$$C = \frac{1}{m} \approx \frac{\pi^2 c^3}{\omega^2 \eta^3 V \Delta\omega} = \frac{\lambda^4}{8\pi\eta^3 V \Delta\lambda}. \tag{41}$$

If we assume a Lorentzian line shape and define the number of modes within the width for the rectangle having the area same as the given Lorenzian,[37] the number m is multiplied by a factor of $\pi/2$, and the C factor becomes

$$C \approx \frac{\lambda^4}{4\pi^2 \eta^3 V \Delta\lambda}. \tag{42}$$

If the mode confinement in the box is not complete, the mode volume V is replaced by V/f_v, where f_v (≤ 1) corresponds to the mode confinement factor in the active material. It should be noted that counting the number of modes is not restricted to a box. We can consider that the space volume V of any kind of shapes has the same number of modes within the same frequency width. Therefore, the present mode counting procedure can be adopted even for an open laser cavity having a mode volume V.

E.   RELATION BETWEEN THE MICROCAVITY LASER RATE EQUATIONS AND THE CONVENTIONAL LASER RATE EQUATIONS

In the derivation of a pair of microcavity laser rate equations, we have started from Fermi's golden rule in order to describe the alteration in the spontaneous emission rate. Thus, it may seem that these equations are not consistent with conventional semiclassical laser theory. However, we here show that the present microcavity laser rate equations are equivalent to the ones derived from the semiclassical theory when the C factor is much smaller than unity.

In the semiclassical laser theory, starting from the plane wave Maxwell-Bloch equations for a single mode standing wave laser,[38] the following rate equations can be derived assuming a homogeneously broadened two-level laser material and central frequency tuning.

$$\dot{N} = P - \frac{\omega T_2 d^2 F_l}{\hbar\varepsilon} NS - \frac{1}{T_1} N, \tag{43}$$

$$\dot{S} = \frac{\omega T_2 d^2 F_l}{\hbar\varepsilon} NS - \gamma S. \tag{44}$$

Here, N is the inverted population density (i.e., number per unit volume), S is the photon density, and P is the pumping density—all in the laser oscillation mode. Also, γ is the photon damping rate, T_1 and T_2 are respectively the longitudinal and transverse relaxation times in the optical Bloch equations, d is the absolute value of the dipole moment, ε is the permittivity, and F_l is the local field correction factor, which is given by

$$F_l = \frac{\eta^2 + 2}{3},$$ (45)

for a refractive index of η.[38] When η is put to unity — the value of vacuum — F_l becomes unity and the permittivity ε takes its vacuum value ε_0. For obtaining the above rate equations, we here again assumed that the lower laser state is quite rapidly depopulated. The second (first) term of the right-hand side of Equations 43 and 44 represents the stimulated emission. The rate $1/T_1$ usually corresponds to the spontaneous emission rate if there are no nonradiative processes. The contribution of spontaneous emission is not included in the photon rate Equation 44, which is a consequence of the semiclassical approach. Therefore, in conventional rate equation analysis, the coupling of the spontaneous emission is artificially added independent from the stimulated emission.

Next, we modify the microcavity laser rate Equations 29 and 30 for the purpose of comparing these with the rate equations derived from the semiclassical theory. The rate Equations 29 and 30, which are for the number of population inversions and for the number of photons inside the cavity, can be converted to density equations by dividing by the volume V as

$$\dot{N} = P - (1 - C)\,AN - CAVSN + CAN,$$ (46)

$$\dot{S} = CAVSN + CAN - \gamma S.$$ (47)

Here, N, S, and P are respectively n/V, s/V, and p/V. Now, the term $CAVSN$ represents the contribution of stimulated emission into the laser oscillation mode. Based on the conventional Fermi golden rule formula, and using Equation 39 for the mode density, the spontaneous emission rate A can be expressed as

$$A = \frac{\pi^2 \omega d^2}{2\hbar\varepsilon}\,g(\omega)F_l = \frac{\omega^3 d^2 \eta^3}{2\hbar\varepsilon c^3}\,F_l.$$ (48)

Using Equation 48 and the expression for C given by Equation 42, and noting that $A = 1/T_1$ and $\Delta\omega = 2/T_2$, Equations 46 and 47 are rewritten as

$$\dot{N} = P - \frac{\omega T_2 d^2 F_l}{\hbar\varepsilon}\,NS - \frac{1}{T_1}N,$$ (49)

$$\dot{S} = \frac{\omega T_2 d^2 F_l}{\hbar \varepsilon} NS + \frac{C}{T_1} N - \gamma S.$$

$$(50)$$

Comparing these two equations with the semiclassical ones, Equations 43 and 44, we can see that the stimulated emission terms are the same in these two sets of rate equations. However, in the photon in Equation 50, the contribution of spontaneous emission is now incorporated properly.

F.   LINEWIDTH AND NOISE OF THE MICROCAVITY LASER

Although discussions about the noise properties of microcavity lasers are not the subject of the present article, we here would like to say a few things regarding this subject. For the laser oscillation linewidth, Björk and Yamamoto gave the expression from an equivalent electric circuit model as[39]

$$\Delta \omega = \gamma - CAn.$$

$$(51)$$

Thus, the linewidth narrowing behavior is quite similar to that of a conventional laser except for the large C value of the microcavity. Furthermore, Yamamoto et al. have shown the possibility of generating photon number squeezed states with very low excitation when the pumping fluctuations are suppressed.[24,26,40] It should be pointed out that the number state could be obtained even in the spontaneous emission regime if the C value is near unity. This is because if regulated excitation is achieved, one can obtain a regulated train of photons in the single cavity mode.

IV.   MICROCAVITY SEMICONDUCTOR LASERS

The main subject of this section is to describe some possible prescriptions for ultra-low-threshold microcavity laser devices. We focus our arguments on the threshold current in microcavity semiconductor lasers based on the rate equation analysis.

A.   BACKGROUND OF THE DEVICE RESEARCH

Before the analytical argument for the laser threshold, we will briefly review the history of ultra-low-threshold semiconductor lasers.

Following the successful room temperature continuous operation of a vertical cavity surface emitting laser (VCSEL) device,[41] significant progress has been made in the last decade for this kind of diode laser.[42-44] Presently, submilliampere threshold devices are obtained in monolithic micropost cavity structures, and a record low threshold of 190 μA has been recently demonstrated.[45] As a new approach for the ultra-low-threshold microcavity laser device, a microdisk laser with whispering gallery (WG) mode operation was fabricated,[46] and a submilliampere threshold was also reported.[47] Furthermore, the performance of the microcylinder diode laser operated with WG modes was examined.[48] Aside from the effort devoted to developing new kinds of microlasers, a marked reduction in the threshold current was also made in the conventional stripe-type laser by reducing the cavity length and incorporating a high reflectivity coating on the cleaved facets. A lasing threshold current of as low as ~250 μA was obtained in this kind of device.[49]

It should be noted that, although the above low threshold current devices are not always developed to be microcavities, *the essential reason for the low threshold is the device-volume reduction*. In other words, as with the microcavity, when the mode volume is reduced, then the C factor is increased. Therefore, from a simple minded viewpoint, a further reduction of the device size may result in tens of microampere threshold laser devices. However, the volume scaling law of the threshold current reduction breaks down in the microcavity semiconductor laser. This is mainly due to the faster nonradiative relaxation of the excited carriers (population inversion) at the surface. In the following, we attempt to determine the lasing threshold current of a microcavity semiconductor laser device and then explore the conditions for realizing ultra-low-threshold lasers.

B.   RATE EQUATIONS OF MICROCAVITY SEMICONDUCTOR LASERS

In Section III, we considered the ideal microcavity laser system. In the operation of a semiconductor microcavity laser device, the influence of the nonnegligible nonradiative depopulation rate (Γ) must be taken into account. Furthermore, band-to-band optical transitions of electrons and holes should be considered instead of an ideally population inverted two-level atomic transition. However, we simplify the latter problem, as previously reported, by introducing the transparency carrier density N_0, and continue to regard the stimulated emission rate as depending linearly upon the carrier density.[39,50] This simplification is also widely used in the analysis of conventional semiconductor lasers. We accordingly modify the rate Equations 29 and 30 as

$$\dot{n} = p - (1 - C)An - CA\left[n + s(n - n_0)\right] - \Gamma n, \qquad (52)$$

$$\dot{s} = CA\left[n + s(n - n_0)\right] - \gamma s. \qquad (53)$$

We take the ratio of the active material volume V_a to the mode volume V to be $f_v = V_a/V$ (this is the so-called confinement factor), giving $n_0 = Vf_vN_0$. Here, we consider the situation that the coupling fraction of spontaneous emission C is less than 0.1, and an approximate value of C is given by Equation 42 in Section III.

If we define the threshold to be the balance of the stimulated emission rate and the photon damping rate, the threshold lasing current I_{th} is given by the steady-state solution of Equations 52 and 53 as

$$\frac{I_{th}}{q} = P_{th} = \left[\frac{\gamma}{C} + \frac{\gamma\Gamma}{CA}\right] + \left[(1-C)A + \Gamma\right]N_0 V f_v, \tag{54}$$

where q is the electron charge. As has already been discussed, a conventional laser analysis uses *density* rate equations of photons and carriers instead of population *number* rate equations, and the inclusion of spontaneous emission is of secondary importance; the stimulated emission rate is the most important factor. Thus, the threshold current is calculated from the injection carrier rate per unit active volume by multiplying by actual volume. On the other hand, in the present number rate equations, the stimulated emission factor is expressed by the spontaneous emission rate A and the C factor. However, in the small C regime, both approaches are equivalent, and should give the same result.

C. LOW THRESHOLD CURRENT PERFORMANCE OF MICROCAVITY SEMICONDUCTOR LASERS

As has been pointed out by other authors,[50] the threshold current given in Equation 54 consists of two components; the first bracket of the right-hand side indicates the component dominated by the cavity structure, and the second bracket is mainly attributed to the number of carriers ($N_0 V f_v$) necessary for transparency. We find that the cavity parameters, which determine the threshold input, are the photon damping rate γ, the mode volume V, and the ratio of the active material volume f_v (thus, the confinement factor). Next, we consider the influence of the nonradiative recombination of minority carriers, which are generated by current injection and contribute to light emission. This is a problem separated from the optical cavity issue, the nature of which depends on the semiconductor materials and device processing technologies. Equation 54 shows that if the nonradiative recombination rate Γ is larger than the spontaneous emission rate, this results in a significant increase in the threshold current.

Based on these parameters, we can analyze the low threshold performance of several types of microcavity lasers; for example, the devices described in Section IV.A. Table 1 summarizes these parameters for several geometries of the microcavity semiconductor laser. Although a very low threshold current of 250 μA was reported in the short cavity conventional stripe laser,[49] the mode

TABLE 1
Several Microcavity Structures for Semiconductor Lasers and Their Properties

Cavity structure	Active volume (μm^3)	Mode volume (μm^3)	Cavity Q	Nonradiative recombination lifetime (ns)
Airpost VCSEL	~10^{-1}	~10	2,000 to 20,000	~1
Microdisk (whispering gallery mode)	~10^{-1}	~1	100 to 1,000	~1
Microcylinder (whispering gallery mode)	~10^{-1}	~10	100 to 1,000	~1
Short Cavity Stripe (LD)	~1	~100	10,000 to 20,000	~10

volume of this laser is one order of magnitude larger than that of a typical micropost VCSEL having a similar value of f_v. This becomes a large handicap in realizing an ultralow threshold operation. Thus, we presume that the present good result is due to the highly established processing technologies for conventional type diode lasers. Concerning the mode volume, the microdisk geometry can be smaller than the microposts and microcylinders by approximately one order of magnitude. However, the actual Q values reported for the microdisk lasers were unexpectedly low — 100 to 200.[46,47,51] The Q values of the microcylinders are almost equal to those of the microdisks.[48] On the other hand, a typical micropost VCSEL can have a Q value of 2,000 to 20,000 (giving a photon lifetime $1/\gamma \doteq Q/\omega$ in the range 1 ps to 10 ps for $\lambda = 1$ μm) depending on the number of $\lambda/4$ layers. Evaluating these figures in total, we find that micropost VCSELs have very high potential as ultra-low-threshold lasers.

Now, we calculate the threshold current of a micropost VCSEL taking into account the nonradiative carrier recombination. As is schematically shown in Figure 8, a typical micropost VCSEL has a pair of $\lambda/4$ stack reflectors several microns thick. The light emitting layer is only a few tens of nanometers thick. In this structure, the nonradiative carrier recombination is dominated by that at the sidewall of the active layers, and the overall nonradiative recombination rate is approximately given by $\Gamma = 2S/r$, where S is the surface nonradiative carrier recombination velocity and r is the micropost radius.[52] For this kind of VCSEL geometry, the mode volume can be expressed as $V = \pi r^2 L_{net}$, where L_{net} is the net cavity length. Thus, the volume ratio $f_v = V_a/V$ simplifies to the ratio of the total active layer thickness to the net cavity length L_{net}. Figure 9 shows the calculated threshold current of a cylindrical VCSEL as a function of the post diameter D ($= 2r$). The C factor is also shown in the upper section of the plot. The parameter is the surface recombination velocity at the etched sidewalls of the QW layers. The other parameters used are the typical for an InGaAs QW VCSEL.

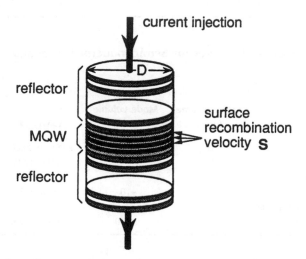

FIGURE 8   Schematic configuration of a cylindrical micropost VCSEL assumed in the threshold current calculation.

It is obvious, from the present calculation, that a large surface recombination velocity seriously increases the threshold current. However, Hamao et al. reported that the surface recombination velocity S is decreased to the order of 10^4 cm·s^{-1} by a thermal annealing surface passivation.[52] If this value is achieved, a sub-100-μA threshold will be obtained rather easily by decreasing the post diameter to less than a few microns.

D.   INPUT-OUTPUT PROPERTIES AND THE EFFECTIVE C FACTOR

In this section, some important features are described regarding the input-output curves and the efficient use of spontaneous emission. If we assume that

FIGURE 9   Calculated threshold current and the C factor of a cylindrical micropost VCSEL as a function of the post radius r. The surface recombination velocity at the etched sidewalls of the QW layers is the parameter. The other parameters used are $A = 10^9$s^{-1}, $\gamma = 10^{12}$ s^{-1}, $N_0 = 10^{18}$ cm^{-3}, $f_v = 0.01$, $\lambda = 1$ μm, $\Delta\lambda = 20$ nm, $L_{net} = 1$ μm, and $\eta = 3.6$.

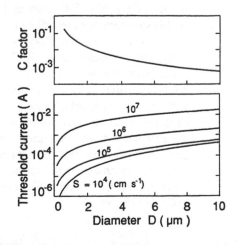

the cavity loss is dominated by the transmission loss, the output photon flux (number per unit time) is given by $S_{out} = \gamma s$. From the static solution of Equations 52 and 53, S_{out} is given by the following expressions for well below threshold ($s \ll 1$, $n \ll n_0$) and well above threshold ($s \gg 1$, and the stimulated emission term dominates the inversion depopulation).

$$S_{out} = \frac{\gamma}{CAn_0 + \gamma} \frac{A}{A + \Gamma} Cp \tag{55}$$

for well below threshold, and

$$S_{out} = p \tag{56}$$

for well above threshold. Therefore, in the logarithmic plot of the input-output curve, downward shift away from the straight line of $\log S_{out} = \log p$ is observed in the spontaneous emission regime. If $\gamma \gg CAn_0$ and $A \gg \Gamma$, the amount of this shift is $\log C$, so we can evaluate the C factor from the input-output properties.[38] However, when the above conditions are not satisfied, the C factor evaluated from an input-output curve is modified by a factor of $\gamma A/[(CAn_0 + \gamma)(A + \Gamma)]$.

When the value of CAn_0 is smaller than γ and Γ is much larger than A, the effective C factor is reduced to $\sim AC/\Gamma$. In this situation, based on Equation 54, it is found that the threshold current is also increased by a factor of $\sim \Gamma/A$. Increases in the threshold current due to large nonradiative recombination rates are seen in the input-output curves in Figure 10. This threshold increase occurs because the excited carriers, and thus the current, are mostly lost as the

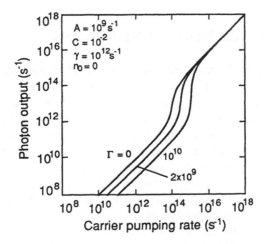

FIGURE 10 Calculated light output S_{out} vs. pumping p of a microcavity semiconductor laser on a logarithmic scale. The parameter is the nonradiative carrier recombination rate Γ. $A = 10^9$ s^{-1}, $\gamma = 10^{12}$ s^{-1}, $C = 10^{-2}$, and $n_0 = N_0 V f_v = 0$.

nonradiative recombination below the laser oscillation threshold. Above the threshold, the intensive stimulated emission dominates the entire carrier recombination process.

On the other hand, when the influence of the nonradiative recombination rate Γ is negligible compared to A, but the value of CAn_0 is much larger than γ, the effective C factor becomes $\sim\gamma/An_0$. This value is independent from the C factor given in Equations 52 and 53. However, it should be noted that the threshold current I_{th} obtained from Equation 54 is kept constant at $\sim qAN_0Vf_v$ in the small γ limit, and is independent of the given values of C and γ. This means that only the transparency carrier number determines the threshold current. Figure 11 shows calculated input-output curves indicating the large influence of the transparency carrier number. It is shown in Figure 11(b) that the effective C factor decreases by a factor of $\sim\gamma/An_0$ with decreasing γ, while the threshold current is fixed at a certain value for a sufficiently small value of γ. The reason for the decrease in the effective C factor at small values of γ, and thus with a high Q cavity, is understood in the following way: when a photon is spontaneously emitted into a cavity resonant mode, the photon does not escape rapidly from the cavity because of the highly reflective mirrors, and this photon will be reabsorbed during the many round trips inside the cavity. If a photon is subsequently emitted into the cavity resonant mode again, the same process will occur. This process can be called "cavity enhanced reabsorption" or "photon recycling".[50] However, the photon can be emitted rather easily into the modes which are not controlled by the cavity. Therefore, the probability of photon emission into a cavity resonant mode is largely decreased after all. The present result indicates an important fact that *the spontaneous emission output will be very small in a semiconductor microcavity designed for low threshold laser operation.*

V.   PROSPECTS FOR DEVICE APPLICATIONS

Starting from the basic principle of the optical microcavity, we have focused on the microcavity laser which can be operated with extremely low input power. Based on the discussion in Section IV.C, we have been led to believe that a sub-100-μA threshold microcavity semiconductor laser can be realized before long with present microfabrication technology. Here, one might ask, "why do we want microcavity devices?" This is an important question motivating the realization of high performance devices for real system applications. An answer is the use in optical interconnection applications for ultrafast parallel data processing of terabits per second throughput. Hayashi has pointed out the following:[53] Making electronic systems of higher density and higher frequency, signal interconnection becomes more difficult and coupling noise increases. This is due to utilizing charged particles (electrons) for signal transmission. On the other hand, photons, which carry energy but have no charge and do not interact among themselves, are ideal for signal interconnection.

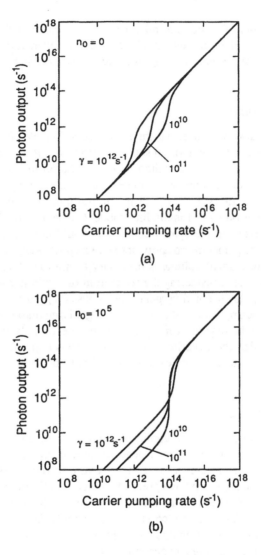

FIGURE 11 Calculated light output S_{out} vs. pumping p of a microcavity semiconductor laser. The cavity photon damping rate γ is the parameter. $A = 10^9$ s^{-1}, $C = 10^{-2}$, $n_0 = N_0 V f_v = 10^{-5}$, and $\Gamma = 0$. (a) $n_0 = N_0 V f_v = 0$, (b) $n_0 = 10^{-5}$.

In optical interconnection systems, high density integration of light-emitting and detecting devices is necessary, where each device should be separated with a very low power consumption, of about 100 µW. With this in mind, consider that the present semiconductor diode lasers in practical use have cavities that are typically several hundred microns long, and consist of tens of milliampere lasing thresholds. These devices are obviously not suitable for large scale integration. Thus, a wavelength sized device, which has microampere operation current and can be integrated on a single chip, is a really desired key element for optical interconnection. Considering microcavity devices from the practical point of view, however, there are several subjects to be considered, in addition to the laser threshold issue. In the following, we briefly discuss these.

A.   SPATIAL COHERENCE

This is also related to the threshold reduction. We have so far assumed single mode excitation in our discussion. However, if a large area — and thus also a large volume — is excited, laser oscillation in multiple cavity modes could occur. For example, when the excitation area of a planar microcavity is large, multiple transverse mode lasing or multiple filament lasing occurs even though there is only one longitudinal mode. Consequently, the lasing threshold would increase and the spatial coherence would deteriorate. Therefore, irrespective of the planar microcavity or the micropost cavity, excitation in a small area and thus a small volume is necessary to improve these properties.

Furthermore, considering the realization of a submicron diameter laser as an example, the beam divergence becomes very large. If we need a low divergence beam, an optical element having a large numerical aperture should be used, and the degree of integration is not necessarily very large. Thus, in order to utilize the benefits of low power consumption of the microcavity device, waveguides for optical coupling should be used.

B.   TEMPORAL COHERENCE

One might consider the application of cavity enhanced and spatially controlled spontaneous emission instead of laser oscillation. In another words, microcavities are usually used as light emitting diodes (LEDs) rather than lasers. Such an approach has a great advantage for eliminating the thresholding behavior, which is a very nonlinear property, or for avoiding a large noise induced by backward reflected laser light. This LED use of course sacrifices the temporal coherence. In this view, the next subject to be taken into account is the utilization of spontaneous emission.

C.   EFFICIENT USE OF SPONTANEOUS EMISSION

For the LED use of microcavities, the microcavity device design concept is different from that of the laser device use. In order to achieve laser oscillation in VCSELs, the reflectivity of each $\lambda/4$ stack is usually taken to be higher than 0.998 so as to compensate for the very small one-pass gain of 0.001 to 0.002 inside the cavity. *In this design, the spontaneous emission cannot be efficiently extracted at the output under low level excitation*; this is due to the cavity-induced strong reabsorption, which is described in Section IV.D. This means

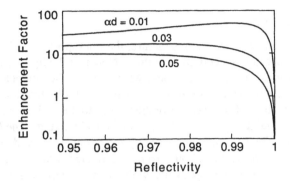

FIGURE 12   Spontaneous emission intensity enhancement factor in the cavity axis direction for a planar FP cavity including an absorptive medium. Equation 22 is plotted as a function of the mirror reflectivity for three different values of the total absorbance. Assuming a QW microcavity, the parameters are taken to be $\alpha = 10^4$ cm^{-1}, and the single QW thickness = 10 nm. Thus, the three curves correspond to 1, 3, and 5 QWs.

that the spontaneous emission coupled to the cavity mode is reabsorbed then sent out to modes uncontrolled by the cavity, or the reabsorbed energy may be transferred to nonradiative channels.

Aside from the laser rate equation discussion in Section IV.D, this important problem can be also discussed utilizing Equation 22 in Section II.C. We now try to estimate the enhancement factor of the spontaneous emission intensity of the planar microcavity in the cavity axis direction. Considering a QW microcavity, the absorption coefficient α is taken to be 10^4 cm^{-1}, and the thickness of the QW is assumed to be 10 nm. In Figure 12, the enhancement factor is plotted as a function of the mirror reflectivity for three different numbers of QWs. As shown in the figure, if the reflectivity is as high as 0.998, the cavity enhancement of spontaneous emission does not occur. This is the case for VCSELs. On the other hand, a slight reduction in the reflectivity results in a large enhancement in the spontaneous emission intensity. In this case, laser oscillation cannot occur. (In reality, the laser gain of semiconductors increases logarithmically rather than linearly with increase in the carrier density, so laser oscillation is not possible even with a very high excitation).

D.   OPERATION SPEED

Even when ignoring restrictions that come from the electric circuit, an operation speed of more than a few gigahertz bandwidth is possible only by laser oscillation. This is because a large increase in the spontaneous emission rate is not expected for a semiconductor microcavity due to the material's very broad emission width,[54] so the operation speed is limited by the spontaneous emission rate of ~10^9 s^{-1}. On the other hand, in laser oscillation, the total emission rate can be very much increased due to the intensive stimulated emission. As a result, a large increase in operation speed is expected.[55]

E.   DEVICE STRUCTURES

A typical microcavity structure, the micropost VCSEL, has a pair of λ/4 stack reflectors several microns thick. Note that the light-emitting layer is only a few tens of nanometer thick. Moreover, if the diameter of a cylindrical micropost is one micron or so, the structure looks like a towerblock. Consequently, there will be serious problems due to the high electrical resistance, the low thermal dissipation, and the low physical strength. The former two effects decrease the efficiency of electricity-light conversion and deteriorate the laser oscillation properties. Thus, from the device structure point of view, the realization of a high Q cavity without thick reflectors is a very important subject. Improvements in the fabrication of such total reflection type cavities as microdisks or microcylinders may solve this problem.

Introducing optical waveguide structures is also expected to be useful for ultralow threshold laser devices. Submicron-scale square, rectangular, or circular semiconductor waveguides, which have a hundred micron length and are surrounded by a low refractive index material, give mode volumes comparable to VCSELs a few microns in diameter. Since the cavity length is on the hundred micron scale, a cleaved facet reflectivity of ~0.3 is sufficient for obtaining laser oscillation.

In the near future, many variations of microcavities described here could be developed, including new methods for current excitation.

VI.   SUMMARY

In summary, we have shown that a microcavity can cause great modifications in such spontaneous emission properties as emission spectrum, spatial emission intensity distribution, and spontaneous emission lifetime. Moreover, these spontaneous emission features also directly result in modification of the laser oscillation properties with a microcavity. Needless to say, controlling spontaneous emission and its application for novel kinds of light sources are deeply concerned with fundamental quantum electrodynamics. However, as has been described in the above sections, the essential point to realize in ultralow power-consumption light-emitting devices is very simple: that is, "make the cavity small." Of course, the present words just symbolically simplify the subjects of optical microcavity device research, because there are several important subjects to be overcome related with semiconductor solid-state physics, crystal growth, and microfabrication technologies, as well as device design technologies. Therefore, in reality, a tight combination of all these highly specialized technologies is quite important for achieving high performance microcavity devices. Thereafter, the door to a new generation of photonic devices will be opened.

ACKNOWLEDGMENTS

The author is grateful to S. D. Brorson and E. P. Ippen for their cooperation and stimulating discussions especially at the Massachusetts Institute of Technology during his visiting research period. Special thanks are due to K. Shimoda, I. Hayashi, R. Lang, T. Kobayashi, K. Ujihara, Y. Yamamoto, G. Björk, and Y. Nambu for their valuable comments and discussions concerning the basic idea, the operating principle, and the applications of optical microcavities.

REFERENCES

1. S. Haroche and D. Kleppner: Cavity Quantum Electrodynamics, *Phys. Today* **42**, 24 (1989).
2. S. E. Morin, Q. Wu, and T. W. Mossberg: Cavity Quantum Electrodynamics at Optical Frequencies, *Opt. Photon. News,* (August 1992) p. 7.
3. S. Haroche and J. Raimond: Cavity Quantum Electrodynamics, Scientific American (April 1993).
4. A. G. Vaidyanathan, W. P. Spencer, and D. Kleppner: Inhibited Absorption of Blackbody Radiation, *Phys. Rev. Lett.* **47**, 1592 (1981).
5. P. Goy, J. M. Raimond, M. Gross, and S. Haroche: Observation of Cavity-Enhanced Single-Atom Spontaneous Emission, *Phys. Rev. Lett.* **50**, 1903 (1983).
6. D. Meschede, H. Walther, and G. Müller: One-Atom Maser, *Phys. Rev. Lett.* **54**, 551 (1985).
7. R. G. Hulet, E. S. Hilfer, and D. Kleppner: Inhibited Spontaneous Emission by a Rydberg Atom, *Phys. Rev. Lett.* **55**, 2137 (1985).
8. G. Rempe and H. Walther: The One Atom Maser and Cavity Quantum Electrodynamics, in Methods of Laser Spectroscopy, Y. Prior et al., Eds., Plenum Press, New York (1986) p. 11.
9. W. Jhe, A. Anderson, E. A. Hinds, D. Meschede, L. Moi, and S. Haroche: Suppression of Spontaneous Decay at Optical Frequencies: Test of Vacuum-Field Anisotropy in Confined Space, *Phys. Rev. Lett.* **58**, 666 (1987).
10. D. J. Heinzen, J. J. Childs, J. E. Thomas, and M. S. Feld: Enhanced and Inhibited Visible Spontaneous Emission by Atoms in a Confocal Resonator, *Phys. Rev. Lett.* **58**, 1320 (1987).
11. D. J. Heinzen and M. S. Feld: Vacuum Radiative Level Shift and Spontaneous-Emission Linewidth of an Atom in an Optical Resonator, *Phys. Rev. Lett.* **59**, 2623 (1987).
12. F. De Martini, G. Innocenti, G. R. Jacobovitz, and P. Mataloni: Anomalous Spontaneous Emission Time in a Microscopic Optical Cavity, *Phys. Rev. Lett.* **59**, 2995 (1987).
13. M. Suzuki, H. Yokoyama, S. D. Brorson, and E. P. Ippen: Observation of Spontaneous Emission Lifetime Change of Dye-Containing Langmuir-Blodgett Films in Optical Microcavities, *Appl. Phys. Lett.* **58**, 998 (1991).

14. T. Tsutsui, C. Adachi, S. Saito, M. Watanabe, and M. Koishi: Effect of Confined Radiation Field on Spontaneous-Emission Lifetime in Vacuum-Deposited Fluorescent Dye Films, *Chem. Phys. Lett.* **182**, 143 (1991).
15. A. J. Campillo, J. D. Eversole, and H-B. Lin: Cavity Quantum Electrodynamic Enhancement of Stimulated Emission in Microdroplets, *Phys. Rev. Lett.* **67**, 437 (1991).
16. E. Yablonovitch, T. J. Gmitter, and R. Bhat: Inhibited and Enhanced Spontaneous Emission from Optically Thin AlGaAs/GaAs Double Heterostructures, *Phys. Rev. Lett.* **61**, 2546 (1988).
17. H. Yokoyama, K. Nishi, T. Anan, and H. Yamada: Enhanced Spontaneous Emission from GaAs QWs with Monolithic Microcavities, paper presented at the Topical Meeting on Quantum Wells for Optics and Optoelectronics, Salt Lake City, UT, March 1989, paper MD4; H. Yokoyama, K. Nishi, T. Anan, H. Yamada, S. D. Brorson, and E. P. Ippen: Enhanced Spontaneous Emission from GaAs Quantum Wells in Monolithic Microcavities, *Appl. Phys. Lett.* **57**, 2814 (1990).
18. Y. Yamamoto, S. Machida, Y. Horikoshi, K. Igeta, and G. Björk: Enhanced and Inhibited Spontaneous Emission of Free Excitons in GaAs Quantum Wells in a Microcavity, *Opt. Commun.* **80**, 337 (1991).
19. T. Tezuka, S, Nunoue, H. Yoshida, and T. Noda: Spontaneous Emission Enhancement in Pillar-Type Microcavities, *Jpn. J. Appl. Phys.* **32**, L54 (1993).
20. T. Kobayashi, T. Segawa, A. Morimoto, and T. Sueta, New Type Lasers, Light Emitting Devices, and Optical Functional Devices with Controlling Spontaneous Emission, Tech. Dig. of 43th Fall Meeting of Japanese Society of Applied Physics, paper 29a-B-6 (Sep. 1982); T. Kobayashi, A. Morimoto, and T. Sueta, Discussions on Closed Microcavity Lasers, Tech. Dig. of 46th Fall Meeting of Japanese Society of Applied Physics, paper 4a-N-1 (Oct. 1985) (both in Japanese).
21. E. Yablonovitch: Inhibited Spontaneous Emission in Solid-State Physics and Electronics, *Phys. Rev. Lett.* **58**, 2059 (1987).
22. H. Yokoyama and S. D. Brorson: Rate Equation Analysis of Microcavity Lasers, *J. Appl. Phys.* **66**, 4801 (1989).
23. S. D. Brorson, H. Yokoyama, and E. P. Ippen: Spontaneous Emission Rate Alteration in Optical Waveguide Structures, *IEEE J. Quantum Electron.* **26**, 1492 (1990).
24. Y. Yamamoto: Controlled Spontaneous Emission in Microcavity Semiconductor Lasers, in Coherence, Amplification, and Quantum Effects in Semiconductor Lasers, Y. Yamamoto, Ed., Wiley-Interscience, New York (1991) p. 561.
25. H. Yokoyama: Physics and Device Applications of Optical Microcavities, *Science* **256**, (3 April 1992) p. 66.
26. Y. Yamamoto, S. Machida, and G. Björk: Microcavity Semiconductor Lasers with Controlled Spontaneous Emission, *Opt. Quantum Electron.* **24**, S215 (1992).
27. H. Yokoyama, K. Nishi, T. Anan, Y. Nambu, S. D. Brorson, E. P. Ippen, and M. Suzuki: Controlling Spontaneous Emission and Threshold-less Laser Oscillation with Optical Microcavities, *Opt. Quantum Electron.* **24**, S245 (1992).
28. K. Shimoda and T. Yajima, Ed.: *Quantum Electronics* (in Japanese), Shokabo, Tokyo (1972), Chap. 1.

29. D. Kleppner: Inhibited Spontaneous Emission, *Phys. Rev. Lett.* **47**, 233 (1981).
30. R. Loudon: *The Quantum Theory of Light (Second Edition)*, Oxford, New York (1983).
31. S. Haroche: Rydberg Atoms and Radiation in a Resonant Cavity, in *Les Houches, Session XXXVIII, 1982–New Trends in Atomic Physics*, G. Grynberg and R. Stora, Eds., Elsevier, Amsterdam (1984), p. 237.
32. E. T. Jaynes and F. W. Cummings: Comparison of Quantum and Semiclassical Radiation Theories with Application to the Beam Maser, *Proc. IEEE* **51**, 89 (1963).
33. K. H. Drexhage: Interaction of Light with Monomolecular Dye Layers, in *Progress in Optics*, Vol. XII, E. Wolf, Ed., North-Holland, Amsterdam (1974), p. 165.
34. H. Yokoyama: A Simplified Analysis of the Optical Bistability of Multiple Quantum Well Étalons, *J. Quantum Electron.* **25**, 1190 (1989).
35. K. Ujihara: Spontaneous Emission and the Concept of Effective Area in a Very Short Optical Cavity with Plane-Parallel Dielectric Mirrors, *Jpn. J. Appl. Phys.* **30**, L901 (1991).
36. F. De Martini and J. R. Jacobovitz: Anomalous Spontaneous-Stimulated-Decay Phase Transition and Zero-Threshold Laser Action in a Microscopic Cavity, *Phys. Rev. Lett.* **60**, 1711 (1988).
37. A. E. Siegman: *Lasers,* University Science, Mill Valley, CA (1986).
38. R. H. Pantell and H. E. Puthoff: *Fundamentals of Quantum Electronics*, John Wiley & Sons, New York (1969), Chap. 3.
39. G. Björk and Y. Yamamoto: Analysis of Semiconductor Microcavity Lasers Using Rate Equations, *IEEE J. Quantum Electron.* **27**, 2386 (1991).
40. Y. Yamamoto, S. Machida, and G. Björk: Microcavity Semiconductor Laser with Enhanced Spontaneous Emission, *Phys. Rev. A* **44**, 657 (1991).
41. F. Koyama, S. Kinoshita, and K. Iga: Room-Temperature Continuous Wave Lasing Characteristics of GaAs Vertical Cavity Surface-Emitting Laser, *Appl. Phys. Lett.* **55**, 1089 (1988).
42. J. L. Jewell, S. L. McCall, Y. H. Lee, A. Schere, A. C. Gossard, and J. H. English: Lasing Characteristics of GaAs Microresonators, *Appl. Phys. Lett.* **54**, 1400 (1989).
43. A. Schere, J. L. Jewell, Y. H. Lee, J. P. Harbison, and L. T. Florez: Fabrication of Microlasers and Microresonator Optical Switches, *Appl. Phys. Lett.* **55**, 2724 (1989).
44. R. S. Geels and L. A. Coldren: Submilliamp Threshold Vertical-Cavity Laser Diodes, *Appl. Phys. Lett.* **57**, 1605 (1990).
45. T. Numai, T. Kawakami, T. Yoshikawa, M. Sugimoto, Y. Sugimoto, H. Yokoyama, K. Kasahara, and K. Asakawa: Record Low Threshold Current in Microcavity Surface-Emitting Laser, *Jpn. J. Appl. Phys.* **32**, L1533 (1993).
46. S. L. McCall, A. F. J. Levi, R. E. Slusher, S. J. Pearton, and R. A. Logan, Whispering-Gallery Mode Microdisk Lasers, *Appl. Phys. Lett.* **60**, 289 (1992).
47. A. F. J. Levi, R. E. Slusher, S. L. McCall, T. Tanbun-Ek, D. L. Coblentz, and S. J. Pearton: Room Temperature Operation of Microdisc Lasers with Submilliamp Threshold Current, *Electron. Lett.* **28**, 1010 (1992).

48. A. F. J. Levi, R. E. Slusher, S. L. McCall, S. J. Pearton, and W. S. Hobson: Room-Temperature Lasing Action in $In_{0.51}Ga_{0.49}P/In_{0.2}Ga_{0.8}As$ Microcylinder Laser Diodes, *Appl. Phys. Lett.* **62**, 2021 (1993).

49. T. R. Chen, B. Zhao, L. Eng, Y. H. Zhuang, A. Yariv, J. E. Unger, and S. Oh: Ultralow Threshold Multi-Quantum Well InGaAs Lasers, *Appl. Phys. Lett.* **60**, 1782 (1992).

50. Y. Yamamoto and G. Björk: Lasers Without Inversion in Microcavities, *Jpn. J. Appl. Phys.* **30**, L 2039 (1991).

51. A. F. J. Levi, R. E. Slusher, S. L. McCall, J. L. Glass, S. J. Pearton, and R. A. Logan: Directional Light Coupling from Microdisk Lasers, *Appl. Phys. Lett.* **62**, 561 (1993).

52. N. Hamao, M. Sugimoto, S. Kohmoto, and H. Yokoyama: Reduction in Sidewall Recombination Velocity by Selective Disordering in GaAs/AlGaAs Quantum Well Mesa Structures, *Appl. Phys. Lett.* **59**, 1488 (1991).

53. I. Hayashi: Optoelectronic Devices and Material Technologies for Photo-Electric Integrated Systems, *Jpn. J. Appl. Phys.* **32**, 266 (1993).

54. T. Baba, T. Hamano, F. Koyama, and K. Iga: Spontaneous Emission Factor of a Microcavity DBR Surface-Emitting Laser, *IEEE J. Quantum Electron.* **27**, 1347 (1991).

55. H. Yokoyama, Y. Nambu, and T. Shimizu: 100 Gbps Response of Microcavity Lasers, in *Ultrafast Phenomena VIII*, J. -L. Mavin et al., Eds., Springer-Verlag, Berlin, p. 220.

9 Recent Progress in Optical Microcavity Experiments

Hiroyuki Yokoyama

TABLE OF CONTENTS

I. INTRODUCTION

In the efforts to control spontaneous emission, many experiments have been carried out using atomic beams in the microwave as well as in the optical spectral regime. In this chapter, however, we survey microcavity experiments in condensed materials from a device physics point of view. Semiconductor microcavities are especially important when we consider device applications because semiconductors can be excited electrically rather than optically. Thus, attention has been mainly focused on semiconductor microcavities. However, we will also describe several experiments utilizing organic dyes because the emission properties of organic dyes resemble those of semiconductors: their emission lifetimes are several nanoseconds and their emission bandwidths are a few tens of nanometers. We discuss optical microcavity experiments dealing with both spontaneous emission and laser oscillation. First, several microcavity configurations are briefly introduced in connection with the experiments. Then, recent experimental studies are described for planar dye microcavities and several kinds of semiconductor microcavities.

II. CAVITY CONFIGURATIONS

Some typical structures are briefly described here. Although many successful cavity quantum electrodynamics (QED) experiments have been carried out in the microwave region,[1-7] experiments in the optical region have proven to be more difficult. The reason for this is that the cavity's dimensions should be on the micron or submicron scale in order to produce strong cavity effects.

A. CONFOCAL AND CONCENTRIC CAVITIES

As discussed in Chapter 8, an FP cavity made by a pair of spherical mirrors gives a rather small mode volume even though the cavity length is much larger than the wavelength of the emitted light. Thus, such a cavity forms an equivalent microcavity configuration. However, the frequency separation of longitudinal cavity modes is rather small due to the large cavity length. Therefore, although this cavity configuration is useful in experiments utilizing atomic beams with very narrow transition widths (so that at most one cavity mode exists within the emission width), it is not very effective for condensed-state-light emitting materials without a significant reduction in the cavity length.

B. PLANAR MICROCAVITIES

As discussed by several authors in different chapters, a planar microcavity can modify the spontaneous emission rate as well as the spatial emission distribution. However, only recently has the mode volume been discussed for this configuration. Ujihara has given a formula for the mode volume of a fundamental planar microcavity mode.[8] The mode volume and mode radius are respectively given by the following expressions:

$$V = \frac{\pi \lambda L^2}{1-R}, \quad r = \sqrt{\frac{\lambda L}{1-R}}. \tag{1}$$

Here, the cavity length L is assumed to be an integer multiple of half a wavelength. This formula shows that the mode volume is proportional to the square of the cavity length for constant mirror reflectivity. Thus, we find that just shortening the planar cavity's length will decrease the threshold population inversion n_{th} because the photon damping rate γ is linearly proportional to the inverse of the cavity length while the mode volume is proportional to square of the cavity length (note that n_{th} is proportinal to $V \cdot \gamma$). It is interesting to see that this one-dimensional microcavity configuration also makes possible the confinement of the field to a small volume.

C. MICROSPHERE CAVITIES

It is well known that microspheres made by liquid or solid dielectric materials act as cavities having high Q values for certain modes. These resonant modes are called whispering gallery (WG) modes, and they propagate azimuthally near the inside surface of a sphere due to total internal reflection. So far, dye-molecule-embedded spheres tens of micron in size have been used for laser oscillation experiments.[9-11] If the sphere size is decreased to the order of the emitted light wavelength, the emission lifetime of molecules coupled to a WG mode is modified.[9] From the application point of view, this cavity structure has a large problem in coupling the light out with high efficiency.

D. MICROPOST CAVITIES

From the device application point of view, these microcavity structures are the most widely used. The microcavity semiconductor surface emitting lasers (VCSELs) usually have square- or cylinder-post structures of dimensions from several microns to a few tens of microns. A pair of upper and bottom reflectors are also made from semiconductor quarter wave stacks. Progress in the crystal growth technologies such as molecular beam epitaxy (MBE) or metal organic chemical vapor deposition (MOCVD), as well as in microfabrication technologies such as reactive ion beam etching (RIBE) enabled us to make these kinds of structures. Figure 1 shows a scanning electron micrograph of such a VCSEL.[12] Baba et al. have calculated the fraction of spontaneous emission coupled into a single cavity mode (the C factor) for such kinds of square-post structures.[13,14] (See Chapter 7 in this volume.) Assuming a good overlap of the transition wavelength and the cavity resonant mode, the C factor can take a value of at most 0.1, even with a sub micron-sized cavity when the emission width is 20 nm wide. If the bandwidth is narrow enough, the value of C approaches 0.5. This means that the fraction of spontaneous emission coupled to the fundamental cavity mode is nearly unity, considering that two orthogonal-polarization modes are degenerate.

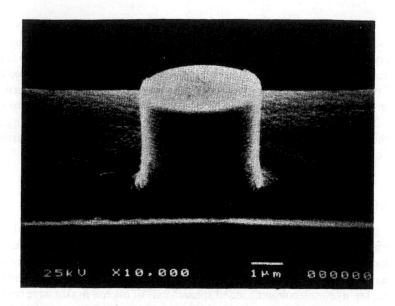

FIGURE 1   Scanning electron microscope photograph of a semiconductor micropost cavity fabricated by MBE and RIBE technologies. The light emitting region consists of InGaAs/GaAs QWs, and a pair of reflectors are λ/4 stacks of GaAs/AlAs.[12]

E.   MICRODISK AND MICROCYLINDER CAVITIES

Microdisk structures are a new class of microcavities based on a large dielectric discontinuity between an optically thin semiconductor layer and air. A typical semiconductor microdisk structure is shown in Figure 2.[15] This cavity configuration was developed also from the standpoint of semiconductor device fabrication. As shown in Figure 1, actual micropost cavities have thick quarter wave stacks in order to obtain high reflectivities. This indicates that it takes a very long time to grow these layers. In theory, a microdisk cavity can achieve a rather high Q value, like a microsphere, since WG modes can propagate inside the structure due to total reflection. Thus, thick reflector structures are removed from the device. As a variation of this cavity, microcylinder WG mode cavities have also been fabricated.[16] A typical structure of the microcylinder WG mode cavity is shown in Figure 3.

III.   ALTERATIONS IN SPONTANEOUS EMISSION PROPERTIES

As discussed in different chapters, in the framework of the golden rule, optical confinement by a microcavity results in a rearrangement of the mode density in two ways simultaneously; one is the nonuniform spatial redistribution, and the other is the spectral redistribution. Thus, in general, a change in

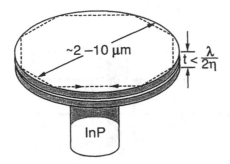

FIGURE 2   Illustration of a microdisk structure reported by McCall et al.[15]

the spatial pattern of the spontaneous emission is accompanied by a change in the spontaneous emission rate. However, as shown for a planar microcavity in the calculations of Brorson, or Ujihara, or Björk and Yamamoto, the emission rate change is not always large. This is especially true if the emission bandwidth of a material is much broader than the cavity resonance width, as is discussed by these authors. Thus, we here summarize principal experiments from these two different viewpoints: spatial emission pattern change and lifetime change.

A.   EMISSION PATTERN ALTERATION

Much pioneering work was done with planar microcavity structures. The first report of altered spontaneous emission in a fabricated microcavity structure was made by Drexhage many years ago.[17] He employed dye-embedded Langmuir-Blodgett (LB) films deposited on metal mirrors. Organic dyes, in spite of having rather large spectral width, are good materials for microcavity experiments because of their high quantum efficiency and low reabsorption. Drexhage could thus successfully observe dramatic modifications in the spatial emission pattern induced by the presence of the microcavities. The features of the spatial emission pattern are described in the calculations by several authors

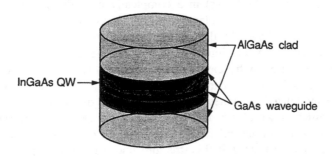

FIGURE 3   Schematic representation of a microcylinder cavity reported by Levi et al.[16]

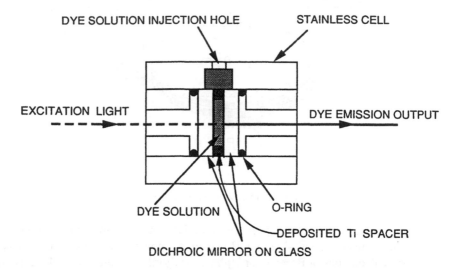

FIGURE 4   Schematic configuration for a dye solution planar microcavity structure.

in several different chapters in this book. In the case of Drexhage's experiment, since the emission width of a dye is usually quite broad, we cannot precisely adapt the monochromatic dipole calculation.

Yokoyama et al. employed a pair of dielectric reflectors to examine the enhancement of the spontaneous emission intensity in the cavity axis. Both LB films[18,19] and a dye solution[18,20] were used in the experiments, and an enhancement factor of 15 to 40 within the cavity resonance peaks were observed. A typical cavity structure used in the dye solution experiment is shown in Figure 4. In order to control the separation between the pair of mirrors, a metal film deposited on one mirror was used as a spacer. For microcavity experiments with dye-embedded LB films, the LB films were directly deposited on the mirror.[17-19]

Similar results were also obtained in our experiment employing a monolithic planar microcavity made exclusively by semiconductors. Initially, we had observed a spontaneous emission alteration (drastic spectral narrowing) of GaAs quantum wells (QWs) in a completely monolithic Fabry-Pérot (FP) étalon structure made by molecular beam epitaxy (MBE). Since the spectral shape does not change by decreasing the excitation intensity to a very low level, this strong modulation of the QWs' PL spectra is due to spontaneous emission modified by the cavity structure rather than laser oscillation. In the initial stage of VCSEL research, there were several reports misinterpreting this kind of modulated spontaneous emission as laser oscillation.

In order to see clearly the influence of the cavity on the spontaneous emission spectrum, the FP cavity employing an external reflector shown in Figure 5 was constructed. In this structure, one reflector was epitaxially grown under the multiple-quantum-well (MQW) layer, and the surface of the MQW layer was antireflection coated. In Figure 6, emission spectra are shown for

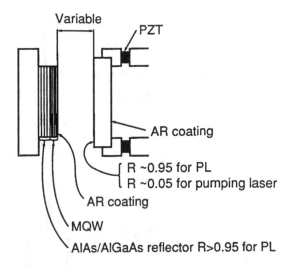

Variable

PZT

AR coating

{ R ~0.95 for PL
  R ~0.05 for pumping laser

AR coating

MQW

AlAs/AlGaAs reflector R>0.95 for PL

FIGURE 5   Schematic illustration of a GaAs MQW Fabry-Pérot cavity, which employs an external reflector in order to vary the cavity length. (From Yokoyama et al., *Opt. Quantum Electron.* **24**, S254, 1992. With permission from Chapman & Hall.)

different cavity lengths. By decreasing the cavity length, it is seen that the emission intensity is concentrated in a few cavity resonance modes. (The heights and widths of the resonance peaks at multiple peaks situation are limited by the spectral resolution of ~0.5 nm.) At the shortest cavity length available in this structure, the spontaneous emission intensity is gathered into a single resonance peak reminiscent of laser oscillation. It should be noted that the spectrally integrated PL intensity is slightly increased in this single mode situation compared to longer cavity cases. This shows the enhancement of spontaneous emission power within a small solid angle around the cavity axis. In an FP cavity structure, averaging over the standing wave effect, the mode density of a resonance peak is enhanced by a factor of $(1 + R)/(1 - R)$ assuming that the absorption loss is negligible. Instead of this, the resonance width is approximately given by the ratio of free spectral range and the above factor (i.e., the number of modes per unit volume is reserved). In other words, the "mode" condenses into a narrow resonance peak. In the situation of a single resonance peak within the material's emission width, the factor η for increase (or decrease) in the spectrally integrated emission intensity in *the cavity axis direction* is roughly given by the following relation:[18]

$$\eta = \frac{P(E_0') \cdot \Delta E}{P(E_0) \cdot \Delta P}.$$

(2)

Here, $P(E)$ is the energy-dependent transition rate at photon energy E, E_0' and E_0 are respectively the photon energies at the cavity resonance peak and

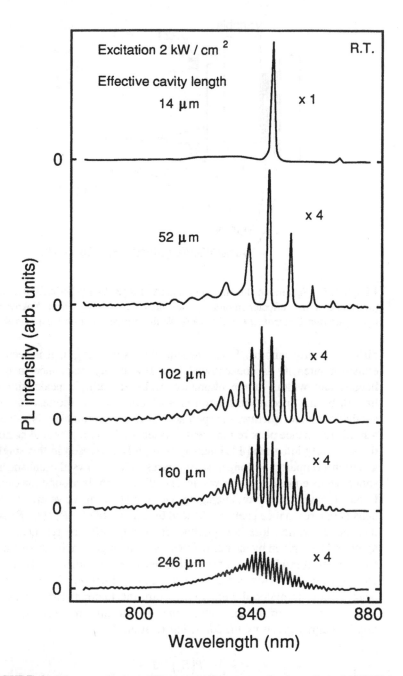

FIGURE 6　Room temperature PL spectra of an external mirror GaAs MQW Fabry-Pérot cavity detected from the cavity axis direction for different cavity lengths. The envelope of the PL spectrum for the longest cavity approximates the free space PL shape. (From Yokoyama et al., *Opt. Quantum Electron.* **24**, S255, 1992. With permission from Chapman & Hall.)

free-space-emission peak, and ΔE and ΔP are respectively the FP mode separation, and the full width at half maximum (FWHM) of $P(E)$. If $E_0' = E_0$, the emission intensity is enhanced by a ratio of $\sim \Delta E / \Delta P$. Note that if $E_0' \neq E_0$, and $P(E_0)/P(E_0') > \Delta E / \Delta P$, the microcavity causes the spontaneous emission into the cavity axis direction to be suppressed instead of enhanced.

In an experiment using an atomic beam, the enhancement and suppression in the cavity axis direction have been clearly shown.[6] Note, however, that if there are multiple cavity resonance peaks within the emission width, enhancement of the spontaneous emission does not occur because the mode density increase caused by the resonance peaks is canceled out by the mode density decrease between the resonance peaks. This situation is described in Chapter 8 of this volume. Thus, a quite short cavity is necessary to observe an enhancement for condensed materials such as semiconductors or organic dyes because of their very broad emission widths.

It is possible to increase the enhancement of spontaneous emission described by Equation 2 by further decreasing the cavity length because of an increase in the cavity mode separation (and thus also in the cavity resonance width). Therefore, we fabricated another microcavity structure, shown in Figure 7. The microcavity consisted of GaAs QWs combined with monolithic layered

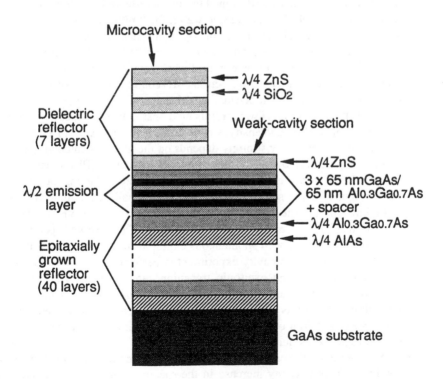

FIGURE 7 Schematic view of a monolithic GaAs MQW planar microcavity structure. (From Yokoyama et al., *Opt. Quantum Electron.* 24, S256, 1992. With permission from Chapman & Hall.)

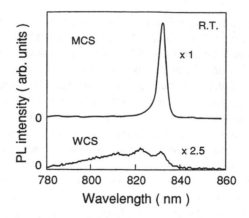

FIGURE 8   Room temperature PL spectra for a monolithic GaAs MQW planar microcavity structure. (From Yokoyama et al., *Opt. Quantum Electron.* **24**, S256, 1992. With permission from Chapman & Hall.)

AlGaAs/AlAs reflectors.[18,21,22] Only the bottom reflector was made by epitaxial growth, and this was designed to yield a reflectivity of 0.98 flat over the transition width. Half of the sample wafer, the "microcavity section" (MCS), was covered by a dielectric reflector having a reflectivity of ~0.9 for emitting light, while showing a reflectivity of less than 0.1 for the excitation wavelength. The other half of the wafer, the "weak-cavity section" (WCS), had only one dielectric layer as an antireflection (AR) coating. There was a weak cavity effect in the WCS section because of the epitaxially grown reflector and the incomplete AR coating. However, this kind of structure is necessary to quantitatively compare the PL intensity with and without the presence of a microcavity because the PL intensity strongly depends on the grown semiconductor wafer quality. Figure 8 shows the static photoluminescence (PL) spectra for the sample. In the measurement, the PL emission is detected along the cavity axis perpendicular to the sample surface within a small solid angle of $\sim 10^{-2}$ π steradian. The PL spectral width for the MCS is ~4 nm FWHM around the lowest quantized electron–heavy hole transition. The observed spectra clearly show the cavity enhanced spontaneous emission in the cavity axis direction.

When a QW microcavity experiment is performed at low temperature, a spectrally narrow band excitonic transition can dominate the spontaneous emission process, instead of the spectrally broad band-to-band, electron-hole recombination. In this situation, an FP cavity resonance curve can cover the entire spectral width of the excitonic emission. Thus, the excitonic emission can be regarded as quasi monochromatic, and the spectrally integrated emission power along the cavity axis is considerably increased in the on-resonance condition. This intensity increase in the spectral domain corresponds in the spatial domain to a concentration of the spontaneous emission power into the cavity axis. These features of an excitonic emission in a monolithic microcavity have been demonstrated experimentally by Yamamoto et al.[23] They also observed changes in the direction and intensity of the radiation depending on the cavity length detuning. Figure 9 shows a part of their experimental results. The observed emission patterns were in agreement with the calculated ones shown in Chapter 6 by Björk and Yamamoto.

B.   SPONTANEOUS EMISSION LIFETIME
CHANGE

Although Drexhage also examined the changes in spontaneous emission lifetime accompanying modifications of the radiation pattern, clear changes were not observed in his experiment.[17] The first report of a spontaneous emission rate change was made by De Martini et al.: they observed the emission from a dye solution confined between a pair of dielectric mirrors and found a factor of ten difference between the shortest and longest lifetimes depending on the cavity detuning.[24] However, in our experiment with a similar structure shown in Figure 4, we could not observe such a clear lifetime change.[18,20] Instead, in our experiment using a dye-embedded LB film instead of a dye solution, a decrease and an increase of the spontaneous emission rate of a factor of at most two were observed depending on the mirror separation.[18,19] Tsutsui et al. followed the one mirror experiment of Drexhage but using amorphous thin films of organic dyes instead of LB films, and observed clear changes of lifetimes by a factor of two to three.[25] A part of their results is shown in Figure 10.

In our planar QW microcavity experiment (shown in Figure 7), we also examined the alteration of the total spontaneous emission rate by the presence of the cavity.[19,21,22] From the time-resolved PL measurement, the nonradiative carrier recombination lifetime (i.e., the inverse of the recombination rate) of the MQW structure was determined to be ~2 ns at room temperature. This indicates that the well/barrier interfacial recombination rate is rather large. The lifetime difference between the MCS and WCS becomes measurable at a rather high excitation condition at which the radiative recombination rate is comparable to the nonradiative rate. From the analysis of the experimental results, we have determined that the radiative lifetime of ~2 ns in the WCS decreased to ~1 ns in the on-resonant MCS.

More recently, Tezuka et al. reported a beautiful result for the lifetime change of excitonic emissions of GaAs QWs confined in micropost cavities.[26] Figure 11 shows emission lifetimes evaluated from the PL decay curves and taking into account carrier diffusion. The lifetime reduction was observed with decreasing the detuning between the cavity resonance wavelength and the peak luminescence wavelength of the active layers. They evaluated that the radiative recombination rate was enhanced by 1.6 to 1.8 times of that of the reference sample having no three-dimensional cavity structure.

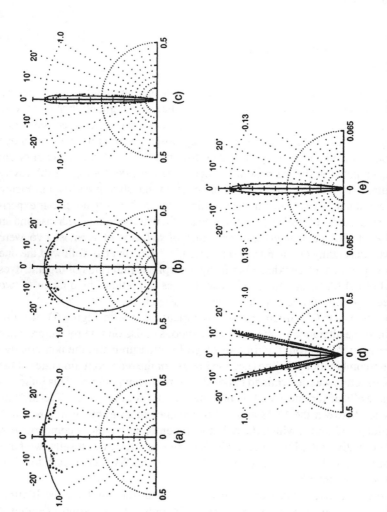

FIGURE 9 (a) and (b) are the experimental and theoretical radiation patterns from a GaAs quantum well in a thick AlGaAs layer: (a) s-wave and (b) p-wave. (c), (d), and (e) are the experimental and theoretical radiation patterns from a GaAs quantum well in a one-wavelength microcavity; the peak wavelength of the excitonic emission is 800 nm, and the cavity resonance wavelengths in the direction perpendicular to the layers are (c) 800 nm, (d) 815 nm, and (e) 790 nm. (From Yamamoto et al., *Opt. Quantum Electron.* **24**, S228, 1992. With permission from Chapman & Hall.)

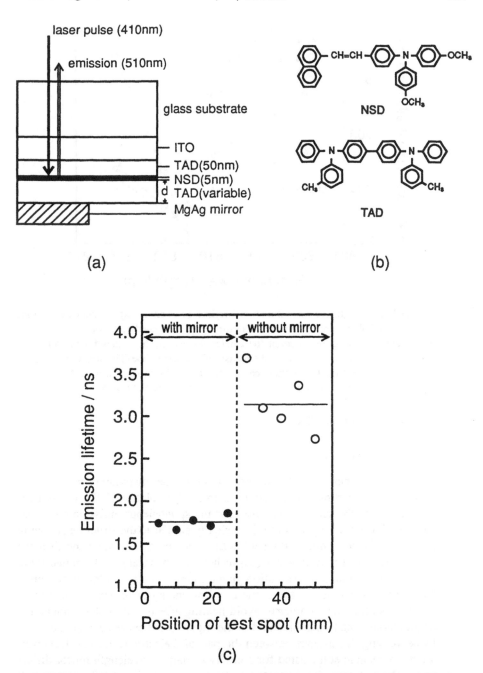

FIGURE 10 The lifetime measurement experiment with a one mirror configuration using amorphous thin films of an organic dye. Sample structure (a), the chemical structure of the dye film (b), and the lifetime measurement result (c). (From Tsutsui et al., *Chem. Phys. Lett.* **182**, 144, 1991. With permission from Elsevier Science.)

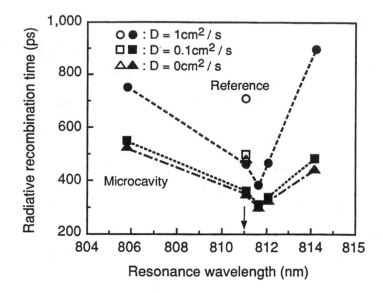

FIGURE 11 Radiative recombination lifetime of excitonic spontaneous emissions from GaAs QW micropost cavities. Lifetimes are evaluated from time-resolved PL measurements taking into account the carrier diffusion. The values for the diffusion constant $D = 0$ cm^{-2} are identical to the measured PL decay time. The arrow in the figure denotes the PL peak of QWs without cavity structures. (From Tezuka et al., *Jpn. J. Appl. Phys.* 32, L56, 1993. With permission from The Japan Society of Applied Physics.)

IV. LASER OSCILLATION

A. DYE MICROCAVITY LASERS

The first experiment of laser oscillation of the microcavity was also reported by De Martini and Jacobovitz using a dye solution.[27] The cavity configuration was the same as that used for the spontaneous emission rate measurement[24] (their cavity structure is basically similar to the structure shown in Figure 4). They reported "thresholdless laser operation" when the distance between a pair of mirrors was equal to half of a wavelength of emitted light.

We also examined dye solution planar microcavities.[20] In this experiment, the planar dye solutions were excited by low energy **picosecond light pulses** at low repetition rate in order to avoid heating effects and photochemical dye degradations. The lasing threshold in the input-output curve became indistinct by decreasing the distance between the pair of dielectric reflectors. However, a fuzzy threshold still existed for a cavity of half a wavelength mirror difference. The C factor was evaluated to be between 0.05 and 0.2 with a small detection solid angle of $\sim 10^{-2}\ \pi$ steradian. The emission lifetime measurement clearly showed a difference between spontaneous and stimulated emission regimes. That is, while the decay time with the lowest excitation energy corresponded to the spontaneous emission lifetime, an increase in excitation energy resulted in faster decay due to stimulated emission.

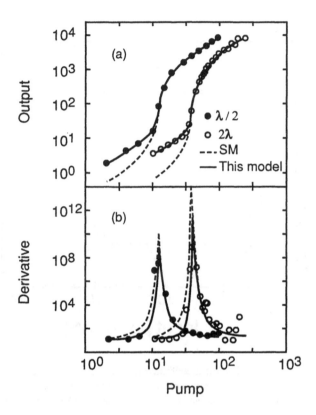

FIGURE 12 Input-output curves for the $\lambda/2$ and 2λ planar microcavities containing a dye solution. (a) Input-output, (b) derivatives of (a). (From Wang et al., *Phys. Rev.* **A47**, R2490, 1993. With permission from The American Physical Society.)

Wang et al. have carried out a precise measurement in order to evaluate the C factor with dye solution microcavities under **quasi steady-state operation**.[28] The input-output curves of optically pumped microcavities having $\lambda/2$ and 2λ resonator sizes are shown in Figure 12. They have observed that the output light became linearly polarized above threshold, and the polarization direction of output followed the excitation laser light polarization. It was also found that the simple rate equation model does not fit the experimental input-output curves well. By positioning the influence of excess spontaneous emission coming from the same excitation spot but not coupling into the laser mode, this discrepancy has been well explained. In this experiment, it has been concluded that the excess spontaneous emission increases the apparent C factor by an order of magnitude over the actual value, even with a small solid angle for the output detection. The actual value has been estimated to be in the order of 10^{-3}, which is not much different from the theoretical estimation given by Björk et al.[29]

Ujihara and Osuge have also carried out the experiment using a dye solution.[30] Based on the beam divergence measurement, they evaluated the mode radius to be 1.8 µm for a 2λ spacing microcavity with a pair of dielectric

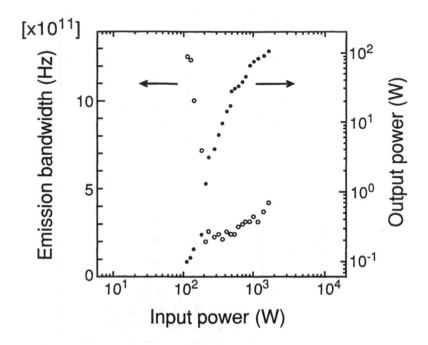

FIGURE 13   The full width at half maximum of the emission spectrum and the output power versus the input power in a planar microcavity dye laser. Open circles: emission bandwidth; filled circles: outout power. (Measured by Osuge and Ujihara.[30])

mirrors of 93% reflectivity. Considering the effective cavity length, they have found that the experimental value of the passive mode radius below lasing threshold is in good agreement with a theoretical estimation obtained from Equation 1. Changes in the emission spectrum have also been measured in detail with varying excitation power. Figure 13 shows the FWHM of the emission spectrum and the output power as functions of the input power. The width decreases sharply as soon as the threshold is reached, but the decrease stops midway in the threshold region and then increases gradually as the input power is further increased. The latter spectral width increase could be attributed to time-dependent frequency shifts due to transient gain depletion or the heating of the solution. Within the limited range corresponding to the sharp spectral width decrease shown in Figure 13, they found a linear dependence of the width on the reciprocal output power in agreement with the improved Schawlow-Townes formula.

    Aside from the planar microcavities, several laser oscillation experiments using microsphere structures have also been carried out.[11-13] Since the sizes of the spheres were larger than several microns in these experiments, multiple cavity mode features were observed in the laser oscillation. Gonokami et al. have reported the switching of laser oscillation modes by injection of external laser light.[11]

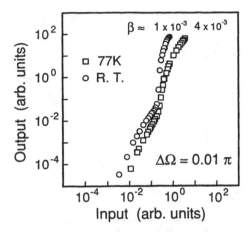

FIGURE 14 Input-output curves for a GaAs QW (3QWs) planar microcavity at room temperature and 77 K.[31] The microcavity structure is similar to the one shown in Figure 7 but with the increased mirror reflectivity.

B. SEMICONDUCTOR MICROCAVITY LASERS WITH OPTICAL EXCITATION

In this subsection, we describe the experiments carried out with optical excitation. It should be noted that almost all the experiments performed to control spontaneous emission have been done with optical excitation. However, many current driven laser devices having wavelength sized cavities have also been fabricated from the surface-emitting-laser point of view. We separately review these current driven devices in Section IV.C.

The lasing behavior of semiconductor microcavities have turned out to be considerably different from those of dye microcavities. Laser oscillation with a fuzzy threshold has not been observed. This feature is common as the input-output curves reported for well-designed VCSELs under optical excitation. We measured the input-output characteristics of InGaAs/AlGaAs QW planar microcavities using a configuration similar to that of the dye microcavity experiments described above.[31] In the experiments, as is shown in Figure 14, clear thresholds were observed in the linear scale input-output curves, and the C factors obtained were on the order of 10^{-3} even at the temperature of liquid nitrogen. At this low temperature the transition width of a QW is much narrower than that of a dye. Accordingly, a larger C factor can be expected compared to a dye microcavity laser having a similar structure under the same detection conditions. However, the experimental results have not fulfilled our expectations.

One possible explanation for the results obtained in semiconductor microcavities is the strong reabsorption of spontaneous emission due to band-to-band transitions. As we have discussed in the previous chapter, if the volume factor of the active medium is not small enough, absorption of spontaneously emitted photons due to band-to-band transition is not negligible. In such a case

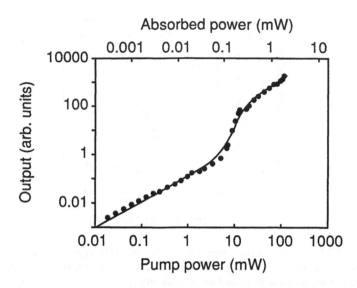

FIGURE 15 Input-output curve measured at the temperature of liquid helium for GaAs QW planar microcavities in which λ spacer layers (including QWs) are sandwiched between λ/4 stacks. A 7-nm single QW is embedded in the spacer layer. (From Horowicz et al., *Appl. Phys. Lett.* **61**, 394, 1992. With permission from American Institute of Physics.)

with a pair of highly reflective mirrors ($R > 0.99$), the observable C factor value decreases orders of magnitude below its value in the absence of absorption (this value is the one given in the rate equation). This is because re-absorbed photons are re-emitted into modes that are not controlled by the highly reflective mirrors. In another words, a high-Q (wall Q) cavity suppresses the efficient spontaneous emission in the laser oscillation mode under a low excitation condition in which net stimulated emission does not occur.

On the other hand, rather large values for the C factor have been reported for the GaAs QW microcavity experiments at the temperature of liquid helium.[32,33] The absorbed pump power for lasing threshold was evaluated to be only 85 μW for a monolithic planar microcavity.[32] The output power as a function of pump power is shown in Figure 15 for a microcavity. The C factor was evaluated to be 10^{-2}. This value is about three orders of magnitude larger than those of conventional semiconductor diode lasers, and it indicates a substantial reduction in the number of cavity modes interacting with the gain medium. This result is due to a reduction in both the mode volume and transition linewidth narrowing due to the very low temperature. A drastic reduction in the transparency carrier density also occurs at a very low temperature, so the reabsorption of spontaneous emission in the cavity may not be important. Furthermore, it should be noted that the radiative band-to-band transition rate increases in a way approximately proportional to the inverse of temperature.

Laser operation with a hemispherical microcavity has also been demonstrated.[33] The cavity consisted of a planar mirror on one side and a curved distributed Bragg reflector on the other side. This resonator had been made intending its beam waist to be located at the QW, yielding further reduction in the mode volume (thus, an increase in the C factor). In this experiment, an InGaAs single QW was used as the active medium, and was located at an antinode position of the standing wave in a 2.5 μm-long cavity configuration. This device also showed a C factor of 10^{-2} at liquid helium temperature.

McCall et al. have demonstrated WG mode laser oscillations in the semiconductor microdisk structure shown in Figure 2.[15] The disk geometry has the advantage of having a smaller total volume compared to the spherical one as well as being easier to fabricate. Lasing modes are formed around the edge of the thin semiconductor microdisk supported below by a semiconductor pedestal. The microdisk consists of InGaAs QW active layers confined in InGaAsP barrier layers, and has a diameter of 3 to 10 μm. The threshold pump power was as low as 50 μW at 77K. This cavity configuration has a problem similar to that of the microsphere: it is difficult to couple light output efficiently in a certain direction. As for improving the directional coupling of light output, they have also demonstrated that patterned asymmetries in the shape of microdisk resonators allowed controlling both the direction and the output power.[34]

C.   SEMICONDUCTOR MICROCAVITY LASERS
WITH ELECTRIC EXCITATION

Great progress has been made in the last decade in developing low threshold current semiconductor diode lasers. This is especially true for the vertical cavity surface emitting laser (VCSEL) device after successful continuous operation at room temperature.[35-37] Submilliampere threshold VCSELs have become rather common.[38,39] On the other hand, a marked reduction in the threshold current was also made in the conventional stripe-type laser by reducing the cavity length incorporated with high reflectivity coating on the cleaved facet. A lasing threshold current of as low as ~250 μA has been obtained in this kind of device.[40] Although these low threshold current semiconductor lasers are not always developed to be microcavities, the threshold reduction is essentially due to the device-volume reduction. In microcavity terminology, it can be said that the mode volume reduction causes an increase in the fractional amount of spontaneous emission (the C factor) coupled to the laser oscillation mode, and this results in the threshold input-power reduction. Therefore, one would simply expect that a further reduction of the device size could realize a laser device having a tens of microamperes threshold. However, the volume scaling law of the threshold current reduction breaks down in the microcavity semiconductor laser. This is mainly due to the onset of fast nonradiative relaxation of the excited carriers (population inversion) at the surface rather than spontaneous emission rate modification. In spite of this problem, a threshold current reduction is still expected upon decreasing the device volume, as was discussed in the last chapter.

From the microcavity device point of view, a current injection version of the microdisk laser was also fabricated, and a submilliampere threshold was observed.[41] Furthermore, the performance of the microcylinder diode laser operated with WG modes was examined.[16] From the device point of view, the latter structure is mechanically stable. Instead of this advantage, the microcylinder has a larger threshold current because of a larger mode volume compared to a microdisk.

To date, the record low threshold of 190 μA has been demonstrated with an airpost microcavity laser.[12] Figure 16 shows a schematic view of the device structure and a current-light output curve for this device. This low threshold value is consistent with the theoretical discussion given in Chapter 8 of this book. Furthermore, realization of this low threshold current has proved that such technologies as MBE and RIEB are presently precise enough for fabricating rather ideal microcavities.

It should be noted that, as discussed in the previous chapter, an efficient use of spontaneous emission as light ouput is not expected in a very low threshold laser device. This is because the spontaneous emission is strongly reabsorbed inside a rather high Q microcavity. Therefore, for efficient use of spontaneous emission, the microcavity light emitting diode (LED) having a rather low Q value has been tried.[42] From the output power point of view for submilliampere devices, the microcavity LED device can be comparable to the microcavity laser. An another advantage of a microcavity LED is that the unintentionally reflected backlight does not cause a serious noise increase because there is no stimulated emission. However, the operation speed is limited by the spontaneous emission rate, whereas, in the laser device, the speed can be increased due to the much faster stimulated emission rate.

V. SUMMARY

We have reviewed here some optical microcavity experiments from a device physics point of view. It should be noted that in contrast to the many cavity QED experiments in the microwave to submillimeter wave regions, in the optical domain, we can directly analyze the output features because of a much higher photon energy compared to the background thermal noise. Clear microcavity effects have been demonstrated showing the modulation in spatial emission patterns accompanying the emission intensity change. On the other hand, changes in the spontaneous emission rate are not as drastic as those in microwave cavity QED experiments. One reason is that although the cavity size is decreased to the wavelength scale in at least one dimension, only a few three-dimensional wavelength sized cavities have been demonstrated. Furthermore, the cavity Q value is not as large as that in a superconducting microwave cavity,[3] and the emission width of condensed state materials are in general

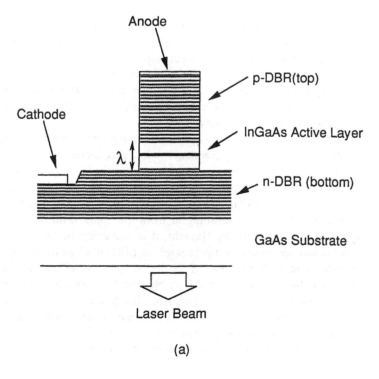

(a)

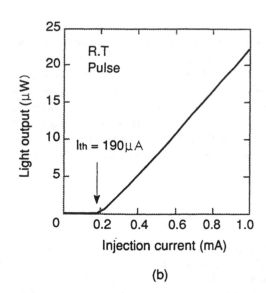

(b)

FIGURE 16 Schematic view (a) and a current-light curve (b) for the airpost microcavity laser which has the record low threshold current of 190 μA. (From Numai et al., *Jpn. J. Appl. Phys.* **32**, L1533, 1993. With permission from The Japan Society of Applied Physics.)

much broader than that of solitary atoms. In many cases, only a fraction of spontaneous emission couples into a single cavity resonant mode. In that sense, we can say that the spontaneous emission properties have been "partly controlled" by the presence of the optical microcavity. If one can confine almost all the spontaneous emission into a single cavity mode in a current driven microcavity device, the generation of number state light with a very low excitation power can be achieved.[23]

In the present microcavity lasers, only a fraction of spontaneous emission couples into a single cavity resonant mode. Thus, increases in the spontaneous emission coupling fraction (C factor) and threshold input reduction are explained simply by reductions in the mode volume, i.e., the scaling law for device volume reduction. However, device size reduction within a simple scaling law is still very meaningful from the application point of view. As discussed in Chapter 10 by Hayashi, it is necessary to have extremely low power consumption microampere laser (or LED) devices for highly integrated optoelectronic systems. Related to this subject, a serious decrease in the effective C factor, due to a strong reabsortion of spontaneous emission inside a semiconductor microcavity, should be clearly verified by experiments.

Consideration for the microcavity geometry, which aids us in power conversion efficiency, reliability, and ease of fabrication, is very important for application of microcavity devices in the real world.

ACKNOWLEDGMENTS

The author is grateful to many colleagues who have cooperated in various aspects of the microcavity research in NEC Corporation. Thanks are due in particular to K. Nishi, M. Suzuki, T. Anan, Y. Nambu, T. Kawakami, T. Numai, T. Yoshikawa, M. Sugimoto, H. Yamada, Y. Sugimoto, H. Saito, K. Kasahara, K. Asakawa, T. Shimizu, R. A. Linke, X. Wang, and G. Devlin. The valuable comments of R. Lang, K. Kobayashi, M. Sakaguchi, K. Onabe, and F. Saito are also gratefully acknowledged. The author also thanks S. D. Brorson, E. P. Ippen, K. Shimoda, I. Hayashi, K. Ujihara, Y. Yamamoto, and G. Björk for their valuable comments and discussions.

REFERENCES

1. The following three review articles cover many experimental works for the Cavity QED: S. Haroche and D. Kleppner: Cavity Quantum Electrodynamics, *Phys. Today* **42**, 24 (1989); S. E. Morin, Q. Wu, and T. W. Mossberg: Cavity Quantum Electrodynamics at Optical Frequencies, *Opt. Photon. News*, (August 1992) p. 7; S. Haroche and J. Raimond: *Cavity Quantum Electrodynamics,* Scientific American (April 1993).
2. A. G. Vaidyanathan, W. P. Spencer, and D. Kleppner: *I*nhibited Absorption of Blackbody Radiation, *Phys. Rev. Lett.* **47**, 1592 (1981).

3. P. Goy, J. M. Raimond, M. Gross, and S. Haroche: Observation of Cavity-Enhanced Single-Atom Spontaneous Emission, *Phys. Rev. Lett.* **50**, 1903 (1983).
4. D. Meschede, H. Walther, and G. Müller: One-Atom Maser, *Phys. Rev. Lett.* **54**, 551 (1985).
5. W. Jhe, A. Anderson, E. A. Hinds, D. Meschede, L. Moi, and S. Haroche: Suppression of Spontaneous Decay at Optical Frequencies: Test of Vacuum-Field Anisotropy in Confined Space, *Phys. Rev. Lett.* **58**, 666 (1987).
6. D. J. Heinzen, J. J. Childs, J. E. Thomas, and M. S. Feld: Enhanced and Inhibited Visible Spontaneous Emission by Atoms in a Confocal Resonator, *Phys. Rev. Lett.* **58**, 1320 (1987).
7. D. J. Heinzen and M. S. Feld: Vacuum Radiative Level Shift and Spontaneous-Emission Linewidth of an Atom in an Optical Resonator, *Phys. Rev. Lett.* **59**, 2623 (1987).
8. K. Ujihara: Spontaneous Emission and the Concept of Effective Area in a Very Short Optical Cavity with Plane-Parallel Dielectric Mirrors, *Jpn. J. Appl. Phys.* **30**, L901 (1991).
9. H.-M. Tzeng, K. F. Wall, M. B. Long, and R. K. Chang: Laser Emission from Individual Droplets at Wavelengths Corresponding to Morphology-Dependent Resonances, *Opt. Lett.* **9**, 499 (1984).
10. A. J. Campillo, J. D. Eversole, and H-B. Lin: Cavity Quantum Electrodynamic Enhancement of Stimulated Emission in Microdroplets, *Phys. Rev. Lett.* **67**, 437 (1991).
11. M. Kuwata-Gonokami, K. Takeda, H. Yasuda, and K. Ema: Laser Emission from Dye-Doped Polystyrene Microsphere, *Jpn. J. Appl. Phys.* **31**, L99 (1992).
12. T. Numai, T. Kawakami, T. Yoshikawa, M. Sugimoto, Y. Sugimoto, H. Yokoyama, K. Kasahara, and K. Asakawa: Record Low Threshold Current in Microcavity Surface-Emitting Laser, *Jpn. J. Appl. Phys.* **32**, L1533 (1993).
13. T. Baba, T. Hamano, F. Koyama, and K. Iga: Spontaneous Emission Factor of a Microcavity DBR Surface-Emitting Laser, *IEEE J. Quantum Electron.* **27**, 1347 (1991).
14. T. Baba, T. Hamano, F. Koyama, and K. Iga: Spontaneous Emission Factor of a Microcavity DBR Surface Emitting Laser (II)—Effects of Electron Quantum Confinements, *IEEE J. Quantum Electron.* **28**, 1310 (1992).
15. S. L. McCall, A. F. J. Levi, R. E. Slusher, S. J. Pearton, and R. A. Logan, Whispering-Gallery Mode Microdisk Lasers, *Appl. Phys. Lett.* **60**, 289 (1992).
16. A. F. J. Levi, R. E. Slusher, S. L. McCall, S. J. Pearton, and W. S. Hobson: Room-Temperature Lasing Action in $In_{0.51}Ga_{0.49}P/In_{0.2}Ga_{0.8}As$ Microcylinder Laser Diodes, *Appl. Phys. Lett.* **62**, 2021 (1993).
17. K. H. Drexhage: Interaction of Light with Monomolecular Dye Layers, in *Progress in Optics*, Vol. XII, E. Wolf, Ed., North Holland, Amsterdam (1974), p. 165.
18. H. Yokoyama, K. Nishi, T. Anan, Y. Nambu, S. D. Brorson, E. P. Ippen, and M. Suzuki: Controlling Spontaneous Emission and Threshold-less Laser Oscillation with Optical Microcavities, *Opt. Quantum Electron.* **24**, S245 (1992).
19. M. Suzuki, H. Yokoyama, S. D. Brorson, and E. P. Ippen: Observation of Spontaneous Emission Lifetime Change of Dye-Containing Langmuir-Blodgett Films in Optical Microcavities, *Appl. Phys. Lett.* **58**, 998 (1991).
20. H. Yokoyama, M. Suzuki, and Y. Nambu: Spontaneous Emission and Laser Oscillation Properties of Microcavities Containing a Dye Solution, *Appl. Phys. Lett.* **58**, 2598 (1991).

21. H. Yokoyama, K. Nishi, T. Anan, and H. Yamada: Enhanced Spontaneous Emission from GaAs QWs with Monolithic Microcavities, paper presented at the Topical Meeting on Quantum Wells for Optics and Optoelectronics, Salt Lake City, UT, March 1989, paper MD4.
22. H. Yokoyama, K. Nishi, T. Anan, H. Yamada, S. D. Brorson, and E. P. Ippen: Enhanced Spontaneous Emission from GaAs Quantum Wells in Monolithic Microcavities, *Appl. Phys. Lett.* **57**, 2814 (1990).
23. Y. Yamamoto, S. Machida, and G. Björk: Micro-Cavity Semiconductor Lasers with Controlled Spontaneous Emission, Opt Quantum Electron. **24**, S215 (1992).
24. F. De Martini, G. Innocenti, G. R. Jacobovitz, and P. Mataloni: Anomalous Spontaneous Emission Time in a Microscopic Optical Cavity, *Phys. Rev. Lett.* **59**, 2995 (1987).
25. T. Tsutsui, C. Adachi, S. Saito, M. Watanabe, and M. Koishi: Effect of Confined Radiation Field on Spontaneous-Emission Lifetime in Vacuum-Deposited Fluorescent Dye Films, *Chem. Phys. Lett.* **182**, 143 (1991).
26. T. Tezuka, S, Nunoue, H. Yoshida, and T. Noda: Spontaneous Emission Enhancement in Pillar-Type Microcavities, *Jpn. J. Appl. Phys.* **32**, L54 (1993).
27. F. De Martini and J. R. Jacobovitz: Anomalous Spontaneous-Stimulated-Decay Phase Transition and Zero-Threshold Laser Action in a Microscopic Cavity, *Phys. Rev. Lett.* **60**, 1711 (1988).
28. X. Wang, R. A. Linke, G. Devlin, and H. Yokoyama: Lasing Threshold Behavior of Microcavities: Observation by Polarization and Spectroscopic Measurements, *Phys. Rev.* **A47**, R2488 (1993).
29. G. Björk, S. Machida, Y. Yamamoto, and K. Igeta: Modification of Spontaneous Emission Rate in Planar Dielectric Microcavity Structures, *Phys. Rev.* **A44**, 669 (1991).
30. M. Osuge and K. Ujihara: Spontaneous Emission and Oscillation in a Planar Microcavity Dye Laser, *J. Appl. Phys.* **76**, 2588 (1994).
31. Y. Nambu and H. Yokoyama, unpublished.
32. R. J. Horowicz, H. Heitmann, Y. Kadota, and Y. Yamamoto: GaAs Microcavity Quantum-Well Laser with Enhanced Coupling of Spontaneous Emission to the Lasing Mode, *Appl. Phys. Lett.* **61**, 393 (1992).
33. F. M. Matinaga, A. Karlson, S. Macida, Y. Yamamoto, T. Suzuki, Y. Kadota, and M. Ikeda: Low-Threshold Operation of Hemisperical Microcavity Single-Quantum-Well Lasers at 4K, *Appl. Phys. Lett.* **62**, 443 (1993).
34. A. F. J. Levi, R. E. Slusher, S. L. McCall, J. L. Glass, S. J. Pearton, and R. A. Logan: Directional Light Coupling from Microdisk Lasers, *Appl. Phys. Lett.* **62**, 561 (1993).
35. F. Koyama, S. Kinoshita, and K. Iga: Room-Temperature Continuous Wave Lasing Characteristics of GaAs Vertical Cavity Surface-Emitting Laser, *Appl. Phys. Lett.* **55**, 1089 (1988).
36. J. L. Jewell, S. L. McCall, Y. H. Lee, A. Schere, A. C. Gossard, and J. H. English: Lasing Characteristics of GaAs Microresonators, *Appl. Phys. Lett.* **54**, 1400 (1989).
37. A. Schere, J. L. Jewell, Y. H. Lee, J. P. Harbison, and L. T. Florez: Fabrication of Microlasers and Microresonator Optical Switches, *Appl. Phys. Lett.* **55**, 2724 (1989).
38. R. S. Geels and L. A. Coldren: Submilliamp Threshold Vertical-Cavity Laser Diodes, *Appl. Phys. Lett.* **57**, 1605 (1990).

39. C. J. Chang-Haanain, Y. A. Wu, G. S. Li, G. Hasnain, K. D. Choquete, C. Ceneau, and L. T. Florez: Low Threshold Buried Heterostructure Vertical Cavity Surface Emitting Laser, *Appl. Phys. Lett.* **63**, 1307 (1993).

40. T. R. Chen, B. Zhao, L. Eng, Y. H. Zhuang, A. Yariv, J. E. Unger, and S. Oh: Ultralow Threshold Multi-Quantum Well InGaAs Lasers, *Appl. Phys. Lett.* **60**, 1782 (1992).

41. A. F. J. Levi, R. E. Slusher, S. L. McCall, T. Tanbun-Ek, D. L. Coblentz, and S. J. Pearton: Room Temperature Operation of Microdisc Lasers with Submilliamp Threshold Current, *Electron. Lett.* **28**, 1010 (1992).

42. E. F. Schubert, Y.-H. Wang, A. Y. Cho, L.-W. Tu, and G. J. Zydzik: Resonant Cavity Light-Emitting Diode, *Appl. Phys. Lett.* **60**, 921 (1992).

10 Application of Microcavities: New Photoelectronic Integrated Systems

Izuo Hayashi

TABLE OF CONTENTS

I. INTRODUCTION

This chapter has been written for the purpose of describing the applications of microcavities in the industrial world which are expected in the next century. I hope to show how important the microcavity is in new photon-electron collaborating systems, which will become the center of information processing systems in a few decades. The microcavity structure will become a key device concept in new *photoelectronic integrated systems,* which will be at the fore-front of the information processing industry for many years to come.

The growth of electronics has been significant during the past few decades. It is still growing at a rather surprising rate, and has penetrated deeper into society. On the other hand, optical communications appeared rather independently and has already covered the whole world, also at a surprising rate. Electronics is taking full advantage of the behavior of electrons, and optical communications is doing the same with photons. Since the fundamental physics of these two elementary particles are very different, it is almost evident that one cannot overwhelm the other. Electrons are ideal for logic operations, because of their large e/m ratio, and, thus their strong interactions. Photons behave extremely well for information transfer, owing to their intrinsic high speed and zero charge. Their characteristics are vastly different, and it is difficult for one to replace the functions of the other. On the other hand, their characteristics are complementary. It is evident that they can help overcome each other's limitations by making use of the strong points of each. If one can build a system in which both particles work in a collaborating way, helping each other, it could be possible to achieve a system having the best total performance. This is the system called the "photoelectronic integrated system", described in this chapter.[1]

Fortunately, we already have photoelectronic (optoelectronic) devices, made of semiconductors, which can convert photons into electrons, and vice versa. They have already been used in optical communications, and computers and have "partially" realized the benefit of photon-electron collaboration. What is described in this article is how one can extend this collaboration fully into entire systems so as to take full advantage of this collaboration. The principle of a photoelectronic integrated system is simple: use electrons for logic operations and use photons for signal interconnections.[1] Theoretically realizing the system is straightforward. In a system the conversion of an electronic signal into a photonic one, and vice versa, must be done wherever and whenever necessary. The photoelectronic integrated system discussed in this chapter is comprised of micron-size devices that are integrated like an LSI.[1] Many small photoelectronic devices having microcavities will be placed in such a system.

Although the basic principles of conventional optoelectronic devices have been established, there are many unsolved problems in these new-generation micron-size devices as well as in material technologies. It will require a large amount of effort to break through these technical barriers. The design and fabrication of a large number of micron-size photoelectronic devices and their integration with microelectronic devices comprise the main parts of the new technologies needed for a photoelectronic integrated system. Here, the microcavity photoelectronic devices will play a key role.

The progress in electronics during recent years has revealed new kinds of problems concerning speed, power, and noise in system behavior, where only electrons are used. In computer systems, signal transfer using conventional metal wires is becoming a bottleneck with the increasing speed of their operation. The new technique of "optical interconnection", which is essentially a short distance optical communication within computer systems, is now being actively developed.[1,2] This indicates limitation of any system using only electrons. This problem will grow along with the progress of computers toward higher speed and higher integration. The problems based on the same basic reason will also grow into systems composed of microelectronics, ICs or LSIs, which are becoming the main body of computers. Possible technological solutions by a photoelectronic integrated system using microcavity photoelectronic devices are described in this chapter. Optical interconnections within LSIs, typically microprocessors, will be very effective, although the distances of the interconnection are very small (on the order of centimeters).[1-3] Three-dimensional circuits will become very attractive by eliminating metal wiring between stacked layers of microelectronics circuits through the use of photonic interconnections. This is a new structure which will be useful for highly parallel processor operation as well as other applications.[1,2,4]

Because of their complementary nature, the photoelectronic integrated systems will also overcome limitations of any photonic systems using mainly photons. Any nonlinear operations, such as amplification or logic operations, can be provided by electronic circuits tightly combined with the photonic parts of the system. Nor will the new systems only supplement deficiencies. A variety of new advantages will be found in the photoelectronic integrated systems. Although some examples are described in this section, it is still not known how many more useful systems will be created which would not be possible without photon-electron collaborations.

Much of the discussion presented in this article is based on studies by the U-OEIC study group, which was formed in 1991. This is a group, comprised of experts from different disciplines: silicon ICs, optoelectronics, materials and computer architecture.[5] Their discussions have focused on the feasibility of optical interconnections at the LSI level. Many new ideas have been born from their discussions, which will appear in this article.

The concept of the OEIC, or optoelectronic integrated circuit, was born and developed in the two national projects for optoelectronics. The first was sponsored by MITI (1979–1986). The second was sponsored by the Key Technology Center (1986–1995).[6] The concept of optically interconnected three-dimensional integrated circuit structures was independently initiated by professor Hirose's group at the Hiroshima University (1984).[7] The main members of these groups have participated in the U-OEIC research group.[5]

An ultra-large-scale integrated circuit, which includes electronic as well as optoelectronic devices, will be called "ultrascale optoelectronic integrated circuits (U-OEIC).

In this chapter, the outlook for studies on photoelectronic integrated systems is given together with the specifications needed for the microcavity photoelectronic devices for these systems.

In the second section, we will describe high speed photoelectronic integrated systems which are extensions of today's microprocessors, using "long distance" optical interconnections across a chip (Section II.B). We also discuss m sively parallel systems incorporating three-dimensional integrated structu s comprising many planar circuits in a stack which are connected vertically with optical beams. Such three-dimensional parallel processors are currently under development at Hiroshima University using optically connected common memories (Section II.C). Microcavity light emitting diodes will be a candidate for massively parallel systems. An image analyzing system using a massively parallel two-dimensional detector array is described (Section II.D). Since individual detector outputs are connected directly to a small processor, high speed image analysis becomes possible.

The operating requirements for micron-size optoelectronic devices are described in the third section (Section III.A). In order to meet the system function needs, microlasers (μ-LDs) must have thresholds that are as low as possible. Tens of microamperes or less are desirable for applications in microprocessors, where hundreds to thousands of these μ-LDs will be used. In massively parallel systems, where millions of light emitters at the μA level are needed, highly efficient microcavity LEDs will be promising. Photodetectors having both small size and high sensitivity are also needed. Photodetectors incorporating microcavities are well suited for such applications.

In Section III.B, fabrication technologies for the design and production of these micron-size photoelectronic devices are discussed. The large numbers and small size of these devices require new design concepts and process procedures, which are very different from those used by today's large, discrete optoelectronic devices. The implementation of microcavity devices will require large amounts of effort in the device design and fabrication processes. The integration of largely lattice-mismatched material systems, such as GaAs devices on silicon substrates, will require a breakthrough. The problem of achieving high yield and high reliability of optoelectronic devices, which is prerequisite for large scale integration, is also a big challenge.

A summary and future prospects of these photoelectronic integrated systems are presented in Section IV. Although it is difficult to foresee future developments, photoelectronic integrated systems will form the groundwork of the information processing systems emerging in the next century, as has been the case for electronic integrated circuits during this century.

II. PHOTOELECTRONIC INTEGRATED SYSTEMS: EXAMPLES OF PHOTON ELECTRON COLLABORATION IN SYSTEMS

In the first part of this section (Section II.A), the general features of optical interconnections are described. It is important to understand the effectiveness of the photonic interconnections, which are completely different from those of electronic interconnections. We then describe a single chip microprocessor with interior optical interconnections in Section II.B. Improvements in speed and power are demonstrated based on computer simulations. We then explain highly parallel processor systems made of stacked layers of electronic circuits, optically interconnected in the vertical direction, thus producing three-dimensional circuits (Section II.C). Many benefits are expected to be obtained by using this type of parallel system. The third application is described in Section II.D. A two-dimensional array of photodetectors is individually connected with small processors. This system is comprised of circuits of a perfectly parallel image processing. An ultra-high speed analysis of input pictures thereby becomes possible.

A. PHOTONIC INTERCONNECTIONS

The largest advantage expected from photonics in electronic systems will come about as the result of signal interconnections. This has already been well demonstrated in fiber optical communications. The minimum amount of power which is required for signal transmission is the power needed to drive the receiver if the loss in the fiber is negligible. Therefore, the minimum amount of energy, which is essentially the minimum number of photons (N_m), is determined by that needed for a sufficient signal to noise ratio at the receiver. It is essentially independent of the distance between the source and the receiver. This is not the case in electronic signal transmission. Whenever electrons move in a metal wire, they consume energy in between the source and the receiver. Power is thus lost in the metal wire as ohmic heat; this part of the power loss is proportional to the distance of signal transmission.

For photonic transmission, the speed of signal transmission is always the velocity of light. For electronic signal transmission, it differs from case to case. If a transmission line is used, the speed is the same as that of photons. On the other hand, in the case of an RC line, the speed is generally much slower. An RC line has some resistivity, and the signal propagates by the charging capacitance of the line through the line resistance (Figure 1). The delay is proportional to $(R \times C)$, and transmission can be much slower than the velocity of light (Figure 2).[3] In electronic systems, transmission lines are used in places where high speed signal transmission is needed. Typical examples are coaxial cables, pairs of parallel wires, and coplanar lines (Figure 3). Although these transmission lines are adequate as far as the speed is concerned, they consume large amounts of power at the terminating registers, which is necessary to avoid reflection at the receiving ends. In addition, the existence of a characteristic impedance creates problems in electronic signal transmission. The photonic interconnection, on the contrary, has no such troubles associated with characteristic impedance.*

---

* The characteristic impedance, or voltage/current ratio of a transmission line is around one hundred ohms. It is determined by the cross-sectional shape of the transmission line. Since it is a logarithmic function of the cross section, it changes little with the geometry. A transmission line is comprised of two isolated metal lines of identical cross sections along the length (Figure 3). They are used in many places in a computer box, and on the boards. Since they require spacing they are difficult to make on small ICs.

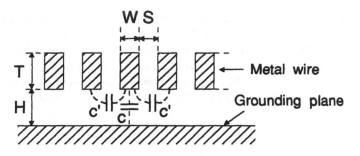

a) Cross section of interconnecting metal wirings on a large scale IC.

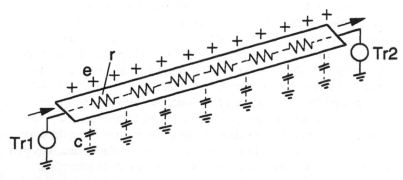

b) A metal wiring.
   (distributed capacitance c
   (distributed resistance   r

FIGURE 1   Electrical signal interconnection in Integrated Circuits (IC). Thin metal wirings have high resistance. Electrical signal shows delay due to capacitance (c) charging through resistance (r).

In electronic ICs, signal transmission is modeled by an RC line. Thin metal wires are patterned on integrated circuits. Therefore their resistance is very high (Figure 1).[3] In large scale ICs, since it is almost one kilohm per centimeter, the propagating speed is low, less than one tenth the speed of light (Figure 2).[3] There are many "long" signal lines in ICs, called global lines, which have lengths comparable to the size of an IC chip, about 1 cm or more.

Not only low speed but also extra power is consumed along on RC line. The power (p) needed to charge up a capacitance (C) to a certain voltage (v) at some frequency (f) is

$$p = C \cdot v^2 \cdot f.$$

Significant amounts of power are lost upon charging and discharging the interconnection wires in an integrated circuit.[3] Global lines consume a large fraction of the power in the ICs lost by this mechanism. Higher integration and

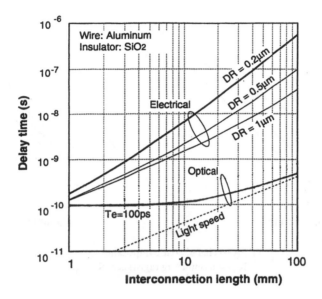

$W = S = $ Design rule (DR)

$H = 1\,\mu m, \quad T = 0.5\,\mu m$

(in Fig. 1, a)

Waveguide

LD driver   Amp

LD : Laser Diode
PD : Photo detector

Te = Delay time of LD, PD
and circuits (driver and amp)

FIGURE 2   Delay time of electrical and optical interconnections. (From Iwata, A., *Optoelectron. Devices Technol.*, **9**, 43, 1994. With permission from Mita Press.)

higher clock frequency ICs will require more power due to this line charging mechanism.

The use of the photonic interconnections in global lines not only increases the signal speed but also significantly decreases the power consumed in these ICs, which is demonstrated later (Section II.B). Photonic interconnections not only offer high speed, low power operation, they also have many other significant features.

Photons propagate in free space. A very important application of free space photon propagation is in a vertical signal interconnection between two planar circuits, each of which exists in a different plane (Figure 4). A high density interconnection between two planes can be obtained, which is difficult using electric interconnections. Electronic circuits can be easily fabricated in a single plane. Three-dimensional circuits can be constructed using vertical photonic signal interconnections between a number of circuits which are stacked vertically. Their importance is explained later (Section II.C.[4]).

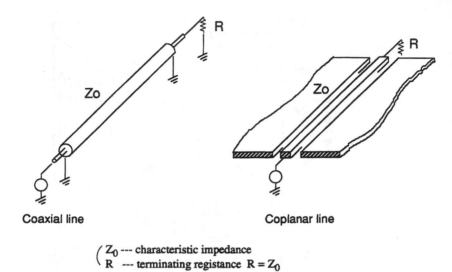

Coaxial line   Coplanar line

$$\left(\begin{array}{l} Z_0 \text{ --- characteristic impedance} \\ R \text{ --- terminating registance } R = Z_0 \end{array}\right.$$

FIGURE 3   Transmission lines.

Significantly lower power is sufficient for high speed photonic signal interconnections. A photon carries an energy of hv, which is about one electron volt for photons used for photonic interconnections.

$$h\nu \approx 1 \text{ electron volt} \approx 10^{-19} \text{ joule, } \lambda \approx 1\mu m.$$

The power carried by the photon flux is essentially determined by the number of photons (n) per second. For 1 μW of power, it is approximately

$$10^{19} \times 10^{-6} = 10^{13} \text{ photon/s.}$$

Therefore, for 1 GHz communication, number of photons per pulse is

$$10^{13}/10^{9} = 10^{4} \text{ photon/pulse.}$$

This is approximately the photon number which is sufficient to obtain a low enough error rate.[8]

B.   PHOTONIC INTERCONNECTIONS IN MICROPROCESSC .S

As discussed in Section II.A, the time delay in an electrical interconnection is significant, even within a single IC having centimeter dimensions. This is because in electrical ICs the interconnections use thin metal strips, and their propagation delay is determined by their RC time constant (Figures 1 and 2).

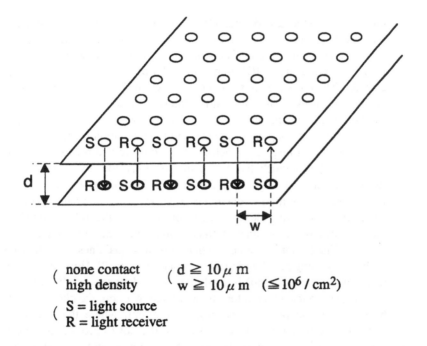

(none contact   (d ≥ 10 μ m
  high density   (w ≥ 10 μ m   (≤ 10⁶ / cm²)

(S = light source
  R = light receiver

FIGURE 4 Vertical photonic interconnections. Face-to-face interconnection between two planar circuits.

A microprocessor is a single chip computer system in which all of the elements of a large size computer are included. Its pattern size of interconnecting wires has already decreased to below one micron, and it will reach 0.1 μm within ten years. A microprocessor is composed of many circuit blocks, processors, memories, and many other elements. Most of the transistors in each circuit block are interconnected within short distances (on the order of microns to tens of microns), so that the interconnection delay is small, even by using conventional metal wiring. However, there are many long metal wires which connect points in different circuit blocks. These include lines which distribute the clock signal and common lines for exchanging signals between distant registers, which are called clock lines and bus lines. Their length extends on the order of the chip size, one to a few centimeters, and their numbers become several thousands in an advanced microprocessor.[3] Delays in these long lines are significant; photonic interconnections show much smaller delays (Figure 2).

In order to implement a photonic interconnection, a light emitter, typically a laser diode, is set so as to convert electrical signals to photonic signals. A photonic signal typically travels through a waveguide and reaches a receiving point (Figure 2). Here, another optoelectronic device, a photodetector (PD), is placed to convert photons into electrons. Amplification will be needed to bring the voltage level of the receiving signal up to that of the electronic circuit. This is the another point which requires innovative thought. The delay time of the photonic interconnection shown in Figure 2, includes delays of electronic circuits (Te) in the LD driver and PD signal amplifier, in addition to delays in the waveguide (Figure 2). The value of Te depends on the speed of the transistors in the driver and in the amplifier. Although it is a few hundred picoseconds when using conventional C-MOS transistors, 100 ps will be achieved for advanced circuits.[3] Therefore the photonic interconnections become effective for interconnection distances of over several millimeters (Figure 2).

Waveguide interconnections are desirable over free space propagation, because of the high coupling efficiency and flexible guiding with patternings. The waveguide patterning can be achieved by conventional photolithographic techniques. In a bus or clock distribution line, photons from an LD are branched into many receivers. Although simplicity of the branching in a photonic interconnection is one of its merits, obtaining equal intensity branching requires some engineering. A waveguide can be made of a line of higher refractive index. Single mode propagation is not an essential requirement, because of the extremely short distance. The photonic interconnection delay in global lines is less than one tenth of the delay using metal wiring interconnects (Figure 2). However, the improvement due to using photons is small over short distances of a millimeter or so.

In order to install a photon signal interconnection, the photon emitter must be placed at the point of the electronic circuit where the source signal is generated, within a millimeter or so, in order not to produce any extra delay (Figure 2). A laser diode (LD) will be suitable as a photon emitter, because of its fast response and high efficiency. The generated photons are fed into an optical waveguide (WG) and then guided to the receiving point.

In order to obtain fast, low power, reliable, and low cost photon signal interconnections, a variety of points must be considered. A laser diode having a low threshold is required for stable operation. The output power changes less with an operating current (I) if I is much larger than the threshold current (I_t), which is a strong function of temperature (Figure 5). In addition, a high electron-photon conversion efficiency and a faster response of the photon output is expected by using a high operating current. Therefore, the low threshold characteristics of a microcavity laser structure are favorable.* For photodiodes (PDs) a fast response and high efficiency are required. Micro-size PDs will have a fast response because of their small capacitance. The use of microcavity PDs should be helpful to obtain high efficiency because the use of a microcavity will increase the photon intensity inside the cavity, which will enhance the photon absorption in the detector material.

Iwata estimated the effect of a photonic interconnection in a C-MOS microprocessor by carrying out a circuit simulation.[3] First, the delay of the whole circuit is the sum of the delays in the individual components, including the electronic part of the circuit. It was found that delays in the electronic part are larger than that of the photonic part. In particular, a photodetector (PD) amplifier will be slow. It also consumes the largest amount of power in the system, because the number of PDs—and thus PD amplifiers—are larger than

---

* Because LEDs have no threshold, they will have stable operation, but slower response and lower external efficiency are drawbacks.

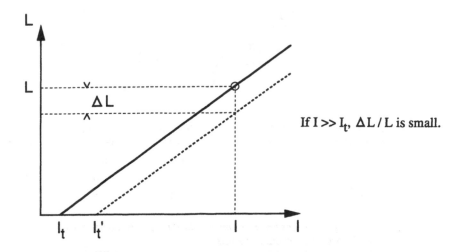

FIGURE 5   Stable operation using a low threshold laser.

that of the LDs in the system. The number of PDs per LD is about 20 in a bus line, and will be on the order of 100 in a clock distribution line. The amount of the minimum light energy to achieve sufficiently low error rate is estimated to be 10^4 photons/pulse at 1 GHz at each of the PDs.[8]

However, the simulation showed that the corresponding output voltage from a PD is about 0.01 volt, which requires a high gain amplifier. According to Iwata's estimate,[3] in a clock distribution system, large numbers of high gain amplifiers increase the total power requirement of the photonic interconnect (Figure 6). Accordingly, increasing the laser power by ten times, from 10^4 to 10^5 photons per pulse in each PD, decreases the required amplifier power by 10, thus yielding a much lower system power requirement (Figure 6).[3] In general, there is a trade-off between the laser power and the PD amplifier power in order to achieve minimum system power. In this example, a conventional C-MOS transistor was assumed as the amplifier.

However, in the long run a more innovative amplifier circuit will be developed. The use of (III-V) transistors, a GaAs HEMT for example, will improve both the speed and power. Photo-detectors made of (III-V)s have better characteristics than those made of silicon. The main part of the PD output capacitance is stray capacitance, mainly due to interconnecting wirings. Therefore, a new device structure, in which a photodetector is directly mounted on an FET gate, will be more effective for lower power as well as higher speed.

A block diagram of clock and bus lines which was used in the simulation are shown in Figure 7.[3] One LD will be sufficient for an entire photonic clock circuit having about 100 PDs. The branching of the waveguide circuit requires careful designing in order to provide equal intensity and equal delay. The difference in the delay between different PDs—skew—can be made practically zero by keeping the length of all the waveguide branches equal. This scheme gives superior performance and is much simpler in structure than an electronic

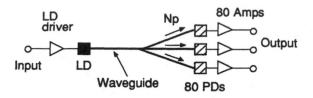

No. of PD receiving photons / pulse	LD driver power	LD output power	80 Amp power	Total system power
(Np)	(mw)	(mw)	(mw)	(mw)
10^4	(0.4)	0.1	80	80.4
10^5	(4)	1.0	6	10

Increasing LD power ten times decreases total power down to 1/8.
One LD drives 80 Photo Detectors.
Total power includes LD driver power and power of 80 PD Amplifier.
Clock frequency 250MHZ, waveguide transmission efficiency 33%.

FIGURE 6   Power dissipation of an optical clock system. (From Iwata, A., *Optoelectron. Devices Technol.*, **9**, 49, 1994. With permission from Mita Press.)

clock distribution circuit, in which additional amplifying stages are necessary (Figure 7). The calculated improvement in speeds is shown in Figure 8.[3]

The use of photonic interconnection delays in bus lines has also decreased significantly, and the overall cycle time will decrease by about one half (Figure 8). This is for the case in which only the clock and bus lines are replaced by photonic devices, and all other circuits remain electronic. Much greater improvements can be achieved by taking full advantage of the delay decrease offered by photonic signal interconnections. However, a completely different circuit architecture must be devised in order to gain the maximum merits.

Iwata has also shown the effect of photonic interconnections on the power consumption of a C-MOS microprocessor (Figure 9).[3] The power in the clock and bus lines has been decreased to less than half by using photonic interconnections. This calculation assumed ten times more laser power, which was necessary for a sufficient error rate (Figure 6).[8] The effect of photonic interconnections will become more significant when improvements in the PD-Amp structure are achieved and the laser power can be lowered.

Microprocessor clock frequencies are increasing every year, and will reach the gigaherz level or more within the next decade. The response times of micron-size LDs and PDs will be in the range of 10 GHz or more because of the small size and high operating current in LDs (Figure 5). Thus, the major speed limitation of photon interconnections will be in the electronic part of the interconnection circuits.

In order to implement optical waveguides for microprocessors, a special independent substrate for the optical components has been proposed by Takeda.[9]

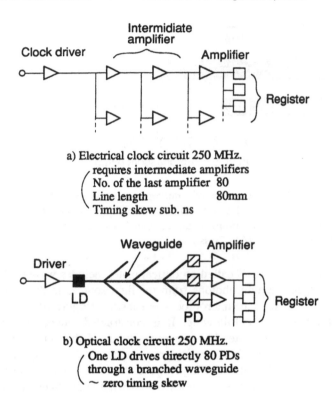

a) Electrical clock circuit 250 MHz.
requires intermediate amplifiers
(No. of the last amplifier 80
   Line length   80mm
   Timing skew sub. ns

b) Optical clock circuit 250 MHz.
(One LD drives directly 80 PDs
   through a branched waveguide
   ~ zero timing skew

FIGURE 7   Comparison of an electrical clock circuit and an equivalent optical clock circuit. Total power of the optical circuit is about one third of that of the electrical circuit. (From Iwata, A., *Optoelectron. Devices Technol.*, **9**, 47, 1994. With permission from Mita Press.)

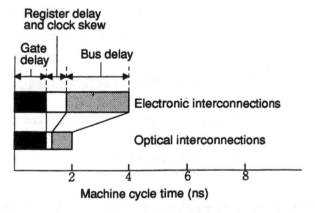

FIGURE 8   Optical interconnections in clock and bus circuits increase the speed of a microprocessor (estimated by a circuit simulation). 0.2 μm C-MOS microprocessor. Clock frequency ∝ (machine cycle time)$^{-1}$. (From Iwata, A., *Optoelectron. Devices Technol.*, **9**, 47, 1994. With permission from Mita Press.)

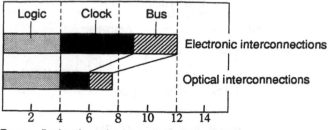

Power dissipation of micro-processors (watt)

FIGURE 9 Power dissipation of a microprocessor decreases by using optical interconnections in clock and bus circuits (estimated by a circuit simulation). 0.2 μm C-MOS microprocessor. (From Iwata, A., *Optoelectron. Devices Technol.*, **9**, 50, 1994. With permission from Mita Press.)

The substrate, which is called an optical plate, includes waveguides, mirrors and lenses for guiding photons from the light sources to the photodetectors which are located on the microprocessor substrate (Figure 10).[3,9] The merits of this structure are the following. It is constructed independently from the microprocessor substrate. The processing steps of the optical plate and the microprocessor do not interfere with one another. Sufficient space is available for the waveguide networks high above the crowded LSI substrate. Even though the width of the waveguides is somewhat larger than the metal wiring (several microns), no problem will result because no practical restriction exists for the thickness of the optical plate. Therefore, a multilayer waveguide structure can be used if necessary.

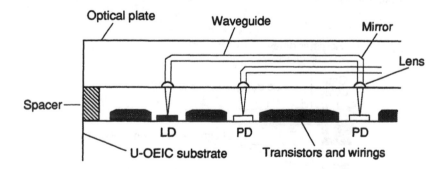

LD = Laser diode
PD = Photo detector

FIGURE 10 Photonic interconnections by using an "optical plate". Waveguide networks, micromirrors, and lenses for photonic interconnections are all accommodated in an optical plate. (From Iwata, A. and Hayashi, I., *IEICE Trans. Electron.*, **E76-C**, 96, 1993. With permission from the Institute of Electronics, Information, and Communication Engineers.)

The input and output parts of the waveguide on the optical plate must be aligned with the light sources and PDs on the IC substrate (Figure 10). A submicron alignment of these corresponding spots is feasible using conventional photolithographic techniques. One positioning step in a horizontal plane should be sufficient for the alignment of all optical contacts between the waveguides on the optical plate and the optoelectronic devices on the substrate. To achieve a high coupling efficiency, micron size lenses will be useful. Small mirrors are also needed for the 90 degree turn in the waveguides (Figure 10).

C. VERTICAL PHOTONIC INTERCONNECTIONS FOR THREE-DIMENSIONAL STRUCTURES

Free space photon propagation can eliminate any "contacts," that are essential in electronic interconnections and require the use of soldering or connectors. This feature will result in a new system of architecture which will only be available through photon-electron collaboration.[4]

The photonic signal interconnections have the unique feature of allowing the construction of three-dimensional integrated systems, which are practically impossible using electronic interconnections alone. Different layers of electronic circuits in a system can be optically interconnected, as shown in Figure 11. Each layer comprises an LSI or a board containing LSI (U-OEIC) chips, which are optically interconnected vertically in many spots with LSIs in the neighboring layers. A high speed system containing massive internal connections can be

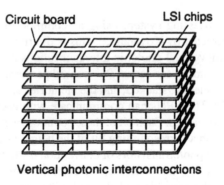

Circuit board   LSI chips

Vertical photonic interconnections

Horizontal photonic interconnections in each circuit board are omitted in the picture.

FIGURE 11   A schematic drawing of a photonically interconnected 3-D photoelectronic integrated system. Three-dimensional circuits can be built by using vertical photonic interconnections. A number of U-OEIC layers are interconnected vertically by photonic beams. Each layer is either a single U-OEIC chip or a circuit board containing U-OEICs. (From Hayashi, I., *Optoelectron. Devices Technol., 9*, 7, 1994. With permission from Mita Press.)

realized by utilizing this type of structure. The effectiveness of vertical photonic interconnections in such three-dimensional (3-D) systems is discussed in this section, which gives examples of several different parallel processing systems as well as image analysis systems.

There are several advantages of 3-D systems:

1. A compact system with high volume density can be realized.
2. High speed systems can be constructed, since the signal path length can be made much shorter than those in a planar two-dimensional structure for the same set of LSIs.
3. Massively parallel systems can be designed most effectively by the 3-D structure[4] (Ae's scheme for example).
4. The fabrication of a 3-D system can be greatly simplified by using vertical photonic interconnections, because an assembly of the 3-D system can be made by a mechanical alignment of LSI layers in the horizontal planes for matching of the vertical photonic interconnections.[4]
5. Higher fabrication yields, and thus a low assembly cost, will result, because the each circuit's planes can be checked out in advance.
6. Since the whole structure can be disassembled if necessary, a flexible reconstruction becomes possible.

Microcavity LEDs will be suitable light emitters for massively parallel 3-D systems using a large number of vertical photonic interconnections.

Electronic circuits in an LSI are intrinsically aligned in a single plane. A stack comprising a few circuit layers might be possible based on special multilayer structures. However, a stack of more than several circuit planes with high density interconnections will be difficult using electrical interconnections.

A parallel processing system using a "common memory" has been proposed and constructed by Koyanagi.[10] This is the first system in which a 3-D structure utilizing vertical photonic interconnections has been applied (Figure 12). Common memories for each CPU (central processing unit) are located on each plane, the planes are stacked, and the memories are photonically interconnected vertically. The CPUs can communicate with each other through these 4 Kbit common memories. Each bit of these common memories are photonically interconnected with a corresponding bit of the neighboring memories. A set of two LEDs and two phototransistors of conventional structure is used to transfer binary data. By using this scheme, all corresponding bits in the common memory contain the same data at the end of the cycle time. The CPUs can communicate with each other through the common memory. The cycle time is 80 ns. Data transfer, itself, takes about 16 ns.[10] The total speed of data transfer is very high, 128 Gbits per second, because of the parallel nature of the data transfer in the memory. This system has actually been built, and its performance has been confirmed.[11]

T. Ae has proposed a new massively parallel processor operation that also uses the common memory scheme with vertical photonic interconnections.

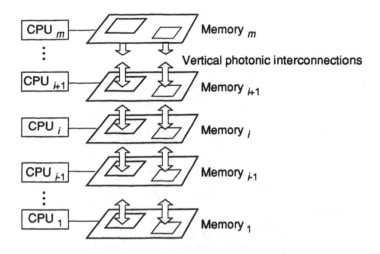

FIGURE 12 A parallel processor system with 3-D optically coupled common (3-D-OCC) memory. The 4-Kbit of common memory is connected with that in the upper and in the lower layers through vertical photonic beams. Pairs of LEDs and phototransistors are used. Memory cells with identical address in all memory layers have the same data after the photonic data transfer. (From Koyanagi, M., *Optoelectron. Devices Technol.*, **9**, 120, 1994. With permission from Mita Press.)

With this scheme, many thousands or even millions of processors can be interconnected.[2,4]

A parallel operation of 12 processors through 8 sets of common memories is shown in Figure 13.[2,4] Eight common memories are located in each of three layers, as shown in Figure 13, and 12 processors are placed in one of these layers. Notice that each processor is connected with its two memories within the same plane. No interconnecting wiring is needed between the layers; interconnection between the layers is achieved photonically. Therefore, the three layers can be fabricated separately. The entire system can be completed by stacking three layers with a proper horizontal alignment. Information from one processor can be transferred to the three neighboring processors directly through the common memory; further transfer can be achieved through two common memories and one processor in two steps. Three steps will be needed to reach the most distant processor.

This scheme can be extended to a massively parallel processor system.[1,2,4] The number of processors (n) is

$$n = N2^{N-1}$$

where N is the number of layers (dimension of the hypercube).

A parallel processor system comprising one million units can be constructed in 17 layers (N = 17). Each processor is located in one of these 17 layers. The adjacent processors are connected through their common memories.

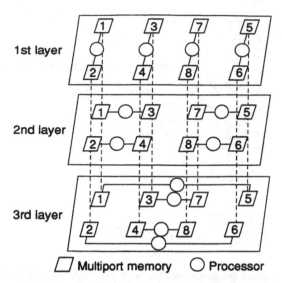

1st layer

2nd layer

3rd layer

☐ Multiport memory   ○ Processor

(a) Multi-layered three-dimensional multiprocessor system connected through
    common memories. Vertical photonic interconnections are used between
    common memories.

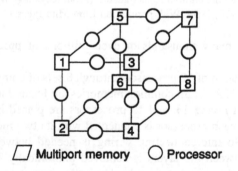

☐ Multiport memory   ○ Processor

(b) Original interconnection architecture

FIGURE 13 A parallel processor architecture connected through common memories.
Vertical photonic interconnections are used to construct a common memory scheme.
(From Ae, T., *Optoelectron. Devices Technol.*, 9, 25, 1994. With permission from Mita
Press.)

No wiring is needed for communications between different layers. Each layered circuit can thus be constructed individually before the total assembly. A
schematic drawing of such a massive parallel system is given in Figure 11. Ae
describes a parallel system comprising 25,000 processors which can be accommodated in 13 layers.[4] If each common memory has 1 Kbit, the total number
of photonic devices necessary for the vertical interconnection is about 10^8

pairs. A rough estimate shows that the total system can be accommodated in a volume of less than $(10 \times 10 \times 10)$ cm.[3] Photonic devices of the smallest possible size, low power, and simple structure are required. The microcavity scheme for LEDs and PDs is particularly attractive for these machines.

According to Ae, a kind of human intelligence machine can be made using this type of a massive parallel system.[4] Such a machine could pick up answers for incoming questions by searching memories to find the closest possible cases recorded, which are made by "teaching," using the experiences of human beings. He has called this scheme *memory based reasoning*.[4] In the future society, there will be strong needs for such intelligent machines—which, for example, can help handicapped persons. Ae's scheme is to use a massive parallel processing system to memorize many incoming cases (e.g., sets of problems and answers) and to sort a proper answer for the incoming questions. According to his estimate, about ten thousand parallel processors will be necessary for a machine for personal use, and about a million for public use. These machines will be able to be designed using his architecture, in which an optically interconnected common memory scheme can be utilized.

D. TWO-DIMENSIONAL IMAGE ANALYZING SYSTEM USING PHOTONIC INTERCONNECTIONS

So far image analysis has been performed using serial scanning of two-dimensional data (TV scheme). The output of a CCD detector is fed to electronic circuits for a data analysis. The scanning speed limits the analyzing speed in such a serial processing scheme. Progress of the microelectronics technology will allow a large number of gates, millions, to be made into a single chip; on the other hand microstructure photonic devices will become available. Thus, a small two-dimensional image analyzing machine with parallel data processing by combining an array of photodetectors and a corresponding set of processing circuits will become feasible, which will operate much faster than the old scanning scheme.

M. Ishikawa has carried out a feasibility experiment of such a parallel image analyzer: two-dimensional photodetector arrays of 64×64 pixels, each of which is connected directly to an electronic processing element (Figure 14).[12] The speed of an image analysis is several orders of magnitude faster than that in the scanning method, because of the parallel nature of the analysis. A compact high speed artificial vision machine will be developed using this scheme (Figure 15).[12] This type of image analyzing machine will be another photoelectronic integrated system for photonic applications using the collaboration of photons and electrons. Microcavity based light receivers and emitters will be suitable for these applications.

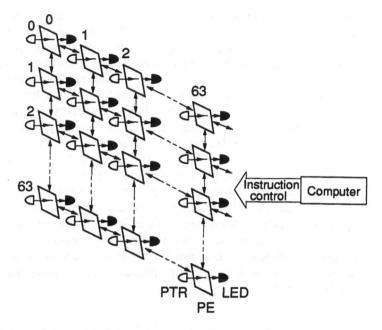

Each process element (PE) has a photo transistor (PTR) in the front, and an LED in the back. Neighboring PEs are interconnected each other.

FIGURE 14 A schematic drawing of a two-dimensional massively parallel image analyzing system. (From Ishikawa, M., *Optoelectron. Devices Technol.*, **9**, 35, 1994. With permission from Mita Press.)

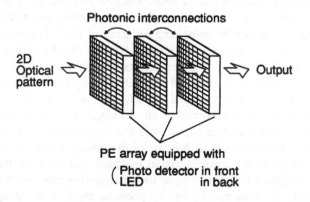

FIGURE 15 Conceptual drawing of an advanced artificial vision machine. (From Ishikawa, M., *Optoelectron. Devices Technol.*, **9**, 30, 1994. With permission from Mita Press.)

III. MICRO-PHOTOELECTRONIC DEVICES

A. NEEDS FOR MICRON-SIZE PHOTOELECTRONIC DEVICES

The effectiveness of photon-electron collaboration has already been demonstrated by "photonic interconnections" in computers, where today's technique of optical communication is useful, due to the larger distances of the interconnections. Needs for photonic interconnections are evident in LSIs, as explained in the previous sections. In order to make photonic interconnections possible within individual integrated circuits, the size of the photoelectronic devices must be as small as possible in order to match the size of the electronic devices in the ICs. As discussed in the present volume, not only lasers, but also LEDs and photon receiving devices (PDs) will have better performances, if they are contained in resonant cavities. We can generally say that the smallest size of any photoelectronic device has dimensions on the order of the wavelength, which is about one micron. The number of photonic devices for interconnections is several orders of magnitude smaller than that of electronic devices in a chip (Section II.B).

The microcavity structure will enhance the characteristics of these photoelectronic devices, as discussed in the present volume. However, in order to realize the performances needed for photoelectronic integrated systems, large numbers of studies will still be needed for device technology. The largest technical barrier to realizing these photoelectronic integrated circuits involves material processing technology. The proposed photonic devices are hundreds of times smaller than those in common use today. In addition, they must be integrated into LSIs together with transistors. These LSIs are made by a well-established technology that has been developed over many years. Transistors are usually based on silicon, compared with the III-Vs used for these photoelectronic devices. Here, a new micro-fabrication technology for micro-photoelectronic devices and an integration technology for dissimilar materials must be performed. In addition, the fabrication process of the Si devices and that for the III-V devices must be compatible—they should not interfere with each other. These are difficult new technologies. Today, fortunately, the progress of material technology is significant, and there is good hope in reaching this goal if sufficient focus can be placed on the target. Performance requirements and the device structure meeting the requirements are described in Section III.B, and materials needed for the fabrication of the structure is described in Section III.C.

B. PERFORMANCE OF MICRON-SIZE PHOTOELECTRONIC DEVICES

Comparing two light emitting diodes with different sizes, the power needed to excite them to the same excited carrier density is proportional to the area of the

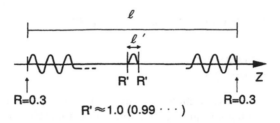

FIGURE 16 Proportional law of semiconductor lasers. Laser thresholds are proportional to the length of the lasers having identical cross sections. In a short laser the mirror reflectivity is increased in order to keep the same excitation density at the same injection current density.

active region of these devices. If a clean heterostructure, such as (AlGaAs/GaAs), is available, LEDs with no extra recombination at the periphery of the active area can be obtained. Thus the linearity assumption of the size versus the power will hold. This is also true for semiconductor lasers. Today, the laser structure has large dimensions of a few hundred microns in one direction (z). Lasers have rather small dimensions of micrometers or less in the other two directions. If one squeezes the long dimension (z), from 100 μm to 1 μm, the power needed to reach the same excitation level can be reduced by a factor of 100 (Figure 16).

If the original large laser has a threshold of 1 mA, which is a possible value for low threshold lasers today, the squeezed one would have that of 10 μA. If the original long cavity has 100 standing waves, for simplicity, the squeezed one would have only one mode, which is a microcavity laser. This simple "proportional law" clearly shows the feasibility of microwatt light-emitting devices.

However, if one looks into the details of this one standing wave laser, there are some conditions which are easily satisfied in the large cavity LD that must also be satisfied in the microcavity one. For instance, the heterointerface of the microlaser must be clean and defectless everywhere in the micron-size cavity. This is difficult to achieve using today's fabrication technology. A highly refined technique must be developed so as to achieve this purpose. The surface recombination velocity of injected carriers in heterointerfaces at the sides of the active volume must be negligible in order to avoid any extra carrier losses. This can be achieved only in the main part of the long laser using today's fabrication technology.

Another new aspect of a microcavity LD is the need for strict dimension control of the cavity. In large cavity LDs, there is no need to have strict cavity dimension control. The LDs will oscillate at a favored wavelength given by the maximum gain, because plenty of modes are available in the cavity. In the microcavity, the only available resonance wavelength is uniquely determined by the dimension of the microcavity. Therefore the cavity dimension must be controlled so as to obtain a resonance wavelength in the range of the gain region. A dimension tolerance of approximately 1 nm will be required for a cavity of 1 μm dimension. Several other dimension controls are also necessary.

Concerning device design, other important aspects are the cost and yield. One must deal with a great many numbers of devices in photoelectronic integrated systems—hundreds, thousands, or even millions. High numbers require high yields. Although a yield of 0.9 is sufficiently high for discrete device applications, a yield of 0.999 is required for 1000 integrated devices. Although it is common practice for microelectronics, it is a completely new world for optoelectronics. In addition, the price for such high technology must be extremely low. The price of discrete photoelectronic devices is on the order of one dollar, even for a cheap item. Integrated ones must have prices many orders of magnitude less, a fraction of a cent per piece, or even less. What I have described concerning the production yield or cost will have basic importance for device design. It is very different from the situation regarding discrete devices. The difference is similar to that for discrete transistors and those in the integrated circuit.

The response-time speed of microcavity devices is, of course, an important characteristic of the photoelectronic devices on integrated circuits. Their smaller size is generally reflected in their faster response. For lasers, decreasing the photon lifetime decreases the response time $\left(\propto \sqrt{\tau} \right)$. The response time of tens of GHz or more (10^{-10} sec or less) will be available with microcavity lasers. Single chip microprocessors will require this high speed. Their clock frequency will approach one gigahertz within the next decade. The slowest part in the circuit will be the electronic amplifier after the photodetector. Therefore, at present, one of the necessary breakthroughs is in the receiving circuit, including the photodetector and the amplifier (Section II.B).

An optical waveguide circuit will be suitable to interconnect LDs and PDs in the system (Section II.B). The advantage of the waveguide scheme has already been explained (Section II.B).

As for 3-D massively parallel systems using vertical photonic interconnections, a large number of light emitters and receivers will be needed—about 10^8 for 25,000 processors with 1 Kbit common memory. Of prime importance here will be simplicity of operation, low power, and low price of the light emitters, rather than high speed. The use of microcavity enhanced LEDs seems to be very attractive. A simple microcavity photodetector is also desirable. The light output intensity of LEDs will be increased ten times by employing a microcavity structure (Figure 17).[13] A response speed of nanoseconds will most likely be sufficient. If one considers the ultraparallel nature of the whole system, the total data-transfer rate will easily exceed a terabit per second (Tb/s), even with a microsecond cycle time. The total power must be as small as possible. One microwatt for the emitter receiver pair will be required.

The paralleled image analysis system described (Section II.D, Figures 14 and 15,[12]) is an example in which an input image is processed by a massively parallel photoelectronic circuit. A micropower receiver emitter pair to be integrated with microelectronics will also be needed.

These examples should explain the importance of such small photonic devices in future society. These systems may not need devices having

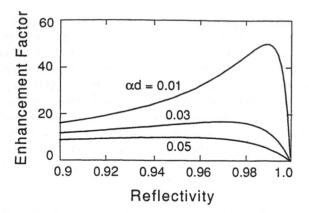

FIGURE 17   Enhancement factor of a microcavity as a function of the reflectivity of reflector. (Calculation by H. Yokoyama.)[13]

sophisticated characteristics, such as picosecond speed or angstrom stability. However millions of these devices must be made with high yield, high reliability, and low cost, and they must be integrated with microelectronics. Although the structure of these micro-photoelectronic devices will make use of microcavity principles, it must be extremely simple and ingenious. Thick multilayer mirrors and other complex structures cannot be used. Clever ideas and sophisticated material engineering will be needed for the necessary breakthroughs.

C.   FABRICATION STUDIES OF PHOTOELECTRONIC DEVICES

The requirements for the structures and performances of micron-size photoelectronic devices were described in the previous section. We can estimate the features of these devices using today's knowledge of photoelectronic devices as well as the detailed studies on microcavity structure described in this book. Their estimated characteristics seem to be sufficient for most of the applications described. However, if one considers a photoelectronic integrated system as a whole, these devices are only a small part of the complete system on an LSI chip. Although photonic devices exist in large numbers on a chip, in the thousands, LSI chips have huge numbers of transistors, sometimes millions. Photonic devices must be placed at designated spots in the LSI circuit, occupy a minimum space on the order of a micron, yet must have nm precision in dimensions. Today nobody knows how to fabricate them with a very high yield of 99.9...% and high reliability. In addition, the processings for their fabrication should not disturb the mother LSI, which is made by silicon microelectronic technology. The coexistence of these completely different devices on a common substrate, which is silicon, is the prime requisite of the photoelectronic integrated systems. Although silicon transistors require high process temperatures near to 1000°C, group III-V for photonic device uses much lower process temperatures of 400 to 600°C. It is likely that they must be processed sequentially: first silicon, then group III-V.

The growth of high quality epitaxial (III-V) material on silicon is the first processing step needed, which has not been established. Micron-size photonic devices with defectless heterointerfaces also need breakthroughs regarding fabrication. In addition the dimensions of these devices must be precisely controlled to meet cavity resonance condition.

In spite of these difficult problems, the recent progress in material technology is encouraging. The controlled epitaxial growth of (III-V) with nanometer precision in thickness has been developed using molecular beam epitaxy (MBE), and epitaxies using gas-phase source materials. The so called atomic layer epitaxy (ALE) must be very useful for making microcavity structures. However, in the lateral direction similar precision is not available. Recent studies aimed at producing quantum wires or quantum dots also need such lateral control. Some of the technologies used for their fabrication will be useful for microcavity fabrication.

In this section, some of the representative process studies, such as the growth of lattice-mismatched heteroepitaxy and studies aimed at defectless interface creation using *in-situ* processing in high vacuum, will be presented. It has been demonstrated that high quality epitaxial growth can be achieved between different semiconductor materials, as long as their lattice constants are closely matched within a few parts in ten thousand. GaAs/AlGaAs is a typical example. Many photoelectronic devices are made using this lattice-matched epitaxial technique. However, once their lattice constants differ by more than 1%, many defects form at the heterointerface. GaAs epitaxy on Si—having 4% mismatch—is a typical example. Many studies have been carried out to improve this epitaxy, since it is a very attractive combination of materials for many device applications. A large number of dislocations originate at the interface, and some propagate upward into the epitaxial layer.[14] The density of the dislocations for a layer of few microns thickness has been decreased to below $10^6/cm^2$. Typically, it was much larger, $10^9/cm^2$ or more. People have been trying many different techniques to break the barrier around the $10^6/cm^2$. Recently, several new approaches show encouraging signs.

If one assumes that a light emitting device must be dislocation-free in order to have long life, the probability of obtaining good device yield is $(1-A \cdot D)$. Here, A is the device active area and D is the dislocation density. The yield is high for smaller devices. However we must have a large number (N) of good devices on a U-OEIC, in order to have a meaningful yield of the whole U-OEIC. The condition will be $(NAD) \ll 1$. If $N = 10^3$ and $A = 10 \ \mu m^2 (= 10^{-7} \ cm^2)$, D must be appreciably less than $10^4/cm^2$. Light-emitting devices are very sensitive to dislocations, because electron-hole recombination creates new defects at the dislocation. Majority carrier devices, such as photodetectors or FETs, are much less sensitive to dislocations. Actually, HEMTs, (heterostructure FETs) that have a high dislocation density of $10^7/cm^2$ are now being produced on GaAs epitaxial layers on silicon. This is very encouraging because it will overcome many technical problems associated with GaAs device production on silicon.

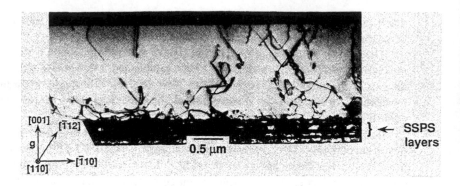

FIGURE 18 An XTEM micrograph showing dislocation structure in a GaAs film grown by a special layer-by-layer growth method (SSPS).[14,15] Dislocations are highly condensed around layer interfaces. (From Tamura, M., *Optoelectron. Devices Technol.*, **9**, 102, 1994. With permission from Mita Press.)

One of the recent techniques yielding high quality epitaxy of GaAs on Si is the layer-by-layer growth of GaAs. After the first few monolayers, GaAs growth tends to occur in islands, and more dislocations may start at coalescing islands. If layer-by-layer growth is forced up to a large thickness using special growth techniques, the dislocation density has decreased (Figure 18).[14,15] Another new approach is to insert buffer layers between the Si substrate and the GaAs epitaxial layer.[14,16] First, a few monolayers of GaSe having a layered crystal were grown on an Si (111) substrate; GaAs was then grown on the GaSe. Under a transmission electron microscope, a GaAs single crystal is seen above the GaSe crystal layers. Because there is no direct bonding between the GaAs and Si, a lattice mismatch doesn't create any problems. Interestingly, the crystal orientation of the GaAs is the same as that of the silicon substrate. Even though the size of the dislocation-free part of the crystal is small at this early stage, this experiment shows the feasibility of this new technique. Several other methods are under study, and progress will be made among these competing studies.

It is possible, even today, to bond completed photoelectronic devices directly onto a silicon substrate. This is the so called "flip-chip" technique. Since positioning of the device can be done accurately, it can be applied to the placement of a relatively small number of photonic devices on an Si chip. Initial OEICs can be made using this flip-chip technique. However, to fabricate many group III-V devices on U-OEICs, heteroepitaxial growth should be used.

As described in the previous section, micron-size photoelectronic devices must have defect-free hetero interfaces at the active region, and must have precisely controlled dimensions. Figure 19 shows a simplified drawing of a light emitter. Using planar epitaxy of the active and cladding layer in succession, the top and bottom interfaces between the GaAs active layer and the sandwiching AlGaAs layer can be made almost defect-free. However, both of the side edges of the active layer facing the AlGaAs sidewalls will have defects, which will be exposed to the outside atmosphere during the process of

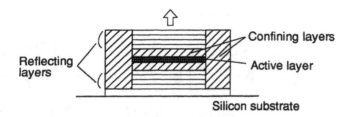

Interface between an active layer and confining layers
must be defect free.

FIGURE 19   A schematic drawing of a light emitter.

etching and regrowth. In general, if the structure is exposed to the outside
environment, immediate oxidation and contamination degrade the exposed
surface of the device. In conventional large devices such degradation at the
device edge is not important. The diffusion length of the injected carriers is
about one micron. Therefore, any nonradiative recombination at the edge has
a minor effect in long devices, but is fatal for micron-size devices, which we
are interested in. Experiments have shown that if one can keep a clean environment during successive process steps, one can obtain clean defectless interfaces. UHV *in-situ* process systems have been developed at the OJL (Optoelectronics Joint Research Laboratory) as well as at the OTL (Optoelectronics
Technology Research Laboratory) (Figure 20). An ultrahigh vacuum, on the
order of 10^{-10} Torr, was maintained during successive processing steps. Studies
are now in progress at OTL, investigating the basic processing sequence of
MBE hetero epitaxy (AlGaAs, GaAs, AlGaAs), etching of a mesalike structure, and regrowth of AlGaAs— all in UHV chambers connected with UHV
pipings. Experiments were performed to confirm the formation of defect-free
interfaces, and studies are now in progress concerning how to control the
dimensions of the mesa structure.[17a-c] A specially developed process involving
in-situ GaAs oxide masking, removal of masking using a focused electron
beam, and Cl_2 gas etching are being investigated.[17a-c]

Many studies must follow these pioneer studies. In order to establish
industrial procedures, a large amount of work remains to be done. For Si LSI
technology, the amount of developmental effort has been huge—it has required
over 20 years and involved a large number of people. The manpower used was
of a magnitude larger than that used for group III-V device development. In
addition group III-V materials are much more difficult than silicon. However,
there is a large amount of accumulated knowledge concerning group III-V
devices and materials, which has not been fully used by industry. This knowledge of group III-V will be useful for the new generation of photoelectronic
integrated systems. The impact of photoelectronic integrated systems on society will be so large that it is certainly worthwhile to make the efforts described
above.

a)

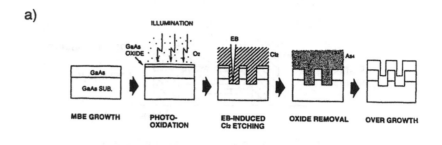

MBE GROWTH   PHOTO-OXIDATION   EB-INDUCED Cl₂ ETCHING   OXIDE REMOVAL   OVER GROWTH

b)

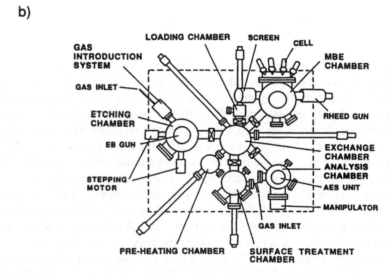

FIGURE 20   Schematic illustration of the UHV multichamber system for *in situ* EB lithography, etching, and MBE. (From Ishikawa et al., *Oyo Buturi* **62**, 107, 1993. With permission from The Japan Society of Applied Physics.)

IV. SUMMARY AND FUTURE PROSPECTS

In this chapter the importance of micro-optoelectronic devices in future systems for information processing has been presented. These systems will be integrated systems comprised of microelectronic and microphotonic devices in order to achieve a close collaboration between photons and electrons in one system. By combining the complementary features of these two particles it should be possible to achieve the highest possible system performance, which would be impossible without this combination. Such systems, in which a large number of both microelectronic and photonic devices are integrated together on a single substrate, will be called *photoelectronic integrated systems*. Corresponding to large scale electronic integrated circuits (LSI), this type of integrated circuit will be called an ultrascale optoelectronic integrated circuit (U-OEIC). Although a U-OEIC doesn't exist today, it is a logical extension of today's technology. The U-OEIC is based on techniques developed for electronic

ICs and photoelectronic devices. Of course, it requires new technologies for microphotonic devices and for the integration of microphotonic devices with microelectronic devices. The functions of the electronic circuits as well as photonic devices which we have today will all be 100 percent utilized. The new systems will expand the functions beyond the limit of today's circuits by the collaboration of two types of particles.

Photonic devices in a U-OEIC will be required to have small size, high speed, low power, and high efficiency. Photonic devices with a microcavity-type structure will be the best candidate for these U-OEICs. As has been shown in this chapter, a photoelectronic integrated system comprising U-OEICs, will be the center of the information processing world in the coming decades. It will therefore comprise the largest market for the microcavity devices. The scale of the market will be comparable, or even larger, than today's microelectronics market. It will be much larger than the photoelectronic device market for optical communications today. Microcavity photonic devices will play a key role in this market.

Examples of photoelectronic integrated systems are described in this chapter. High speed microprocessors show how one can extend the performance beyond the limit of purely electronic microprocessors by employing an optical interconnections scheme using U-OEIC circuits. The high performance has been estimated by a circuit simulation (Section II.B).[3]

Not only will there be improvements in performances of the system (Section II.B), but a completely new type of circuit will be achieved by the electron-photon combination. Three-dimensional circuit structures will be constructed by the combination, which are impossible to build using only electronic circuits (see Section II.C). A feasibility experiment using a six-processor system is in progress, and the basic function has been confirmed.[11] Using this scheme, architectural designs of ultraparallel processing systems are being studied. A 25,000-processor parallel system can be built using a 13-layer structure.[4] Each processor is connected to its neighbors in different layers through common memories which are located in all the layers and are interconnected by vertical photonic beams. The situation is similar to those in Figure 13. The system will be useful as artificial brains for handicapped people, for example.

The high speed analysis of two-dimensional (64×64) pictures using 4096 completely parallel processor elements connected to an array of photo detectors has been designed and a feasibility experiment was performed (Section II.D). An operation four orders of magnitude faster than that of the normal TV scanning scheme has been demonstrated, using a large scale model circuit.[12] A monolithic version of this system will be used for a robot's eye having the size of a single chip. Further the use of three-dimensional structures will open the way toward more advanced artificial vision systems (Figure 15).[12]

In order to fabricate such photoelectronic integrated circuits, technical barriers for microphotonic devices and their integration onto U-OEICs must be overcome. Basic material studies aiming at these breakthroughs are in progress. Microfabrication techniques for defectless heterointerfaces,[17] and studies aiming at defectless lattice-mismatched heterostructure are introduced.[14-16]

Optical interconnections on the board level will be developed one step before the chip-level interconnections described in this chapter. Although they will have optical interconnections of tens of centimeters using more conventional optoelectronic technologies, they will also require that of the chip level in the advanced stage. The technological problems described in this chapter will become important. Although developmental efforts have already started using today's technology, it is important to prepare for the next stage of the chip-level interconnections described here. This will give us a long-range target. Some basic studies must be started in advance.

It is still too early to predict what kind of new applications will result from using photoelectronic integrated systems. They will utilize not only the benefits of high speed and low power but also many other benefits, such as less mutual coupling and low noise, and the coupling-free nature of crossing beams, free of DC voltage matching. As for the microcavity structure not only can the sensitivity and directionality be increased, a wavelength selective feature can be achieved for LDs, LEDs, and PDs. A multiwavelength interconnection system can be built which is unique for photonic interconnections. Extremely simple multicolor communications will result in new systems using photoelectronic integrated schemes.

In this section, I have been discussing the performance of microcavity photoelectronic devices in the first generation of photoelectronic integrated systems. These systems use one-to-one photonic signal interconnections similar to today's optical communication. Even with simple one-to-one interconnections, the potential benefits of photonic signal transmission are extremely large compared with metal wire interconnections, as was previously explained. In these systems logical operations, switching, and nonlinear operation are all achieved by electronic devices. Because of the recent progress of transistors in LSIs, their operation can be performed by a small device of 1 μm or less, and at a negligible price per operation. It is almost meaningless to use photonic devices for logical operations in these systems. The basic philosophy of photoelectronic integrated systems is to utilize photoelectronic devices for photon-electron or electron-photon conversion at any locations and at any instances required, and to use photons or electrons for the most efficient purpose for the system.

However, microcavity devices will have further advanced features, such as wavelength selection, and will have applications in advanced systems. In microcavity devices the wavelength of operation is determined by the resonant condition, which is controlled by the mechanical dimensions. Precise control of the dimension, on the order of nanometers, is necessary for any light emitting or detecting devices. Light emitters of different wavelengths (λ) will be made from the same material by tuning the cavity dimension. By adjusting the material or by changing the dimension of the quantum wells, lasers in a further enlarged wavelength range will result. Noncoherent light emitting devices will have an even larger wavelength range. Today, there is a wide wavelength range of LEDs from infrared to ultraviolet.

Today's photodetectors have only a cut-off wavelength, beyond which the PD has no selective sensitivity. By using a microcavity structure, the PD will have a sensitive wavelength range. The cavity quality factor (Q) will determine the bandwidth.* Therefore, PDs with almost any wavelengths can be prepared, corresponding to LDs or LEDs. In this way, multiwavelength signal interconnection systems can be fabricated, which will be simple, and thus, mass producible. They will have applications in many systems, such as, multicolor bus lines for parallel processor systems or neurosystems that require (n to m) interconnections.

Switching is another feature that can be realized by a microcavity structure. Two cavities appropriately coupled will have two stable oscillating frequencies. A laser structure that has such two coupled cavities can oscillate in either one of these two wavelengths. Switching can be achieved by changing the excitation. The separation of wavelengths can be made much larger than those in a large cavity laser, which was done, for example, by changing the refractive index induced by carrier injection. The speed of switching can be as fast as the response speed of a laser. The wavelength switching can be converted into beam deflection, which has been desired for years regarding many system applications.

There are many unique features of microcavities, using either a single mode or multimodes, which can be used in many systems even today. Some of the requirements of optical fiber communication systems that are difficult to realize using today's technology will be solved using microcavity engineering.

It is known that silicon can be used as a detector material. It can absorb photons having wavelengths shorter than about a micrometer. However, the absorption length is large at longer wavelengths. It is about 30 μm at the GaAs laser wavelength, and decreases quickly to 3 μm at $\lambda = 0.6$ μm. The possibility of increasing the absorption by using a microcavity is attractive, because a silicon detector is the best detector material to be integrated on an Si LSI. There will be applications using many detectors but few light emitters in the system.

The availability of either electronic or photonic signals for input and output will be a very convenient method for interconnecting any exotic systems to the traditional photoelectronic systems. For starting operation in any exotic system, the photoelectronic integrated system will provide the most appropriate base system.

In conclusion, the photoelectronic integrated system using micro-photoelectronic devices will open a new era of information processing in the next century. A close collaboration between experts of different disciplines in the related fields is essential for the invisioned development.

ACKNOWLEDGMENTS

A special thank you is offered to the members of the U-OEIC research group for years of constructive discussions on photoelectronic integrated systems which have formed the basis of this chapter. In particular, T. Ae's idea on

---

*A microcavity also increases the sensitivity in PDs. In very small PDs, lack of width comparable with the wavelength of light may decrease the sensitivity, which will be recovered by the cavity operation.

massively parallel 3-D systems stimulated my thinking on using microcavity LEDs in such highly integrated systems. Discussions in the group also indicated the importance of applying the microcavity structures in light-emitting as well as light-receiving devices for photoelectronic integrated systems. Appreciation is expressed to H. Yokoyama for the many discussions with him in the past that been useful in developing the concept of microcavity performance and structure; to E. Yablonovitch, for stimulating discussions; to Y. Yamamoto, for discussions on microcavity structures; and to S. D. Brorson for his in-depth reading and many constructive comments on the rewriting and finalizing of this chapter.

REFERENCES AND NOTES

1. I. Hayashi, Photo-Electronic Integrated Systems — Basic Concepts of U-OEIC and Their Feasibilities, *Optoelectron-Devices Technol.* (OP-DET). **9**, 1 (1994).*

2. I. Hayashi, T. Ae, and M. Koyanagi, Optical Interconnection, *J. IEICE.* **75**, 951 (1993) (in Japanese).

3. A. Iwata, Optical Interconnection for ULSI Technology Innovation, *Optoelectron.-Devices Techn.* (OP-DET). **9**, 39 (1994).

4. T. Ae, Optical Interconnection for Parallel Data Processing, *Optoelectron.-Devices Technol.* (OP-DET). **9**, 15 (1994).

5. U-OEIC Research Group: The office is in the Optoelectronic Industry and Technology Development Association, Tokyo, Japan (Tel 81 3 5632 7721), Members: H. Hirose, M. Koyanagi, M. Abe, O. Wada, H. Ohno, M. Kashiwagi, I. Hayashi, Y. Katayama, W. Suzaki, M. Takeda, T. Nakamura, E. Arai, A. Iwata, M. Ishikawa, E. Sano, T. Nakada, M. Nakamura, H. Yonezu, K. Yamashita, H. Ishida.

6. I. Hayashi, M. Hirano, and Y. Katayama, Collaborative Semiconductor Research in Japan, *Proc. IEEE.* **77**, 1430 (1989).

7a. T. Ae and R. Aihara, *J. Inf. Process.* **26**, 1145 (1985).

7b. M. Koyanagi, H. Takata, and M. Hirose, Optically Coupled Three-Dimensional Common Memory, *Optoelectron.-Devices Technol.* (OP-DET). **3**, 83 (1988).

8. H. Ohno, Minimum Light Power for Optical Interconnection in Integrated Circuits, *Optoelectron.-Devices and Technol.* (OP-DET). **9**, 131 (1994).

9. M. Takeda, Stack-Type Optical Buslines Based on a Modified Dragone Star Coupler: A Concept, *Optoelectron.-Devices Technol.* (OP-DET). **9**, 137 (1994).

10. M. Koyanagi, K. Miyake, H. Kurotaki, S. Yokoyama, Y. Horiike, and M. Hirose, Fundamental Characteristics of Optically Coupled Three-Dimensional Common Memory, *Optoelectron.-Devices Technol.* (OP-DET). **9**, 119 (1994).

Optoelectronics—Devices and Technologies (OP-DET) is a technical journal published by Mita Press, Ochanomizu Center Building 2-12, Bunkyo-ku, Tokyho 113, Japan. Tel 03-3818-1011.

11. K. Miyake, T. Namba, K. Hashimoto, H. Sakaue, S. Miyazaki, Y. Horiike, S. Yokoyama, M. Koyanagi, and M. Hirose, Fabrication and Evaluation of Three-Dimensional Optically-coupled Common Memory, Extended Abstracts of the 1994 International Conference on Solid State Devices and Materials, Yokohama, 965 (1994).

12. M. Ishikawa, System Architecture for Integrated Optoelectronic Computing, *Optoelectron.-Devices Technol.* (OP-DET). **9**, 29 (1994).

13. H. Yokoyama, Spontaneous and Stimulated Emissions in the Microcavity Laser, Figure 12, this volume, chapter 8.

14. M. Tamura, Recent Progress in GaAs on Si Technology — Concerning the Reduction of Threading Dislocation, *Optoelectron.-Devices Technol.* (OP-DET). **9**, 95 (1994).

15. Y. Takagi, H. Yonezu, T. Kawai, K. Hayashida, N. Ohshima, and K. Pak, Suppression of Threading Dislocation Generation in GaAs-on-Si with Strained Short-Period Superlattices, 8th Int'l Conf. on MBE, Aug. 29–Sep. 2, 1994, Osaka, Japan, A10-16, 254.

16. J. E. Palmer, T. Saitoh, T. Yodo, and M. Tamura, Growth and Characterization of GaSe and GaAs/GaSe on As-Passivated Si (111) Substrates, *J. Appl. Phys.* **74**, 7211 (1993).

17a. H. Akita, Y. Sugimoto, M. Taneya, Y. Hiratani, Y. Ohki, H. Kawanishi, and Y. Katayama, Pattern etching and selective growth of GaAs by *in situ* electron beam lithography using an oxidized thin layer, in Advanced Techniques for Integrated Circuits Processing, Santa Clara, *Proc. Soc. Photo-Opt. Instrum. Eng.* **1392**, 576 (1990).

17b. T. Ishikawa, H. Kawanishi, N. Tanaka, and Y. Sugimoto, *In-situ* UHV Process for Compound Semiconductors, *Ohyo Butusri.* **62**, 102 (1993)(in Japanese).

17c. T. Ishikawa, N. Tanaka, M. López, and I. Matsuyama, Electron Beam Lithography Using GaAs Oxidized Resist for GaAs/AlGaAs Ultrafine Structure Fabrication, *J. Photopolymer Sci. Tech.* **7**, 595 (1994).

INDEX

Milton Keynes UK
Ingram Content Group UK Ltd.
UKHW020015071024
449327UK00031B/2800